ANNUAL EDITIONS

Environment

08/09

Twenty-Seventh Edition

EDITOR

Zachary Sharp

University of New Mexico

Zachary Sharp is a professor in the Department of Earth and Planetary Sciences at the University of New Mexico. He received a B.S. in Geology from U.C. Berkeley and a Masters and Ph.D. from the University of Michigan. He was a postdoctoral fellow at the Geophysical Laboratory, Carnegie Institution of Washington, and then spent eight years at the Université de Lausanne in Switzerland, before moving to New Mexico. He has over 100 publications, including a book on stable isotope geochemistry, and has been teaching a course on environmental science for the last eight years. His passion is educating students about environmental issues.

Boston   Burr Ridge, IL   Dubuque, IA   New York   San Francisco   St. Louis
Bangkok   Bogotá   Caracas   Kuala Lumpur   Lisbon   London   Madrid   Mexico City
Milan   Montreal   New Delhi   Santiago   Seoul   Singapore   Sydney   Taipei   Toronto

ANNUAL EDITIONS: ENVIRONMENT, TWENTY-SEVENTH EDITION

1 2 3 4 5 6 7 8 9 0 QPD/QPD 0 9 8

ISBN 978–0–07–351548–9
MHID 0–07–351548–5
ISSN 0272–9008w

Managing Editor: *Larry Loeppke*
Managing Editor *: Faye Schilling*
Developmental Editor: *Dave Welsh*
Editorial Assistant: *Nancy Meissner*
Production Service Assistant: *Rita Hingtgen*
Permissions Coordinator: *Shirley Lanners*
Senior Marketing Manager: *Julie Keck*
Marketing Communications Specialist: *Mary Klein*
Marketing Coordinator: *Alice Link*
Project Manager: *Jean Smith*
Senior Administrative Assistant: *DeAnna Dausener*
Senior Production Supervisor: *Laura Fuller*
Cover Graphics: *Tara McDermott*

Compositor: Laserwords Private Limited
Cover Image: © Goodshot/Almay (background), © IMS Communication LTD., All Rights Reserved (foreground).

Library in Congress Cataloging-in-Publication Data
Main entry under title: Annual Editions: Environment. 2008/2009.
1. Environment—Periodicals. I. Sharp, Zachary, *comp.* II. Title: Environment.
658'.05

Editors/Advisory Board

Members of the Advisory Board are instrumental in the final selection of articles for each edition of ANNUAL EDITIONS. Their review of articles for content, level, currentness, and appropriateness provides critical direction to the editor and staff. We think that you will find their careful consideration well reflected in this volume.

EDITOR

Zachary Sharp
University of New Mexico

ADVISORY BOARD

Daniel Agley
Towson University

John Aliff
Georgia Perimeter College

Matthew R. Auer
Indiana University

Robert V. Bartlett
Purdue University

Susan W. Beatty
University of Colorado, Boulder

Kelly Cartwright
College of Lake County

Michelle Cawthorn
Georgia Southern University

William P. Cunningham
University of Minnesota—St. Paul

Rosa Guedes
Philadelphia University

Gian Gupta
University of Maryland

M. Iqbal
University of Northern Iowa

Vishnu R. Khade
Eastern Connecticut State University

Steve Malcolm
Western Michigan University

Leslie McFadden
University of New Mexico

John Milmore
Westchester Community College

Adil Najam
Tufts University

David J. Nemeth
University of Toledo

Shannon O'Lear
University of Kansas

David Padgett
Tennessee State University

Colette Palamar
Antioch College

John H. Parker
Florida International University

Melanie Rathburn
Boston University

Bradley F. Smith
Western Washington University

Nicholas J. Smith-Sebasto
Montclair State University

Chris Sutton
Western Illinois University

Robert A. Taylor
Ohio State University

Robert Taylor
Montclair State University

Mica Tomkiewicz
Brooklyn College of SUNY

Karen Warren
Randolph College

Ben E. Wodi
SUNY at Cortland

Preface

The term "environmentalism" in the modern sense was only defined in the 1970s. Prior to that, the term referred to the 'nature vs. nurture' debate. The modern definition reflects our recognition that resources are not infinite, and that humankind can produce long-lasting changes to the world around us. In the 1960s and 70s we realized that environmental degradation could no longer be ignored, and as a result, a great number of positive advances took place. The first Earth Day was held in 1970. In the same year, an executive order created the Environmental Protection Agency. The Clean Air Act, originally passed by Congress in 1963, was extended and amended in the 1970s. The Clean Water Act was passed in 1972. The Corporate Average Fuel Economy (CAFÉ) regulations, increasing fuel efficiency standards, were passed in 1975. Catalytic converters, which reduced carbon monoxide emissions by 96%, were mandated in the same year. These rulings and others led to dramatic improvements in environmental quality.

But our Nation's concern for the environment was short-lived. In one of his first acts as President, Ronald Reagan removed the solar panels from the White House that Jimmy Carter had erected only the year before. (The panels were transferred to and used at Unity College in Maine for over 20 years). Tax credits for solar panels lapsed, and the budget for the Solar Energy Research Institute was slashed. Efforts were made to gut the Clean Air Act and Clean Water Act, and eliminate the Superfund, a Congressional Act originally created to protect families and communities from toxic waste. Whereas environmentalism had been politically neutral (Nixon created the EPA and many Republican congressmen were environmentalists), in the 1980s a linkage between political affiliation and environmentalism was created. A shift in American attitudes tilted in favor of business, profits, and less government regulations. Sadly, the big loser was the environment. Organizations and groups started working actively to slow or reverse years of progress in environmental protection.

Today we face a new challenge, one that affects us all, crosses all political boundaries, and forces us to sit up and take notice. The environmental ramifications of global warming are staggering. But so are the costs. While pro-environmental groups work hard to bring about change, powerful industry groups push back—determined to continue extracting fossil fuels from the Earth at breakneck speed, to continue producing cars based on the internal combustion engine, and to make sure that we continue our consumption of manufactured goods with little concern for the environment. Today, the burning of fossil fuels (oil, coal, natural gas) add over 50 trillion pounds of CO_2 to our air annually. The concentration of atmospheric CO_2 has increased from the pre-industrial level of 280 ppm to over 380 ppm today.

CO_2 is a greenhouse gas, meaning it traps outgoing infrared radiation. Scientists predicted long ago that the increased CO_2 levels would raise the temperature of the planet, and yet their arguments were at first ignored and then refuted by groups opposing any action to reverse the trend. The arguments have changed over the years, but continue unabated. First, we heard from the climate change skeptics that the planet was cooling, not warming. Then we heard that the scientific evidence was not exact, and that there was considerable doubt about the long-term effects from CO_2 emissions. Finally, in light of overwhelming evidence that the Earth is indeed warming, the debate has shifted once again. Now the skeptics tell us that warming is a natural phenomenon, having nothing to do with burning fossil fuels. We even heard recently from the White House Press Secretary that "there are public health benefits to climate change" and that "many people die from cold-related deaths every winter." Message: 'Yes, it's warming, but no cause for concern.'

And so, today we find ourselves at a crossroads. The environmental degradation we are experiencing could have been avoided had we chosen to act sooner. As far back as the early 1970s, major renewable energy initiatives were proposed by Jimmy Carter, but never adopted. Instead, we face irreparable damage to our planet at many levels. Drought is a problem across much of the country, and on a global scale, droughts in impoverished regions of the world are causing widespread suffering and death. Heat waves are becoming more common and extreme (the August 2003 heat wave killed over 15,000 people in France alone). Global temperatures are warming. "Eleven of the last twelve years (1995–2006) rank among the twelve warmest years in the instrumental record of global surface temperature (since 1850)" (see summery of Policymakers, IPCC report, 2007). Glaciers, which supply needed water to many low-lying populated regions, are melting at a dramatic rate. Mt. Kilimanjaro, which has been covered with glaciers for the last 10,000 years, will be ice-free in the next several years!

In spite of the gloomy picture outlined above, there is cause for hope. For one, our awareness of the environment is growing. Unfortunately, our newfound awareness is tied to the unavoidable realization that our planet is sick. But with this recognition, there is a newfound drive for change. And there are many ways in which we can begin to turn things around. It won't be easy. People in developed countries enjoy the luxuries granted to them from cheap and abundant energy. Emerging nations consume energy at prodigious rates as a path to economic strength. Change requires sacrifice. We need to decide that we care equally about the future as the present. In spite of the magnitude of the task at hand, we have the resources to change course. And our voice is being heard. Politicians are beginning to come together on the need to work to save the planet. We are now beginning to see that economic opportunities and 'Green Technology' are not incompatible. There are numerous alternative energies that exist that could reduce our reliance on fossil fuels. Solar, wind, biomass, geothermal, and nuclear energy all have advantages and disadvantages. Some are costly, intermittent in their supply, dangerous, or limited in geographic extent. But intelligent utilization of alternative energies is imperative if we hope to minimize the effects of climate change.

Weaning ourselves off fossil fuels also reduces our reliance on foreign petroleum sources. We presently import over 500 million barrels of oil per year. As oil prices climb to $100/barrel, this translates to a trade deficit of $50 billion/year. And this staggering figure ignores the costs of insuring the continuity of these supplies. We are at war in Iraq, and have extremely hostile relations with other major oil producers, such as Iran and Venezuela. Fully one third of the world's petroleum is held in countries with high political risk.

Technology is only part of the answer. Our future path must be guided by intelligent decisions. Conservation is important, and can be implemented immediately, and while it may be slightly inconvenient to our accepted daily practices, conservation can have widespread economic benefits. Special interests must not dictate our future, rather it must be decided by what is best for the people and our world. We need a sustainable economy, one that grows, but one that also conserves our natural resources for future generations. The public must be aware of environmental needs and must be educated about our actions and what benefits and detriments each holds. Change will only occur when citizens recognize the problems and demand action from their elected officials.

With the need for a broad understanding of complex environmental issues, *Annual Editions: Environment 08/09* has been organized to address a wide array of topics. The compilation of published articles has been organized into five categories, each dealing with a subset of the environment. There is no organized overall outline of all environmental problems such as would be taught in an introductory environmental science class. Instead, ideas focused on specific issues are presented. The goal is that each article will lead to different ideas and concerns that can be addressed and debated in a classroom setting. Awareness is critical. Time is not on our side. But by raising awareness of environmental issues, we can make a difference. Ultimately, we can and must change to save the planet.

Zachary Sharp
Editor

Contents

UNIT 1
The Global Environment: An Emerging World View

UNIT 2
Population, Policy, and Economy

The concepts in bold italics are developed in the article. For further expansion, please refer to the Topic Guide.

UNIT 3
Energy: Present and Future Problems

The concepts in bold italics are developed in the article. For further expansion, please refer to the Topic Guide.

UNIT 4
Biosphere: Endangered Species

The concepts in bold italics are developed in the article. For further expansion, please refer to the Topic Guide.

UNIT 5
Resources: Land and Water

UNIT 6
The Politics of Climate Change

The concepts in bold italics are developed in the article. For further expansion, please refer to the Topic Guide.

Correlation Guide

The *Annual Editions* series provides students with convenient, inexpensive access to current, carefully selected articles from the public press. **Annual Editions: Environment 08/09** is an easy-to-use reader that presents articles on important topics such as *the global environment, population policy, energy use and options, the biosphere, natural resources, climate change politics,* and many more. For more information on *Annual Editions* and other *McGraw-Hill Contemporary Learning Series* titles visit www.mhcls.com.

This convenient guide matches the units in **Annual Editions: Environment 08/09** with the corresponding chapters in three of our best-selling McGraw-Hill Environmental Science textbooks by Kaufmann/Cleveland and Cunningham/Cunningham.

Annual Editions: Environment 08/09	Environmental Science by Kaufmann/Cleveland	Principles of Environmental Science: Inquiry & Applications, 4/e by Cunningham/Cunningham	Environmental Science: A Global Concern, 10/e by Cunningham/Cunningham
Unit 1: The Global Environment: An Emerging World View	**Chapter 1:** Environment and Society: A Sustainable Partnership? **Chapter 13:** Global Climate Change: A Warming Planet	**Chapter 1:** Understanding Our Environment **Chapter 14:** Sustainability and Human Development	**Chapter 1:** Understanding Our Environment
Unit 2: Population, Policy, and Economy	**Chapter 9:** Carrying Capacity: How Large a Population? **Chapter 11:** The Driving Forces of Environmental Change	**Chapter 4:** Human Populations **Chapter 14:** Sustainability and Human Development	**Chapter 7:** Human Populations **Chapter 12:** Land Use: Forests and Grasslands **Chapter 13:** Preserving and Restoring Nature **Chapter 22:** Urbanization and Sustainable Cities **Chapter 23:** Ecological Economics **Chapter 24:** Environmental Policy, Law, and Planning
Unit 3: Energy: Present and Future Problems	**Chapter 2:** The Laws of Energy and Matter **Chapter 20:** Fossil Fuels: The Lifeblood of the Global Economy **Chapter 21:** Nuclear Power **Chapter 22:** Renewable Energy and Energy Efficiency **Chapter 23:** Materials, Society, and the Environment **Chapter 24:** A Sustainable Future: Will Business as Usual Get Us There?	**Chapter 12:** Energy **Chapter 13:** Solid and Hazardous Waste	**Chapter 3:** Matter, Energy, and Life **Chapter 19:** Conventional Energy **Chapter 20:** Sustainable Energy **Chapter 21:** Solid, Toxic, and Hazardous Waste
Unit 4: Biosphere: Endangered Species	**Chapter 7:** Biomes: Where Do Plants and Animals Live? **Chapter 12:** Biodiversity: Species and So Much More	**Chapter 3:** Populations, Communities, and Species Interaction **Chapter 5:** Biomes and Biodiversity	**Chapter 4:** Evolution, Biological Communities, and Species Interactions **Chapter 5:** Biomes: Global Patterns of Life **Chapter 11:** Biodiversity
Unit 5: Resources: Land and Water	**Chapter 15:** Soil: A Potentially Sustainable Resource **Chapter 18:** Water Resources	**Chapter 2:** Principles of Ecology: Matter, Energy, and Life **Chapter 10:** Water: Resources and Pollution **Chapter 11:** Environmental Geology and Earth Resources	**Chapter 12:** Land Use: Forests and Grasslands **Chapter 14:** Geology and Earth Resources **Chapter 17:** Water Use and Management **Chapter 18:** Water Pollution
Unit 6: The Politics of Climate Change	**Chapter 1:** Environment and Society: A Sustainable Partnership? **Chapter 3:** Systems: Why Are Environmental Problems So Difficult to Solve? **Chapter 24:** A Sustainable Future: Will Business as Usual Get Us There?	**Chapter 15:** Environmental Science and Policy	**Chapter 24:** Environmental Policy, Law, and Planning **Chapter 25:** What Then Shall We Do?

Topic Guide

This topic guide suggests how the selections in this book relate to the subjects covered in your course. You may want to use the topics listed on these pages to search the Web more easily.

On the following pages a number of Web sites have been gathered specifically for this book. They are arranged to reflect the units of this *Annual Edition*. You can link to these sites by going to the student online support site at *http://www.mhcls.com/online/*.

ALL THE ARTICLES THAT RELATE TO EACH TOPIC ARE LISTED BELOW THE BOLD-FACED TERM.

Internet References

The following Internet sites have been carefully researched and selected to support the articles found in this reader. The easiest way to access these selected sites is to go to our student online support site at *http://www.mhcls.com/online/*.

AE: Environment 08/09

The following sites were available at the time of publication. Visit our Web site—we update our student online support site regularly to reflect any changes.

General Sources

Britannica's Internet Guide
http://www.britannica.com

This site presents extensive links to material on world geography and culture, encompassing material on wildlife, human lifestyles, and the environment.

CIA Factbook
http://www.cia.gov/cia/publications/factbook/

This site is the U.S. government's official source for data on the population, production, resources, geography, political systems, and other important characteristics of each of the world's countries.

CO2 Calculator
http://actonco2.direct.gov.uk/index.html

This site is a fun link to illustrate where we use energy in our daily lives and how we can reduce our CO2 footprint. Graphical interface.

EnviroLink
http://www.envirolink.org/

One of the world's largest environmental information clearinghouses, EnviroLink is a grassroots nonprofit organization that unites organizations and volunteers around the world and provides up-to-date information and resources.

IPPC (International Panel on Climate Change)
http://www.ipcc.ch/

Link to the Nobel Prize winning report on climate change. It's an incredible resource and is very detailed, but with a short informative Summary for Policy Makers.

Library of Congress
http://www.loc.gov

Examine this extensive Web site to learn about resource tools, library services/resources, exhibitions, and databases in many different subfields of environmental studies.

The New York Times
http://www.nytimes.com

Browsing through the archives of the New York Times will provide a wide array of articles and information related to the different subfields of the environment.

SocioSite: Sociological Subject Areas
http://www.pscw.uva.nl/sociosite/TOPICS/

This huge sociological site from the University of Amsterdam provides many discussions and references of interest to students of the environment, such as the links to information on ecology and consumerism.

U.S. Geological Survey
http://www.usgs.gov

This site and its many links are replete with information and resources in environmental studies, from explanations of El Niño to discussion of concerns about water resources.

UNIT 1: The Global Environment: An Emerging World View

Alternative Energy Institute (AEI)
http://www.altenergy.org

The AEI will continue to monitor the transition from today's energy forms to the future in a "surprising journey of twists and turns." This site is the beginning of an incredible journey.

Earth Science Enterprise
http://www.earth.nasa.gov

Information about NASA's Mission to Planet Earth program and its Science of the Earth System can be found here. Surf to learn about satellites, El Niño, and even "strategic visions" of interest to environmentalists.

IISDnet
http://www.iisd.org

The International Institute for Sustainable Development, a Canadian organization, presents information through gateways entitled Business, Climate Change, Measurement and Assessment, and Natural Resources. IISD Linkages is its multimedia resource for environment and development policy makers.

National Geographic Society
http://www.nationalgeographic.com

Links to National Geographic's huge archive are provided here. There is a great deal of material related to the atmosphere, the oceans, and other environmental topics.

Research and Reference (Library of Congress)
http://lcweb.loc.gov/rr/

This research and reference site of the Library of Congress will lead to invaluable information on different countries. It provides links to numerous publications, bibliographies, and guides in area studies that can be of great help to environmentalists.

Solstice: Documents and Databases
http://solstice.crest.org/index.html

In this online source for sustainable energy information, the Center for Renewable Energy and Sustainable Technology (CREST) offers documents and databases on renewable energy, energy efficiency, and sustainable living. The site also offers related Web sites, case studies, and policy issues.

United Nations
http://www.unsystem.org

Visit this official Web site Locator for the United Nations System of Organizations to get a sense of the scope of international environmental inquiry today. Various UN organizations concern themselves with everything from maritime law to habitat protection to agriculture.

United Nations Environment Programme (UNEP)
http://www.unep.ch

Consult this home page of UNEP for links to critical topics of concern to environmentalists, including desertification, migratory species, and the impact of trade on the environment. The site will direct you to useful databases and global resource information.

World Resources Institute (WRI)
http://www.wri.org/

The World Resources Institute is committed to change for a sustainable world and believes that change in human behavior is urgently needed to halt the accelerating rate of environmental deterioration in some areas. It sponsors not only the general Web site above, but also The Environmental Information Portal (www.earthtrends.wri.org), which provides a rich database on the interaction between human disease, pollution, and large-scale environmental, development, and demographic issues.

UNIT 2: Population, Policy, and Economy

The Hunger Project
http://www.thp.org

Browse through this nonprofit organization's site to explore the ways that it attempts to achieve its goal: the sustainable end to global hunger through leadership at all levels of society. The Hunger Project contends that the persistence of hunger is at the heart of the major security issues that are threatening our planet.

Poverty Mapping
http://www.povertymap.net

Poverty maps can quickly provide information on the spatial distribution of poverty. This site provides maps, graphics, data, publications, news, and links that provide the public with poverty mapping from the global to the subnational level.

World Health Organization
http://www.who.int

The home page of the World Health Organization provides links to a wealth of statistical and analytical information about health and the environment in the developing world.

World Population and Demographic Data
http://geography.about.com/cs/worldpopulation/

On this site, information about world population and additional demographic data for all the countries of the world are provided.

WWW Virtual Library: Demography & Population Studies
http://demography.anu.edu.au/VirtualLibrary/

This is a definitive guide to demography and population studies. A multitude of important links to information about global poverty and hunger can be found here.

UNIT 3: Energy: Present and Future Problems

Alliance for Global Sustainability (AGS)
http://globalsustainability.org/

The AGS is a cooperative venture seeking solutions to today's urgent and complex environmental problems. Research teams from four universities study large-scale, multidisciplinary

environmental problems that are faced by the world's ecosystems, economies, and societies.

Alternative Energy Institute, Inc.
http://www.altenergy.org

On this site created by a nonprofit organization, discover how the use of conventional fuels affects the environment. Also learn about research work on new forms of energy.

Energy and the Environment: Resources for a Networked World
http://zebu.uoregon.edu/energy.html

An extensive array of materials having to do with energy sources—both renewable and nonrenewable—as well as other topics of interest to students of the environment is found on this site.

Institute for Global Communication/EcoNet
http://www.igc.org/

This environmentally friendly site provides links to dozens of governmental, organizational, and commercial sites having to do with energy sources. Resources address energy efficiency, renewable generating sources, global warming, and more.

Nuclear Power Introduction
http://library.thinkquest.org/17658/pdfs/nucintro.pdf

Information regarding alternative energy forms can be accessed here. There is a brief introduction to nuclear power and a link to maps that show where nuclear power plants exist.

U.S. Department of Energy
http://www.energy.gov

Scrolling through the links provided by this Department of Energy home page will lead to information about fossil fuels and a variety of sustainable/renewable energy sources.

UNIT 4: Biosphere: Endangered Species

Endangered Species
http://www.endangeredspecie.com/

This site provides a wealth of information on endangered species anywhere in the world. Links providing data on the causes, interesting facts, law issues, case studies, and other issues on endangered species are available.

Friends of the Earth
http://www.foe.co.uk/index.html

Friends of the Earth, a nonprofit organization based in the United Kingdom, pursues a number of campaigns to protect the Earth and its living creatures. This site has links to many important environmental sites, covering such broad topics as ozone depletion, soil erosion, and biodiversity.

Natural Resources Defense Council
http://nrdc.org

The Natural Resources Defense Council (NRDC) uses law, science, and the support of more than one million members and activists to protect the planet's wildlife, plants, water, soils, and other resources. The site provides abundant information on global issues and political responses.

Smithsonian Institution Web Site
http://www.si.edu

Looking through this site, which provides access to many of the enormous resources of the Smithsonian, one gets a sense of the biological diversity that is threatened by humans' unsound environmental policies and practices.

World Wildlife Federation (WWF)
http://www.wwf.org

This home page of the WWF leads to an extensive array of information links about endangered species, wildlife management and preservation, and more. It provides many suggestions for how to take an active part in protecting the biosphere.

UNIT 5: Resources: Land and Water

Global Climate Change
http://www.puc.state.oh.us/consumer/gcc/index.html

The goal of this PUCO (Public Utilities Commission of Ohio) site is to serve as a clearinghouse of information related to global climate change. Its extensive links provide an explanation of the science and chronology of global climate change, acronyms, definitions, and more.

National Oceanic and Atmospheric Administration (NOAA)
http://www.noaa.gov

Through this home page of NOAA, you can find information about coastal issues, fisheries, climate, and more.

National Operational Hydrologic Remote Sensing Center (NOHRSC)
http://www.nohrsc.nws.gov

Flood images are available at this site of the NOHRSC, which works with the U.S. National Weather Service to track weather-related information.

Terrestrial Sciences
http://www.cgd.ucar.edu/tss/

The Terrestrial Sciences Section (TSS) is part of the Climate and Global Dynamics (CGD) Division at the National Center for Atmospheric Research (NCAR) in Boulder, Colorado. Scientists in the section study land-atmosphere interactions; in particular, surface forcing of the atmosphere, through model development, application, and observational analyses. Here, you'll find a link to VEMAP, The Vegetation/Ecosystem Modeling and Analysis Project.

UNIT 6: The Hazards of Growth: Pollution and Climate Change

Persistent Organic Pollutants (POP)
http://www.chem.unep.ch/pops/

Visit this site to learn more about persistent organic pollutants (POPs) and the issues and concerns surrounding them.

School of Labor and Industrial Relations (SLIR): Hot Links
http://www.lir.msu.edu/hotlinks/

Michigan State University's SLIR page connects to industrial relations sites throughout the world. It has links to U.S. government statistics, newspapers and libraries, international intergovernmental organizations, and more.

Space Research Institute
http://arc.iki.rssi.ru/eng/index.htm

For a change of pace, browse through this home page of Russia's Space Research Institute for information on its Environment Monitoring Information Systems, the IKI Satellite Situation Center, and its Data Archive.

Worldwatch Institute
http://www.worldwatch.org

The Worldwatch Institute, dedicated to fostering the evolution of an environmentally sustainable society, presents this site with access to World Watch Magazine and State of the World 2000. Click on In the News and Press Releases for discussions of current problems.

We highly recommend that you review our Web site for expanded information and our other product lines. We are continually updating and adding links to our Web site in order to offer you the most usable and useful information that will support and expand the value of your Annual Editions. You can reach us at: *http://www.mhcls.com/annualeditions/*.

UNIT 1

The Global Environment: An Emerging World View

Unit Selections

1. **Climate Change 2007,** Intergovernmental Panel on Climate Change
2. **How Many Planets?** *The Economist*
3. **Five Meta-Trends Changing the World,** David Pearce Snyder
4. **Globalization's Effects on the Environment,** Jo Kwong
5. **Do Global Attitudes and Behaviors Support Sustainable Development?,** Anthony A. Leiserowitz, Robert W. Kates, and Thomas M. Parris

Key Points to Consider

- What is the consensus of the Scientific Community? What do we know with certainty about climate change and what questions are still debated? How serious are the consequences of climate change and how far should we be willing to go to mitigate its effects?

- What are the connections between the attempts to develop sustainable systems and the quantity and quality of environmental data? Are there also relationships between data and the role of technology and economic systems in shaping the environmental future?

- What are some of the key "meta-trends" produced by increasing globalization of economic and other human systems? How can human societies and cultures adapt to such trends in order to prevent significant environmental disruption?

- How has the process of "globalization" altered the cultural and environmental patterns of the world? What kinds of changes brought about by an increasingly global economy have been unforeseen?

- What is the relationship between human attitudes and behavior and the attempts to develop systems of economic development that can be environmentally sustainable?

Student Web Site

www.mhcls.com/online

Internet References

Further information regarding these Web sites may be found in this book's preface or online.

Alternative Energy Institute (AEI)
http://www.altenergy.org

Earth Science Enterprise
http://www.earth.nasa.gov

IISDnet
http://www.iisd.org

National Geographic Society
http://www.nationalgeographic.com

Research and Reference (Library of Congress)
http://lcweb.loc.gov/rr/

Solstice: Documents and Databases
http://solstice.crest.org/index.html

United Nations
http://www.unsystem.org

United Nations Environment Programme (UNEP)
http://www.unep.ch

World Resources Institute (WRI)
http://www.wri.org/

The Industrial Revolution changed the world forever, raising the standard of living for a vast number of people, increasing productivity, lowering mortality rates, and dramatically increasing the population of our planet. The key to the Industrial Revolution was the harnessing of the energy from fossil fuels, allowing a single factory worker to produce what would otherwise require dozens of individuals using only the power of their own hands and that of domesticated animals. Agricultural output increased dramatically, and at the same time, fewer laborers were needed in the fields, leading to a migration to the factories of the cities. Manufactured goods were readily available to the vast majority of people in developed countries. But the explosive industrial growth in the last two hundred years has also created a host of unforeseen consequences that negatively impact the residents of our planet. The toxic waste from power plants and urban factories had to be disposed of, and the most expedient and least expensive methods were simply to dump them into rivers or expel them into the air. The most egregious actions were quickly recognized by affected citizens, and governments quickly enacted regulations to ameliorate the harmful practices. But other examples of air and water pollution, while equally harmful, were often insidious and difficult to identify. At the top of this list is the phenomenon of global warming, where CO_2 generated from fossil fuel burning is increasing the intensity of the Greenhouse Effect and warming our planet. The consequences of this warming have, to this point, been subtle, yet the predicted effects, outlined in the first article, "Climate Change 2007" are extremely serious, leading to consequences that are truly frightening. To make matters worse, the solutions to lowering greenhouse gas emissions require sacrifices and difficult, expensive changes from our manufacturing sector. Powerful interest groups oppose such changes, so that progress in reversing our course have been pitifully slow.

Pollution cleanup is not cheap and will not be done without government regulation. There is no impetus for the CEO of a company to pour money into environment remediation. Voluntary cleanup is actually against the very goals vested in the CEO, which are to maximize investors' profits. Only with government regulation can any meaningful pollution controls be put into place. Obvious harmful pollution practices, such as egregious dumping of toxic waste into a river, are easily recognized and quickly dealt with at the local level. Other problems are regional or global in extent and require an organized effort amongst communities or nations. Global warming is the most serious affront on our environment, but remedies are costly and would require a complete paradigm shift in the way that our modern industrialized world conducts business. Wholesale changes in the way we produce and use energy must be adopted. At present, there is no magic bullet to reversing the course of global warming. Conservation, improving automobile efficiency, changing to

Chad Baker/Getty Images

renewable energy sources are all viable actions that help reduce CO_2 emissions, but all are expensive modifications that require sacrifice from all. Those who implore our leaders to address this problem are countered by strong lobbying efforts from powerful business interests that want to continue the 'business as usual' model of energy consumption. But as evidence mounts and more and more politicians and business leaders are resigned to accepting the fact that global warming is upon us and a serious concern, we can hope that change is possible before it is too late—before we reach the 'tipping point' of no return.

Al Gore and the IPCC received the Nobel Prize this year for their work highlighting the concerns of global warming. Nevertheless, the importance of environmental issues is not foremost on most people's minds. As the 2008 election approaches, presidential candidates speak of many things—the economy, immigration, health care, the war in Iraq. Rarely do we hear politicians speak forcefully about climate change, and the issue is relegated to the back burner. Change will only come when one of two conditions is reached: the first is that there is a general consensus among the vast majority of people that something needs to be done to avoid environmental damage. The second, and this would be the worst case scenario, is if we wait until the environmental damage is done, and we've reached a point of no return.

The articles in this section bring together some of the concerns and suggest possibilities to deal with our major environmental problems. The "Climate Change 2007" written by the Intergovernmental Panel on Climate Change, summarizes the climatic changes expected in the next decades and provides the level of confidence for each consideration (note that this report is followed by another IPCC report on expected outcomes, which can be found at http://www.ipcc.ch/pdf/assessment-report/ar4/syr/ar4_syr_spm.pdf). In order to achieve consensus amongst all members, the conclusions are very conservative, with most climate scientists expecting even more dramatic changes to occur. Nevertheless, the results are sobering, to say the least. The summary states *"Warming of the climate system is unequivocal, as is now evident from observations of increases in global average air and ocean temperatures, widespread melting of snow and ice, and rising global mean sea level,"* with 11 of the last 12 years being the warmest since modern records began in 1850. And *"Anthropogenic warming and sea level rise would continue for centuries due to the timescales associated with climate processes and feedbacks, even if greenhouse gas concentrations were to be stabilized".* The IPCC report is alarming, and will hopefully awaken us to action.

In "How Many Planets?" the editors of the Economist magazine discuss what needs to be done to develop a sustainable economy, one where our consumption of resources is matched by their regeneration. They then ask how technology can address the climate change crisis. The long-term policy must be to change to a low-carbon energy system. The remaining articles deal with globalization and its effects on the environment and return to the question of sustainable development.

Climate Change 2007
The Physical Science Basis, Summary for Policymakers

INTERGOVERNMENTAL PANEL ON CLIMATE CHANGE

Introduction

The Working Group I contribution to the IPCC Fourth Assessment Report describes progress in understanding of the human and natural drivers of climate change[1], observed climate change, climate processes and attribution, and estimates of projected future climate change. It builds upon past IPCC assessments and incorporates new findings from the past six years of research. Scientific progress since the TAR is based upon large amounts of new and more comprehensive data, more sophisticated analyses of data, improvements in understanding of processes and their simulation in models, and more extensive exploration of uncertainty ranges.

The basis for substantive paragraphs in this Summary for Policymakers can be found in the chapter sections specified in curly brackets.

Human and Natural Drivers of Climate Change

Changes in the atmospheric abundance of greenhouse gases and aerosols, in solar radiation and in land surface properties alter the energy balance of the climate system. These changes are expressed in terms of radiative forcing[2], which is used to compare how a range of human and natural factors drive warming or cooling influences on global climate. Since the Third Assessment Report (TAR), new observations and related modelling of greenhouse gases, solar activity, land surface properties and some aspects of aerosols have led to improvements in the quantitative estimates of radiative forcing.

- Carbon dioxide is the most important anthropogenic greenhouse gas (see Figure SPM-2). The global

> Global atmospheric concentrations of carbon dioxide, methane and nitrous oxide have increased markedly as a result of human activities since 1750 and now far exceed pre-industrial values determined from ice cores spanning many thousands of years (see Figure SPM-1). The global increases in carbon dioxide concentration are due primarily to fossil fuel use and land-use change, while those of methane and nitrous oxide are primarily due to agriculture. {2.3, 6.4, 7.3}

atmospheric concentration of carbon dioxide has increased from a pre-industrial value of about 280 ppm to 379 ppm[3] in 2005. The atmospheric concentration of carbon dioxide in 2005 exceeds by far the natural range over the last 650,000 years (180 to 300 ppm) as determined from ice cores. The annual carbon dioxide concentration growth-rate was larger during the last 10 years (1995–2005 average: 1.9 ppm per year), than it has been since the beginning of continuous direct atmospheric measurements (1960–2005 average: 1.4 ppm per year) although there is year-to-year variability in growth rates. {2.3, 7.3}

- The primary source of the increased atmospheric concentration of carbon dioxide since the pre-industrial period results from fossil fuel use, with land use change providing another significant but smaller contribution. Annual fossil carbon dioxide emissions[4] increased from an average of 6.4 [6.0 to 6.8][5] GtC (23.5 [22.0 to 25.0] $GtCO_2$) per year in the 1990s, to 7.2 [6.9 to 7.5] GtC (26.4 [25.3 to 27.5] $GtCO_2$) per year in 2000–2005 (2004 and 2005 data are interim estimates). Carbon dioxide emissions associated with land-use change are estimated to be 1.6 [0.5 to 2.7] GtC (5.9 [1.8 to 9.9] $GtCO_2$) per year over the 1990s, although these estimates have a large uncertainty. {7.3}

- The global atmospheric concentration of methane has increased from a pre-industrial value of about 715 ppb to 1732 ppb in the early 1990s, and is 1774 ppb in 2005. The atmospheric concentration of methane in 2005 exceeds by far the natural range of the last 650,000 years (320 to 790 ppb) as determined from ice cores. Growth rates have declined since the early 1990s, consistent with total emissions (sum of anthropogenic and natural sources) being nearly constant during this period. It is *very likely*[6] that the observed increase in methane concentration is due to anthropogenic activities, predominantly agriculture and fossil fuel use, but relative contributions from different source types are not well determined. {2.3, 7.4}

- The global atmospheric nitrous oxide concentration increased from a pre-industrial value of about 270 ppb to 319 ppb in 2005. The growth rate has been approximately

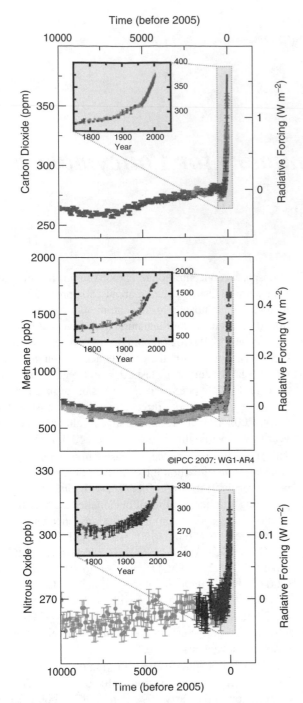

FIGURE SPM-1 Changes in greenhouse gases from ice-Core and modern data. Atmospheric concentrations of carbon dioxide, methane and nitrous oxide over the last 10,000 years (large panels) and since 1750 (inset panels). {Figure 6.4}

constant since 1980. More than a third of all nitrous oxide emissions are anthropogenic and are primarily due to agriculture. {2.3, 7.4}

- The combined radiative forcing due to increases in carbon dioxide, methane, and nitrous oxide is +2.30 [+2.07 to +2.53] W m^{-2}, and its rate of increase during the industrial era is *very likely* to have been unprecedented in more than 10,000 years (see Figures

The understanding of anthropogenic warming and cooling influences on climate has improved since the Third Assessment Report (TAR), leading to *very high confidence*[7] that the globally averaged net effect of human activities since 1750 has been one of warming, with a radiative forcing of +1.6 [+0.6 to +2.4] W m^{-2}. (see Figure SPM-2). {2.3. 6.5, 2.9}

SPM-1 and SPM-2). The carbon dioxide radiative forcing increased by 20% from 1995 to 2005, the largest change for any decade in at least the last 200 years. {2.3, 6.4}

- Anthropogenic contributions to aerosols (primarily sulphate, organic carbon, black carbon, nitrate and dust) together produce a cooling effect, with a total direct radiative forcing of −0.5 [−0.9 to −0.1] W m^{-2} and an indirect cloud albedo forcing of −0.7 [−1.8 to −0.3] W m^{-2}. These forcings are now better understood than at the time of the TAR due to improved *in situ,* satellite and ground-based measurements and more comprehensive modelling, but remain the dominant uncertainty in radiative forcing. Aerosols also influence cloud lifetime and precipitation. {2.4, 2.9, 7.5}

- Significant anthropogenic contributions to radiative forcing come from several other sources. Tropospheric ozone changes due to emissions of ozone-forming chemicals (nitrogen oxides, carbon monoxide, and hydrocarbons) contribute +0.35 [+0.25 to +0.65] W m^{-2}. The direct radiative forcing due to changes in halocarbons[8] is +0.34 [+0.31 to +0.37] W m^{-2}. Changes in surface albedo, due to land-cover changes and deposition of black carbon aerosols on snow, exert respective forcings of −0.2 [−0.4 to 0.0] and +0.1 [0.0 to +0.2] W m^{-2}. Additional terms smaller than ∓ 0.1 W m^{-2} are shown in Figure SPM-2. {2.3, 2.5, 7.2}

- Changes in solar irradiance since 1750 are estimated to cause a radiative forcing of +0.12 [+0.06 to +0.30] W m^{-2}, which is less than half the estimate given in the TAR. {2.7}

Direct Observations Of Recent Climate Change

Since the TAR, progress in understanding how climate is changing in space and in time has been gained through improvements and extensions of numerous datasets and data analyses, broader geographical coverage, better understanding of uncertainties, and a wider variety of measurements. Increasingly comprehensive observations are available for glaciers and snow cover since the 1960s, and for sea level and ice sheets since about the past decade. However, data coverage remains limited in some regions.

- Eleven of the last twelve years (1995–2006) rank among the 12 warmest years in the instrumental record of global surface temperature[9] (since 1850).

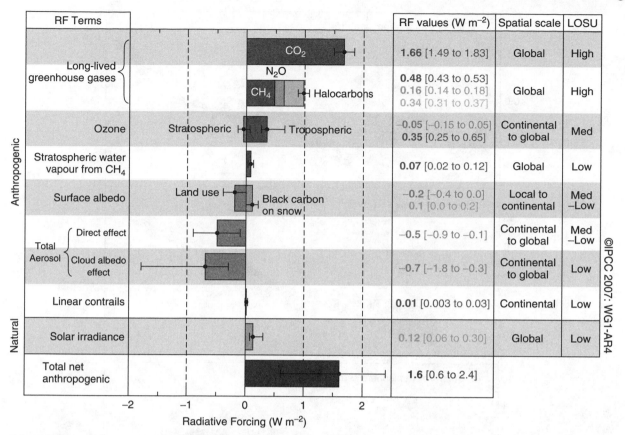

FIGURE SPM-2 Radiative forcing components. Global-average radiative forcing (RF) estimates and ranges in 2005 for anthropogenic carbon dioxide (CO_2), methane (CH_4), nitrous oxide (N_2O) and other important agents and mechanisms, together with the typical geographical extent (spatial scale) of the forcing and the assessed level of scientific understanding (LOSU). The net anthropogenic radiative forcing and its range are also shown. These require summing asymmetric uncertainty estimates from the component terms, and cannot be obtained by simple addition. Additional forcing factors not included here are considered to have a very low LOSU. Volcanic aerosols contribute an additional natural forcing but are not included in this figure due to their episodic nature. Range for linear contrails does not include other possible effects of aviation on cloudiness. {2.9, Figure 2.20}

Warming of the climate system is unequivocal, as is now evident from observations of increases in global average air and ocean temperatures, widespread melting of snow and ice, and rising global average sea level (see Figure SPM-3). {3.2, 4.2, 5.5}

The updated 100-year linear trend (1906–2005) of 0.74 [0.56 to 0.92]°C is therefore larger than the corresponding trend for 1901–2000 given in the TAR of 0.6 [0.4 to 0.8] °C. The linear warming trend over the last 50 years (0.13 [0.10 to 0.16] °C per decade) is nearly twice that for the last 100 years. The total temperature increase from 1850–1899 to 2001–2005 is 0.76 [0.57 to 0.95] °C. Urban heat island effects are real but local, and have a negligible influence (less than 0.006 °C per decade over land and zero over the oceans) on these values. {3.2}

- New analyses of balloon-borne and satellite measurements of lower- and mid-tropospheric temperature show warming rates that are similar to those of the surface temperature record and are consistent within their respective uncertainties, largely reconciling a discrepancy noted in the TAR. {3.2, 3.4}

- The average atmospheric water vapour content has increased since at least the 1980s over land and ocean as well as in the upper troposphere. The increase is broadly consistent with the extra water vapour that warmer air can hold. {3.4}

- Observations since 1961 show that the average temperature of the global ocean has increased to depths of at least 3000 m and that the ocean has been absorbing more than 80% of the heat added to the climate system. Such warming causes seawater to expand, contributing to sea level rise (see Table SPM-1). {5.2, 5.5}

- Mountain glaciers and snow cover have declined on average in both hemispheres. Widespread decreases in glaciers and ice caps have contributed to sea level rise (ice caps do not include contributions from the Greenland and Antarctic ice sheets). (See Table SPM-1.) {4.6, 4.7, 4.8, 5.5}

- New data since the TAR now show that losses from the ice sheets of Greenland and Antarctica have *very likely* contributed to sea level rise over 1993 to 2003

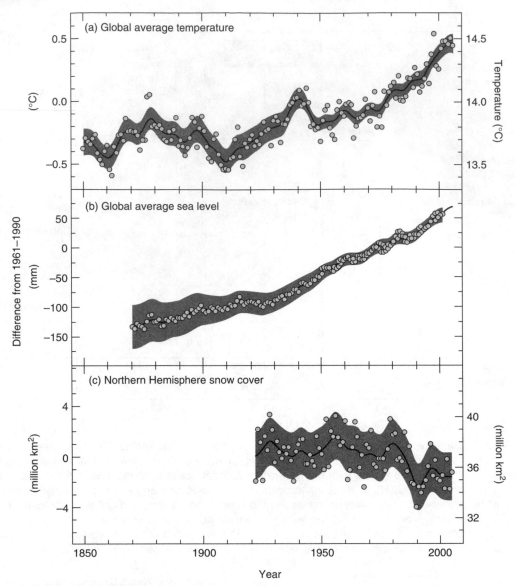

FIGURE SPM-3 Changes in temperature, sea level and northern hemisphere snow cover. The shaded areas are the uncertainty intervals estimated from a comprehensive analysis of known uncertainties (a and b) and from the time series (c). {FAQ 3.1, Figure 1, Figure 4.2 and Figure 5.13}

(see Table SPM-1). Flow speed has increased for some Greenland and Antarctic outlet glaciers, which drain ice from the interior of the ice sheets. The corresponding increased ice sheet mass loss has often followed thinning, reduction or loss of ice shelves or loss of floating glacier tongues. Such dynamical ice loss is sufficient to explain most of the Antarctic net mass loss and approximately half of the Greenland net mass loss. The remainder of the ice loss from Greenland has occurred because losses due to melting have exceeded accumulation due to snowfall. {4.6, 4.8, 5.5}

- Global average sea level rose at an average rate of 1.8 [1.3 to 2.3] mm per year over 1961 to 2003. The rate was faster over 1993 to 2003, about 3.1 [2.4 to 3.8] mm per year. Whether the faster rate for 1993 to 2003 reflects decadal variability or an increase in the longer-term

trend is unclear. There is *high confidence* that the rate of observed sea level rise increased from the 19th to the 20th century. The total 20th century rise is estimated to be 0.17 [0.12 to 0.22] m. {5.5}

- For 1993–2003, the sum of the climate contributions is consistent within uncertainties with the total sea level rise that is directly observed (see Table SPM-1). These estimates are based on improved satellite and *in-situ* data now available. For the period of 1961 to 2003, the sum of climate contributions is estimated to be smaller than the observed sea level rise. The TAR reported a similar discrepancy for 1910 to 1990. {5.5}

- Average Arctic temperatures increased at almost twice the global average rate in the past 100 years. Arctic temperatures have high decadal variability, and a warm period was also observed from 1925 to 1945. {3.2}

Table SPM-1   Observed rate of sea level rise and estimated contributions from different sources. {5.5, Table 5.3}

Source of sea level rise	Rate of sea level rise (mm per year)	
	1961–2003	1993–2003
Thermal expansion	0.42 ∓ 0.12	1.6 ∓ 0.5
Glaciers and ice caps	0.50 ∓ 0.18	0.77 ∓ 0.22
Greenland ice sheet	0.05 ∓ 0.12	0.21 ∓ 0.07
Antarctic ice sheet	0.14 ∓ 0.41	0.21 ∓ 0.35
Sum of individual climate contributions to sea level rise	1.1 ∓ 0.5	2.8 ∓ 0.7
Observed total sea level rise	1.8 ∓ 0.5[a]	3.1 ∓ 0.7[a]
Difference (Observed minus sum of estimated climate contributions)	0.7 ∓ 0.7	0.3 ∓ 1.0

[a]Data prior to 1993 are from tide gauges and after 1993 are from satellite altimetry.

At continental, regional, and ocean basin scales, numerous long-term changes in climate have been observed. These include changes in Arctic temperatures and ice, widespread changes in precipitation amounts, ocean salinity, wind patterns and aspects of extreme weather including droughts, heavy precipitation, heat waves and the intensity of tropical cyclones[10]. {3.2, 3.3, 3.4, 3.5, 3.6, 5.2}

- Satellite data since 1978 show that annual average Arctic sea ice extent has shrunk by 2.7 [2.1 to 3.3]% per decade, with larger decreases in summer of 7.4 [5.0 to 9.8]% per decade. These values are consistent with those reported in the TAR. {4.4}

- Temperatures at the top of the permafrost layer have generally increased since the 1980s in the Arctic (by up to 3 °C). The maximum area covered by seasonally frozen ground has decreased by about 7% in the Northern Hemisphere since 1900, with a decrease in spring of up to 15% {4.7}

- Long-term trends from 1900 to 2005 have been observed in precipitation amount over many large regions[11]. Significantly increased precipitation has been observed in eastern parts of North and South America, northern Europe and northern and central Asia. Drying has been observed in the Sahel, the Mediterranean, southern Africa and parts of southern Asia. Precipitation is highly variable spatially and temporally, and data are limited in some regions. Long-term trends have not been observed for the other large regions assessed[11]. {3.3, 3.9}

- Changes in precipitation and evaporation over the oceans are suggested by freshening of mid and high latitude waters together with increased salinity in low latitude waters. {5.2}

- Mid-latitude westerly winds have strengthened in both hemispheres since the 1960s. {3.5}

- More intense and longer droughts have been observed over wider areas since the 1970s, particularly in the tropics and subtropics. Increased drying linked with higher temperatures and decreased precipitation have contributed to changes in drought. Changes in sea surface temperatures (SST), wind patterns, and decreased snowpack and snow cover have also been linked to droughts. {3.3}

- The frequency of heavy precipitation events has increased over most land areas, consistent with warming and observed increases of atmospheric water vapour. {3.8, 3.9}

- Widespread changes in extreme temperatures have been observed over the last 50 years. Cold days, cold nights and frost have become less frequent, while hot days, hot nights, and heat waves have become more frequent (see Table SPM-2). {3.8}

- There is observational evidence for an increase of intense tropical cyclone activity in the North Atlantic since about 1970, correlated with increases of tropical sea surface temperatures. There are also suggestions of increased intense tropical cyclone activity in some other regions where concerns over data quality are greater. Multi-decadal variability and the quality of the tropical cyclone records prior to routine satellite observations in about 1970 complicate the detection of long-term trends in tropical cyclone activity. There is no clear trend in the annual numbers of tropical cyclones. {3.8}

- A decrease in diurnal temperature range (DTR) was reported in the TAR, but the data available then extended only from 1950 to 1993. Updated observations reveal that DTR has not changed from 1979 to 2004 as both day- and night-time temperature have risen at about the same rate. The trends are highly variable from one region to another. {3.2}

- Antarctic sea ice extent continues to show inter-annual variability and localized changes but no statistically

Table SPM-2   Recent trends, assessment of human influence on the trend, and projections for extreme weather events for which there is an observed late 20th century trend. {Tables 3.7, 3.8, 9.4, Sections 3.8, 5.5, 9.7, 11.2–11.9}

Phenomenon [a] and direction of trend	Likelihood that trend occurred in late 20th century (typically post 1960)	Likelihood of a human contribution to observed trend [b]	Likelihood of future trends based on projections for 21st century using SRES scenarios
Warmer and fewer cold days and nights over most land areas	Very likely [c]	Likely [d]	Virtually certain [d]
Warmer and more frequent hot days and nights over most land areas	Very likely [e]	Likely (nights) [d]	Virtually certain [d]
Warm spells / heat waves. Frequency increases over most land areas	Likely	More likely than not [f]	Very likely
Heavy precipitation events. Frequency (or proportion of total rainfall from heavy falls) increases over most areas	Likely	More likely than not [f]	Very likely
Area affected by droughts increases	Likely in many regions since 1970s	More likely than not	Likely
Intense tropical cyclone activity increases	Likely in some regions since 1970	More likely than not [f]	Likely
Increased incidence of extreme high sea level (excludes tsunamis) [g]	Likely	More likely than not [f,h]	Likely [i]

[a] See Table 3.7 for further details regarding definitions.

[b] See Table TS-4, Box TS.3.4 and Table 9.4.

[c] Decreased frequency of cold days and nights (coldest 10%).

[d] Warming of the most extreme days and nights each year.

[e] Increased frequency of hot days and nights (hottest 10%).

[f] Magnitude of anthropogenic contributions not assessed. Attribution for these phenomena based on expert judgement rather than formal attribution studies.

[g] Extreme high sea level depends on average sea level and on regional weather systems. It is defined here as the highest 1% of hourly values of observed sea level at a station for a given reference period.

[h] Changes in observed extreme high sea level closely follow the changes in average sea level {5.5.2.6}. It is *very likely* that anthropogenic activity contributed to a rise in average sea level. {9.5.2}

[i] In all scenarios, the projected global average sea level at 2100 is higher than in the reference period {10.6}. The effect of changes in regional weather systems on sea level extremes has not been assessed.

> Some aspects of climate have not been observed to change. {3.2, 3.8, 4.4, 5.3}

significant average trends, consistent with the lack of warming reflected in atmospheric temperatures averaged across the region. {3.2, 4.4}

- There is insufficient evidence to determine whether trends exist in the meridional overturning circulation of the global ocean or in small scale phenomena such as tornadoes, hail, lightning and dust-storms. {3.8, 5.3}

A Paleoclimatic Perspective

Paleoclimatic studies use changes in climatically sensitive indicators to infer past changes in global climate on time scales ranging from decades to millions of years. Such proxy data (e.g., tree ring width) may be influenced by both local temperature and other factors such as precipitation, and are often representative of particular seasons rather than full years. Studies since the TAR draw increased confidence from additional data showing coherent behaviour across multiple indicators in different parts of the world. However, uncertainties generally increase with time into the past due to increasingly limited spatial coverage.

Paleoclimate information supports the interpretation that the warmth of the last half century is unusual in at least the previous 1300 years. The last time the polar regions were significantly warmer than present for an extended period (about 125,000 years ago), reductions in polar ice volume led to 4 to 6 metres of sea level rise. {6.4, 6.6}

- Average Northern Hemisphere temperatures during the second half of the 20th century were *very likely* higher than during any other 50-year period in the last 500 years and *likely* the highest in at least the past 1300 years. Some recent studies indicate greater variability in Northern Hemisphere temperatures than suggested in the TAR, particularly finding that cooler periods existed in the 12 to 14th, 17th, and 19th centuries. Warmer periods prior to the 20th century are within the uncertainty range given in the TAR. {6.6}

- Global average sea level in the last interglacial period (about 125,000 years ago) was *likely* 4 to 6 m higher than during the 20th century, mainly due to the retreat of polar ice. Ice core data indicate that average polar temperatures at that time were 3 to 5 °C higher than present, because of differences in the Earth's orbit. The Greenland ice sheet and other Arctic ice fields *likely* contributed no more than 4 m of the observed sea level rise. There may also have been a contribution from Antarctica. {6.4}

Understanding and Attributing Climate Change

This Assessment considers longer and improved records, an expanded range of observations, and improvements in the simulation of many aspects of climate and its variability based on studies since the TAR. It also considers the results of new attribution studies that have evaluated whether observed changes are quantitatively consistent with the expected response to external forcings and inconsistent with alternative physically plausible explanations.

Most of the observed increase in globally averaged temperatures since the mid-20th century is *very likely* due to the observed increase in anthropogenic greenhouse gas concentrations[12]. This is an advance since the TAR's conclusion that "most of the observed warming over the last 50 years is *likely* to have been due to the increase in greenhouse gas concentrations". Discernible human influences now extend to other aspects of climate, including ocean warming, continental-average temperatures, temperature extremes and wind patterns (see Figure SPM-4 and Table SPM-2). {9.4, 9.5}

- It is *likely* that increases in greenhouse gas concentrations alone would have caused more warming than observed because volcanic and anthropogenic aerosols have offset some warming that would otherwise have taken place. {2.9, 7.5, 9.4}

- The observed widespread warming of the atmosphere and ocean, together with ice mass loss, support the conclusion that it is *extremely unlikely* that global climate change of the past fifty years can be explained without external forcing, and *very likely* that it is not due to known natural causes alone. {4.8, 5.2, 9.4, 9.5, 9.7}

- Warming of the climate system has been detected in changes of surface and atmospheric temperatures, temperatures in the upper several hundred metres of the ocean and in contributions to sea level rise. Attribution studies have established anthropogenic contributions to all of these changes. The observed pattern of tropospheric warming and stratospheric cooling is *very likely* due to the combined influences of greenhouse gas increases and stratospheric ozone depletion. {3.2, 3.4, 9.4, 9.5}

- It is *likely* that there has been significant anthropogenic warming over the past 50 years averaged over each continent except Antarctica (see Figure SPM-4). The observed patterns of warming, including greater warming over land than over the ocean, and their changes over time, are only simulated by models that include anthropogenic forcing. The ability of coupled climate models to simulate the observed temperature evolution on each of six continents provides stronger evidence of human influence on climate than was available in the TAR. {3.2, 9.4}

- Difficulties remain in reliably simulating and attributing observed temperature changes at smaller scales. On these scales, natural climate variability is relatively larger making it harder to distinguish changes expected due to external forcings. Uncertainties in local forcings and feedbacks also make it difficult to estimate the contribution of greenhouse gas increases to observed small-scale temperature changes. {8.3, 9.4}

- Anthropogenic forcing is *likely* to have contributed to changes in wind patterns[13], affecting extra-tropical storm tracks and temperature patterns in both hemispheres. However, the observed changes in the Northern Hemisphere circulation are larger than simulated in response to 20th century forcing change. {3.5, 3.6, 9.5, 10.3}

- Temperatures of the most extreme hot nights, cold nights and cold days are *likely* to have increased due to anthropogenic forcing. It is *more likely than not* that anthropogenic forcing has increased the risk of heat waves (see Table SPM-2). {9.4}

- The equilibrium climate sensitivity is a measure of the climate system response to sustained radiative forcing. It is not a projection but is defined as the global average surface warming following a doubling of carbon dioxide concentrations. It is *likely* to be in the range 2 to 4.5 °C with a best estimate of about 3 °C, and is *very unlikely*

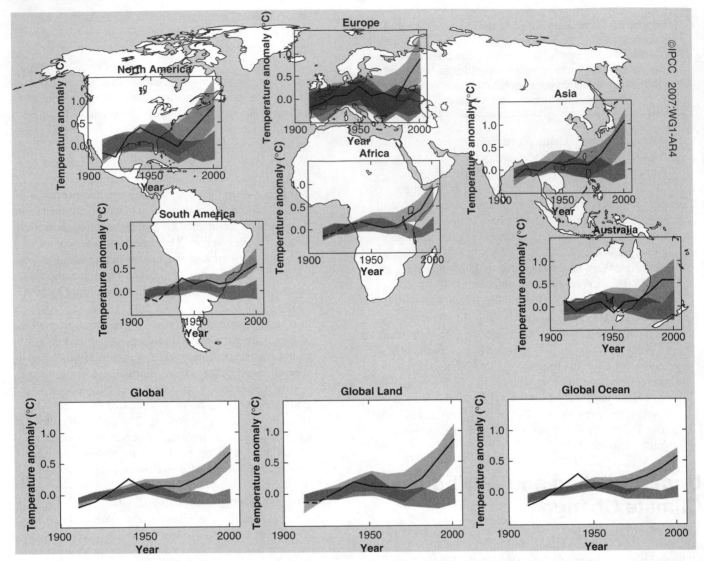

FIGURE SPM-4. Global and continental temperature change. Comparison of observed continental- and global-scale changes in surface temperature with results simulated by climate models using natural and anthropogenic forcings. Decadal averages of observations are shown for the period 1906–2005 (black line) plotted against the centre of the decade and relative to the corresponding average for 1901–1950. Lines are dashed where spatial coverage is less than 50%. Light gray shaded bands show the 5–95% range for 19 simulations from 5 climate models using only the natural forcings due to solar activity and volcanoes. Medium gray shaded bands show the 5–95% range for 58 simulations from 14 climate models using both natural and anthropogenic forcings. {FAQ 9.2, Figure 1}

Analysis of climate models together with constraints from observations enables an assessed *likely* range to be given for climate sensitivity for the first time and provides increased confidence in the understanding of the climate system response to radiative forcing. {6.6, 8.6, 9.6, Box 10.2}

to be less than 1.5 °C. Values substantially higher than 4.5 °C cannot be excluded, but agreement of models with observations is not as good for those values. Water vapour changes represent the largest feedback affecting climate sensitivity and are now better understood than in the TAR. Cloud feedbacks remain the largest source of uncertainty. {8.6, 9.6, Box 10.2}

- It is *very unlikely* that climate changes of at least the seven centuries prior to 1950 were due to variability generated within the climate system alone. A significant fraction of the reconstructed Northern Hemisphere interdecadal temperature variability over those centuries is *very likely* attributable to volcanic eruptions and changes in solar irradiance, and it is *likely* that anthropogenic forcing contributed to the early 20th century warming evident in these records. {2.7, 2.8, 6.6, 9.3}

Projections of Future Changes in Climate

A major advance of this assessment of climate change projections compared with the TAR is the large number of simulations available from a broader range of models. Taken together with additional information from observations, these provide a quantitative basis for estimating likelihoods for many aspects of future climate change. Model simulations cover a range of possible futures including idealised emission or concentration assumptions. These include SRES[14] illustrative marker scenarios for the 2000–2100 period and model experiments with greenhouse gases and aerosol concentrations held constant after year 2000 or 2100.

> For the next two decades a warming of about 0.2 °C per decade is projected for a range of SRES emission scenarios. Even if the concentrations of all greenhouse gases and aerosols had been kept constant at year 2000 levels, a further warming of about 0.1 °C per decade would be expected. {10.3, 10.7}

- Since IPCC's first report in 1990, assessed projections have suggested global averaged temperature increases between about 0.15 and 0.3 °C per decade for 1990 to 2005. This can now be compared with observed values of about 0.2 °C per decade, strengthening confidence in near-term projections. {1.2, 3.2}

- Model experiments show that even if all radiative forcing agents are held constant at year 2000 levels, a further warming trend would occur in the next two decades at a rate of about 0.1 °C per decade, due mainly to the slow response of the oceans. About twice as much warming (0.2 °C per decade) would be expected if emissions are within the range of the SRES scenarios. Best-estimate projections from models indicate that decadal-average warming over each inhabited continent by 2030 is insensitive to the choice among SRES scenarios and is *very likely* to be at least twice as large as the corresponding model-estimated natural variability during the 20th century. {9.4, 10.3, 10.5, 11.2–11.7, Figure TS-29}

- Advances in climate change modelling now enable best estimates and *likely* assessed uncertainty ranges to be given for projected warming for different emission scenarios. Results for different emission scenarios are provided explicitly in this report to avoid loss of this policy-relevant information. Projected globally-averaged surface warmings for the end of the 21st century (2090–2099) relative to 1980–1999 are shown in Table SPM-3. These illustrate the differences between lower to higher SRES emission scenarios and the projected warming uncertainty associated with these scenarios. {10.5}

> Continued greenhouse gas emissions at or above current rates would cause further warming and induce many changes in the global climate system during the 21st century that would *very likely* be larger than those observed during the 20th century. {10.3}

- Best estimates and *likely* ranges for globally average surface air warming for six SRES emissions marker scenarios are given in this assessment and are shown in Table SPM-3. For example, the best estimate for the low scenario (B1) is 1.8 °C (*likely* range is 1.1 °C to 2.9 °C), and the best estimate for the high scenario (A1FI) is 4.0 °C (*likely* range is 2.4 °C to 6.4 °C). Although these projections are broadly consistent with the span quoted in the TAR (1.4 to 5.8 °C), they are not directly comparable (see Figure SPM-5). The AR4 is more advanced as it provides best estimates and an assessed likelihood range for each of the marker scenarios. The new assessment of the *likely* ranges now relies on a larger number of climate models of increasing complexity and realism, as well as new information regarding the nature of feedbacks from the carbon cycle and constraints on climate response from observations. {10.5}

- Warming tends to reduce land and ocean uptake of atmospheric carbon dioxide, increasing the fraction of anthropogenic emissions that remains in the atmosphere. For the A2 scenario, for example, the climate-carbon cycle feedback increases the corresponding global average warming at 2100 by more than 1 °C. Assessed upper ranges for temperature projections are larger than in the TAR (see Table SPM-3) mainly because the broader range of models now available suggests stronger climate-carbon cycle feedbacks. {7.3, 10.5}

- Model-based projections of global average sea level rise at the end of the 21st century (2090–2099) are shown in Table SPM-3. For each scenario, the midpoint of the range in Table SPM-3 is within 10% of the TAR model average for 2090–2099. The ranges are narrower than in the TAR mainly because of improved information about some uncertainties in the projected contributions[15]. {10.6}

- Models used to date do not include uncertainties in climate-carbon cycle feedback nor do they include the full effects of changes in ice sheet flow, because a basis in published literature is lacking. The projections include a contribution due to increased ice flow from Greenland and Antarctica at the rates observed for 1993–2003, but these flow rates could increase or decrease in the future. For example, if this contribution were to grow linearly with global average temperature change, the upper ranges of sea level rise for SRES scenarios shown in Table SPM-3 would increase by 0.1 m to 0.2 m. Larger values cannot be excluded, but understanding of these effects is too limited to assess

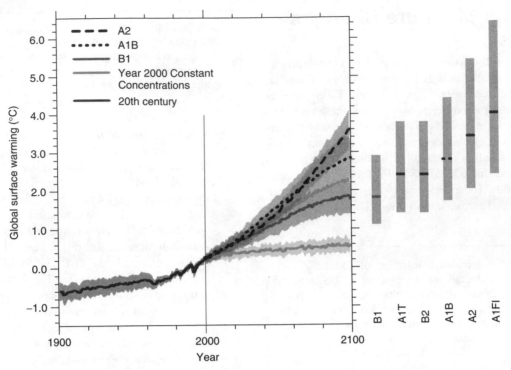

Figure Spm-5 Multi-model averages and assessed ranges for surface warming. Solid lines are multi-model global averages of surface warming (relative to 1980–99) for the scenarios A2, A1B and B1, shown as continuations of the 20th century simulations. Shading denotes the plus/minus one standard deviation range of individual model annual averages. The orange line is for the experiment where concentrations were held constant at year 2000 values. The gray bars at right indicate the best estimate (solid line within each bar) and the *likely* range assessed for the six SRES marker scenarios. The assessment of the best estimate and *likely* ranges in the gray bars includes the AOGCMs in the left part of the figure, as well as results from a hierarchy of independent models and observational constraints. {Figures 10.4 and 10.29}

Table SPM-3   Projected globally averaged surface warming and sea level rise at the end of the 21st century. {10.5, 10.6, Table 10.7}

Case	Temperature Change (°C at 2090–2099 relative to 1980–1999) [a]		Sea Level Rise (m at 2090–2099 relative to 1980–1999)
	Best estimate	*Likely* range	Model-based range excluding future rapid dynamical changes in ice flow
Constant Year 2000 concentrations [b]	0.6	0.3–0.9	NA
B1 scenario	1.8	1.1–2.9	0.18–0.38
A1T scenario	2.4	1.4–3.8	0.20–0.45
B2 scenario	2.4	1.4–3.8	0.20–0.43
A1B scenario	2.8	1.7–4.4	0.21–0.48
A2 scenario	3.4	2.0–5.4	0.23–0.51
A1FI scenario	4.0	2.4–6.4	0.26–0.59

[a]These estimates are assessed from a hierarchy of models that encompass a simple climate model, several Earth Models of Intermediate Complexity (EMICs), and a large number of Atmosphere-Ocean Global Circulaion Models (AOGCMs).

[b]Year 2000 constant composition is derived from AOGCMs only.

their likelihood or provide a best estimate or an upper bound for sea level rise. {10.6}

- Increasing atmospheric carbon dioxide concentrations leads to increasing acidification of the ocean. Projections based on SRES scenarios give reductions in average global surface ocean pH[16] of between 0.14 and 0.35 units over the 21st century, adding to the present decrease of 0.1 units since pre-industrial times. {5.4, Box 7.3, 10.4}

> There is now higher confidence in projected patterns of warming and other regional-scale features, including changes in wind patterns, precipitation, and some aspects of extremes and of ice. {8.2, 8.3, 8.4, 8.5, 9.4, 9.5, 10.3, 11.1}

- Projected warming in the 21st century shows scenario-independent geographical patterns similar to those observed over the past several decades. Warming is expected to be greatest over land and at most high northern latitudes, and least over the Southern Ocean and parts of the North Atlantic ocean (see Figure SPM-6). {10.3}
- Snow cover is projected to contract. Widespread increases in thaw depth are projected over most permafrost regions. {10.3, 10.6}
- Sea ice is projected to shrink in both the Arctic and Antarctic under all SRES scenarios. In some projections, Arctic late-summer sea ice disappears almost entirely by the latter part of the 21st century. {10.3}
- It is *very likely* that hot extremes, heat waves, and heavy precipitation events will continue to become more frequent. {10.3}
- Based on a range of models, it is *likely* that future tropical cyclones (typhoons and hurricanes) will become more intense, with larger peak wind speeds and more heavy precipitation associated with ongoing increases of tropical SSTs. There is less confidence in projections of a global decrease in numbers of tropical cyclones. The apparent increase in the proportion of very intense storms since 1970 in some regions is much larger than simulated by current models for that period. {9.5, 10.3, 3.8}
- Extra-tropical storm tracks are projected to move poleward, with consequent changes in wind, precipitation, and temperature patterns, continuing the broad pattern of observed trends over the last half-century. {3.6, 10.3}
- Since the TAR there is an improving understanding of projected patterns of precipitation. Increases in the amount of precipitation are *very likely* in high-latitudes, while decreases are *likely* in most subtropical land regions (by as much as about 20% in the A1B scenario in 2100, see Figure SPM-7), continuing

observed patterns in recent trends. {3.3, 8.3, 9.5, 10.3, 11.2 to 11.9}

- Based on current model simulations, it is *very likely* that the meridional overturning circulation (MOC) of the Atlantic Ocean will slow down during the 21st century. The multi-model average reduction by 2100 is 25% (range from zero to about 50%) for SRES emission scenario A1B. Temperatures in the Atlantic region are projected to increase despite such changes due to the much larger warming associated with projected increases of greenhouse gases. It is *very unlikely* that the MOC will undergo a large abrupt transition during the 21st century. Longer-term changes in the MOC cannot be assessed with confidence. {10.3, 10.7}

> Anthropogenic warming and sea level rise would continue for centuries due to the timescales associated with climate processes and feedbacks, even if greenhouse gas concentrations were to be stabilized. {10.4, 10.5, 10.7}

- Climate carbon cycle coupling is expected to add carbon dioxide to the atmosphere as the climate system warms, but the magnitude of this feedback is uncertain. This increases the uncertainty in the trajectory of carbon dioxide emissions required to achieve a particular stabilisation level of atmospheric carbon dioxide concentration. Based on current understanding of climate carbon cycle feedback, model studies suggest that to stabilise at 450 ppm carbon dioxide, could require that cumulative emissions over the 21st century be reduced from an average of approximately 670 [630 to 710] GtC (2460 [2310 to 2600] GtCO$_2$) to approximately 490 [375 to 600] GtC (1800 [1370 to 2200] GtCO$_2$). Similarly, to stabilise at 1000 ppm this feedback could require that cumulative emissions be reduced from a model average of approximately 1415 [1340 to 1490] GtC (5190 [4910 to 5460] GtCO$_2$) to approximately 1100 [980 to 1250] GtC (4030 [3590 to 4580] GtCO$_2$). {7.3, 10.4}
- If radiative forcing were to be stabilized in 2100 at B1 or A1B levels[11] a further increase in global average temperature of about 0.5 °C would still be expected, mostly by 2200. {10.7}
- If radiative forcing were to be stabilized in 2100 at A1B levels[11], thermal expansion alone would lead to 0.3 to 0.8 m of sea level rise by 2300 (relative to 1980–1999). Thermal expansion would continue for many centuries, due to the time required to transport heat into the deep ocean. {10.7}
- Contraction of the Greenland ice sheet is projected to continue to contribute to sea level rise after 2100. Current models suggest ice mass losses increase with temperature more rapidly than gains due to precipitation and that the surface mass balance becomes negative

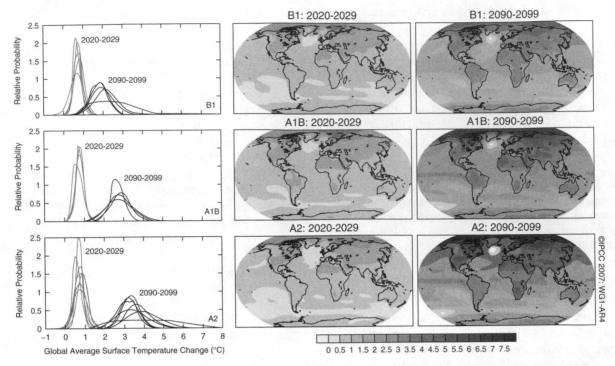

Figure SPM-6 AOGCM projections of surface temperatures.   Projected surface temperature changes for the early and late 21st century relative to the period 1980–1999. The central and right panels show the Atmosphere-Ocean General Circulation multi-Model average projections for the B1 (top), A1B (middle) and A2 (bottom) SRES scenarios averaged over decades 2020–2029 (center) and 2090–2099 (right). The left panel shows corresponding uncertainties as the relative probabilities of estimated global average warming from several different AOGCM and EMICs studies for the same periods. Some studies present results only for a subset of the SRES scenarios, or for various model versions. Therefore the difference in the number of curves, shown in the left-hand panels, is due only to differences in the availability of results. {Figures 10.8 and 10.28}

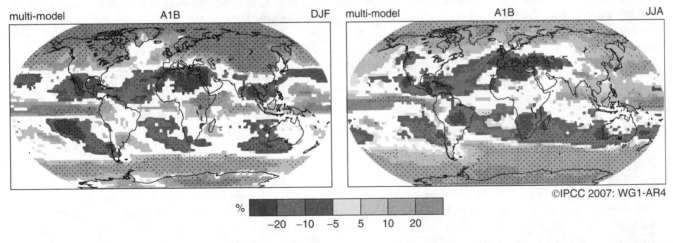

Figure SPM-7 Projected patterns of precipitation changes.   Relative changes in precipitation (in percent) for the period 2090–2099, relative to 1980–1999. Values are multi-model averages based on the SRES A1B scenario for December to February (left) and June to August (right). White areas are where less than 66% of the models agree in the sign of the change and stippled areas are where more than 90% of the models agree in the sign of the change. {Figure 10.9}

at a global average warming (relative to pre-industrial values) in excess of 1.9 to 4.6 °C. If a negative surface mass balance were sustained for millennia, that would lead to virtually complete elimination of the Greenland ice sheet and a resulting contribution to sea level rise

of about 7 m. The corresponding future temperatures in Greenland are comparable to those inferred for the last interglacial period 125,000 years ago, when paleoclimatic information suggests reductions of polar land ice extent and 4 to 6 m of sea level rise. {6.4, 10.7}

The Emission Scenarios of the IPCC Special Report on Emission Scenarios (SRES) [17]

A1. The A1 storyline and scenario family describes a future world of very rapid economic growth, global population that peaks in mid-century and declines thereafter, and the rapid introduction of new and more efficient technologies. Major underlying themes are convergence among regions, capacity building and increased cultural and social interactions, with a substantial reduction in regional differences in per capita income. The A1 scenario family develops into three groups that describe alternative directions of technological change in the energy system. The three A1 groups are distinguished by their technological emphasis: fossil intensive (A1FI), non-fossil energy sources (A1T), or a balance across all sources (A1B) (where balanced is defined as not relying too heavily on one particular energy source, on the assumption that similar improvement rates apply to all energy supply and end use technologies).

A2. The A2 storyline and scenario family describes a very heterogeneous world. The underlying theme is self reliance and preservation of local identities. Fertility patterns across regions converge very slowly, which results in continuously increasing population. Economic development is primarily regionally oriented and per capita economic growth and technological change more fragmented and slower than other storylines.

B1. The B1 storyline and scenario family describes a convergent world with the same global population, that peaks in mid-century and declines thereafter, as in the A1 storyline, but with rapid change in economic structures toward a service and information economy, with reductions in material intensity and the introduction of clean and resource efficient technologies. The emphasis is on global solutions to economic, social and environmental sustainability, including improved equity, but without additional climate initiatives.

B2. The B2 storyline and scenario family describes a world in which the emphasis is on local solutions to economic, social and environmental sustainability. It is a world with continuously increasing global population, at a rate lower than A2, intermediate levels of economic development, and less rapid and more diverse technological change than in the B1 and A1 storylines. While the scenario is also oriented towards environmental protection and social equity, it focuses on local and regional levels.

An illustrative scenario was chosen for each of the six scenario groups A1B, A1FI, A1T, A2, B1 and B2. All should be considered equally sound.

The SRES scenarios do not include additional climate initiatives, which means that no scenarios are included that explicitly assume implementation of the United Nations Framework Convention on Climate Change or the emissions targets of the Kyoto Protocol.

- Dynamical processes related to ice flow not included in current models but suggested by recent observations could increase the vulnerability of the ice sheets to warming, increasing future sea level rise. Understanding of these processes is limited and there is no consensus on their magnitude. {4.6, 10.7}
- Current global model studies project that the Antarctic ice sheet will remain too cold for widespread surface melting and is expected to gain in mass due to increased snowfall. However, net loss of ice mass could occur if dynamical ice discharge dominates the ice sheet mass balance. {10.7}
- Both past and future anthropogenic carbon dioxide emissions will continue to contribute to warming and sea level rise for more than a millennium, due to the timescales required for removal of this gas from the atmosphere. {7.3, 10.3}

Notes

1. *Climate change* in IPCC usage refers to any change in climate over time, whether due to natural variability or as a result of human activity. This usage differs from that in the Framework Convention on Climate Change, where climate change refers to a change of climate that is attributed directly or indirectly to human activity that alters the composition of the global atmosphere and that is in addition to natural climate variability observed over comparable time periods.

2. *Radiative forcing* is a measure of the influence that a factor has in altering the balance of incoming and outgoing energy in the Earth-atmosphere system and is an index of the importance of the factor as a potential climate change mechanism. Positive forcing tends to warm the surface while negative forcing tends to cool it. In this report radiative forcing values are for 2005 relative to pre-industrial conditions defined at 1750 and are expressed in watts per square metre ($W\ m^{-2}$). See Glossary and Section 2.2 for further details.

3. ppm (parts per million) or ppb (parts per billion, 1 billion = 1,000 million) is the ratio of the number of greenhouse gas molecules to the total number of molecules of dry air. For example: 300 ppm means 300 molecules of a greenhouse gas per million molecules of dry air.

4. Fossil carbon dioxide emissions include those from the production, distribution and consumption of fossil fuels and as a by-product from cement production. An emission of 1 GtC corresponds to 3.67 $GtCO_2$.

5. In general, uncertainty ranges for results given in this Summary for Policymakers are 90% uncertainty intervals unless stated otherwise, i.e., there is an estimated 5% likelihood that the

value could be above the range given in square brackets and 5% likelihood that the value could be below that range. Best estimates are given where available. Assessed uncertainty intervals are not always symmetric about the corresponding best estimate. Note that a number of uncertainty ranges in the Working Group I TAR corresponded to 2-sigma (95%), often using expert judgement.

6. In this Summary for Policymakers, the following terms have been used to indicate the assessed likelihood, using expert judgement, of an outcome or a result: *Virtually certain* > 99% probability of occurrence, *Extremely likely* > 95%, *Very likely* > 90%, *Likely* > 66%, *More likely than not* > 50%, *Unlikely* < 33%, *Very unlikely* < 10%, *Extremely unlikely* < 5%. (See Box TS.1.1 for more details).

7. In this Summary for Policymakers the following levels of confidence have been used to express expert judgments on the correctness of the underlying science: *very high confidence* at least a 9 out of 10 chance of being correct; *high confidence* about an 8 out of 10 chance of being correct. (See Box TS.1.1)

8. Halocarbon radiative forcing has been recently assessed in detail in IPCC's Special Report on Safeguarding the Ozone Layer and the Global Climate System (2005).

9. The average of near surface air temperature over land, and sea surface temperature.

10. Tropical cyclones include hurricanes and typhoons.

11. The assessed regions are those considered in the regional projections Chapter of the TAR and in Chapter 11 of this Report.

12. Consideration of remaining uncertainty is based on current methodologies.

13. In particular, the Southern and Northern Annular Modes and related changes in the North Atlantic Oscillation. {3.6, 9.5, Box TS.3.1

14. SRES refers to the IPCC Special Report on Emission Scenarios (2000). The SRES scenario families and illustrative cases, which did not include additional climate initiatives, are summarized in a box at the end of this Summary for Policymakers. Approximate CO_2 equivalent concentrations corresponding to the computed radiative forcing due to anthropogenic greenhouse gases and aerosols in 2100 (see p. 823 of the TAR) for the SRES B1, A1T, B2, A1B, A2 and A1FI illustrative marker scenarios are about 600, 700, 800, 850, 1250 and 1550 ppm respectively. Scenarios B1, A1B, and A2 have been the focus of model inter-comparison studies and many of those results are assessed in this report.

15. TAR projections were made for 2100, whereas projections in this Report are for 2090–2099. The TAR would have had similar ranges to those in Table SPM-2 if it had treated the uncertainties in the same way.

16. Decreases in pH correspond to increases in acidity of a solution. See Glossary for further details.

17. Emission scenarios are not assessed in this Working Group One report of the IPCC. This box summarizing the SRES scenarios is taken from the TAR and has been subject to prior line by line approval by the Panel.

How Many Planets?
A Survey of the Global Environment
The Great Race

Growth need not be the enemy of greenery. But much more effort is required to make the two compatible, says Vijay Vaitheeswaran.

Sustainable development is a dangerously slippery concept. Who could possibly be against something that invokes such alluring images of untouched wildernesses and happy creatures? The difficulty comes in trying to reconcile the "development" with the "sustainable" bit: look more closely, and you will notice that there are no people in the picture.

That seems unlikely to stop a contingent of some of 60,000 world leaders, businessmen, activists, bureaucrats and journalists from travelling to South Africa next month for the UN-sponsored World Summit on Sustainable Development in Johannesburg. Whether the summit achieves anything remains to be seen, but at least it is asking the right questions. This survey will argue that sustainable development cuts to the heart of mankind's relationship with nature—or, as Paul Portney of Resources for the Future, an American think-tank, puts it, "the great race between development and degradation". It will also explain why there is reason for hope about the planet's future.

The best way known to help the poor today—economic growth—has to be handled with care, or it can leave a degraded or even devastated natural environment for the future. That explains why ecologists and economists have long held diametrically opposed views on development. The difficult part is to work out what we owe future generations, and how to reconcile that moral obligation with what we owe the poorest among us today.

It is worth recalling some of the arguments fielded in the run-up to the big Earth Summit in Rio de Janeiro a decade ago. A publication from UNESCO, a United Nations agency, offered the following vision of the future: "Every generation should leave water, air and soil resources as pure and unpolluted as when it came on earth. Each generation should leave undiminished all the species of animals it found existing on earth." Man, that suggests, is but a strand in the web of life, and the natural order is fixed and supreme. Put earth first, it seems to say.

Robert Solow, an economist at the Massachusetts Institute of Technology, replied at the time that this was "fundamentally the wrong way to go", arguing that the obligation to the future is "not to leave the world as we found it in detail, but rather to leave the option or the capacity to be as well off as we are." Implicit in that argument is the seemingly hard-hearted notion of "fungibility": that natural resources, whether petroleum or giant pandas, are substitutable.

Rio's Fatal Flaw

Champions of development and defenders of the environment have been locked in battle ever since a UN summit in Stockholm launched the sustainable-development debate three decades ago. Over the years, this debate often pitted indignant politicians and social activists from the poor world against equally indignant politicians and greens from the rich world. But by the time the Rio summit came along, it seemed they had reached a truce. With the help of a committee of grandees led by Gro Harlem Brundtland, a former Norwegian prime minister, the interested parties struck a deal in 1987: development and the environment, they declared, were inextricably linked. That compromise generated a good deal of euphoria. Green groups grew concerned over poverty, and development charities waxed lyrical about greenery. Even the World Bank joined in. Its World Development Report in 1992 gushed about "win-win" strategies, such as ending environmentally harmful subsidies, that would help both the economy and the environment.

By nearly universal agreement, those grand aspirations have fallen flat in the decade since that summit. Little headway has been made with environmental problems such as climate change and loss of biodiversity. Such progress as has been achieved has been largely due to three factors that this survey will explore in later sections: more decision-making at local level, technological innovation, and the rise of market forces in environmental matters.

The main explanation for the disappointment—and the chief lesson for those about to gather in South Africa—is that Rio overreached itself. Its participants were so anxious to reach a political consensus that they agreed to the Brundtland definition of sustainable development, which Daniel Esty of Yale University thinks has turned into "a buzz-word largely devoid of content". The biggest mistake, he reckons, is that it slides over the difficult trade-offs between environment and development in the real world. He is careful to note that there are plenty of cases where those goals are linked—but also many where they are not: "Environmental and economic policy goals are distinct, and the actions needed to achieve them are not the same."

No Such Thing as Win-Win

To insist that the two are "impossible to separate", as the Brundtland commission claimed, is nonsense. Even the World Bank now accepts that its much-trumpeted 1992 report was much too optimistic. Kristalina Georgieva, the Bank's director for the environment, echoes comments from various colleagues when she says: "I've never seen a real win-win in my life. There's always somebody, usually an elite group grabbing rents, that loses. And we've learned in the past decade that those losers fight hard to make sure that technically elegant win-win policies do not get very fat."

So would it be better to ditch the concept of sustainable development altogether? Probably not. Even people with their feet firmly planted on the ground think one aspect of it is worth salvaging: the emphasis on the future.

Nobody would accuse John Graham of jumping on green bandwagons. As an official in President George Bush's Office of Management and Budget, and previously as head of Harvard University's Centre for Risk Analysis, he has built a reputation for evidence-based policymaking. Yet he insists sustainable development is a worthwhile concept: "It's good therapy for the tunnel vision common in government ministries, as it forces integrated policymaking. In practical terms, it means that you have to take economic cost-benefit trade-offs into account in environmental laws, and keep environmental trade-offs in mind with economic development."

Jose Maria Figueres, a former president of Costa Rica, takes a similar view. "As a politician, I saw at first hand how often policies were dictated by short-term considerations such as elections or partisan pressure. Sustainability is a useful template to align short-term policies with medium- to long-term goals."

It is not only politicians who see value in saving the sensible aspects of sustainable development. Achim Steiner, head of the International Union for the Conservation of Nature, the world's biggest conservation group, puts it this way: "Let's be honest: greens and businesses do not have the same objective, but they can find common ground. We look for pragmatic ways to save species. From our own work on the ground on poverty, our members—be they bird watchers or passionate ecologists—have learned that 'sustainable use' is a better way to conserve."

Sir Robert Wilson, boss of Rio Tinto, a mining giant, agrees. He and other business leaders say it forces hard choices about the future out into the open: "I like this concept because it frames the trade-offs inherent in a business like ours. It means that single-issue activism is simply not as viable."

Kenneth Arrow and Larry Goulder, two economists at Stanford University, suggest that the old ideological enemies are converging: "Many economists now accept the idea that natural capital has to be valued, and that we need to account for ecosystem services. Many ecologists now accept that prohibiting everything in the name of protecting nature is not useful, and so are being selective." They think the debate is narrowing to the more empirical question of how far it is possible to substitute natural capital with the man-made sort, and specific forms of natural capital for one another.

The Job for Johannesburg

So what can the Johannesburg summit contribute? The prospects are limited. There are no big, set-piece political treaties to be signed as there were at Rio. America's acrimonious departure from the Kyoto Protocol, a UN treaty on climate change, has left a bitter taste in many mouths. And the final pre-summit gathering, held in early June in Indonesia, broke up in disarray. Still, the gathered worthies could usefully concentrate on a handful of areas where international co-operation can help deal with environmental problems. Those include improving access for the poor to cleaner energy and to safe drinking water, two areas where concerns about human health and the environmental overlap. If rich countries want to make progress, they must agree on firm targets and offer the money needed to meet them. Only if they do so will poor countries be willing to cooperate on problems such as global warming that rich countries care about.

That seems like a modest goal, but it just might get the world thinking seriously about sustainability once again. If the Johannesburg summit helps rebuild a bit of faith in international environmental cooperation, then it will have been worthwhile. Minimising the harm that future economic growth does to the environment will require the rich world to work hand in glove with the poor world—which seems nearly unimaginable in today's atmosphere poisoned by the shortcomings of Rio and Kyoto.

To understand why this matters, recall that great race between development and degradation. Mankind has stayed comfortably ahead in that race so far, but can it go on doing so? The sheer magnitude of the economic growth that is hoped for in the coming decades makes it seem inevitable that the clashes between mankind and nature will grow worse. Some are now asking whether all this economic growth is really necessary or useful in the first place, citing past advocates of the simple life.

"God forbid that India should ever take to industrialism after the manner of the West . . . It took Britain half the resources of the planet to achieve this prosperity. How many planets will a country like India require?", Mahatma Gandhi asked half a century ago. That question encapsulated the bundle of worries that haunts the sustainable-development debate to this day. Today, the vast majority of Gandhi's countrymen are still living the simple life—full of simple misery, malnourishment and material want. Grinding poverty, it turns out, is pretty sustainable.

If Gandhi were alive today, he might look at China next door and find that the country, once as poor as India, has been transformed beyond recognition by two decades of roaring economic growth. Vast numbers of people have been lifted out of poverty and into middle-class comfort. That could prompt him to reframe his question: how many planets will it take to satisfy China's needs if it ever achieves profligate America's affluence? One green group reckons the answer is three. The next section looks at the environmental data that might underpin such claims. It makes for alarming reading—though not for the reason that first springs to mind.

Flying Blind

It comes as a shock to discover how little information there is on the environment.

What is the true state of the planet? It depends from which side you are peering at it. "Things are really looking up," comes the cry from one corner (usually overflowing with economists and technologists), pointing to a set of rosy statistics. "Disaster is nigh," shouts the other corner (usually full of ecologists and environmental lobbyists), holding up a rival set of troubling indicators.

According to the optimists, the 20th century marked a period of unprecedented economic growth that lifted masses of people out of abject poverty. It also brought technological innovations such as vaccines and other advances in public health that tackled many preventable diseases. The result has been a breath-taking enhancement of human welfare and longer, better lives for people everywhere on earth.

At this point, the pessimists interject: "Ah, but at what ecological cost?" They note that the economic growth which made all these gains possible sprang from the rapid spread of industrialisation and its resource-guzzling cousins, urbanisation, motorisation and electrification. The earth provided the necessary raw materials, ranging from coal to pulp to iron. Its ecosystems—rivers, seas, the atmosphere—also absorbed much of the noxious fallout from that process. The sheer magnitude of ecological change resulting directly from the past century's economic activity is remarkable.

To answer that Gandhian question about how many planets it would take if everybody lived like the West, we need to know how much—or how little—damage the West's transformation from poverty to plenty has done to the planet to date. Economists point to the remarkable improvement in local air and water pollution in the rich world in recent decades. "It's Getting Better All the Time", a cheerful tract co-written by the late Julian Simon, insists that: "One of the greatest trends of the past 100 years has been the astonishing rate of

progress in reducing almost every form of pollution." The conclusion seems unavoidable: "Relax! If we keep growing as usual, we'll inevitably grow greener."

The ecologically minded crowd takes a different view. "GEO3", a new report from the United Nations Environment Programme, looks back at the past few decades and sees much reason for concern. Its thoughtful boss, Klaus Töpfer (a former German environment minister), insists that his report is not "a document of doom and gloom". Yet, in summing it up, UNEP decries "the declining environmental quality of planet earth", and wags a finger at economic prosperity: "Currently, one-fifth of the world's population enjoys high, some would say excessive, levels of affluence." The conclusion seems unavoidable: "Panic! If we keep growing as usual, we'll inevitably choke the planet to death."

"People and Ecosystems", a collaboration between the World Resources Institute, the World Bank and the United Nations, tried to gauge the condition of ecosystems by examining the goods and services they produce—food, fibre, clean water, carbon storage and so on—and their capacity to continue producing them. The authors explain why ecosystems matter: half of all jobs worldwide are in agriculture, forestry and fishing, and the output from those three commodity businesses still dominates the economies of a quarter of the world's countries.

The report reached two chief conclusions after surveying the best available environmental data. First, a number of ecosystems are "fraying" under the impact of human activity. Second, ecosystems in future will be less able than in the past to deliver the goods and services human life depends upon, which points to unsustainability. But it took care to say: "It's hard, of course, to know what will be truly sustainable." The reason this collection of leading experts could not reach a firm conclusion was that, remarkably, much of the information they needed was incomplete or missing altogether: "Our knowledge of ecosystems has increased dramatically, but it simply has not kept pace with our ability to alter them."

Another group of experts, this time organised by the World Economic Forum, found itself similarly frustrated. The leader of that project, Daniel Esty of Yale, exclaims, throwing his arms in the air: "Why hasn't anyone done careful environmental measurement before? Businessmen always say, 'what matters gets measured.' Social scientists started quantitative measurement 30 years ago, and even political science turned to hard numbers 15 years ago. Yet look at environmental policy, and the data are lousy."

Gaping Holes

At long last, efforts are under way to improve environmental data collection. The most ambitious of these is the Millennium Ecosystem Assessment, a joint effort among leading development agencies and environmental groups. This four-year effort is billed as an attempt to establish systematic data sets on all environmental matters across the world. But one of the researchers involved grouses that it "has very, very little new money to collect or analyse new data". It seems astonishing that governments have been making sweeping decisions on environmental policy for decades without such a baseline in the first place.

One positive sign is the growing interest of the private sector in collecting environmental data. It seems plain that leaving the task to the public sector has not worked. Information on the environment comes far lower on the bureaucratic pecking order than data on education or social affairs, which tend to be overseen by ministries with bigger budgets and more political clout. A number of countries, ranging from New Zealand to Austria, are now looking to the private sector to help collect and manage data in areas such as climate. Development banks are also considering using private contractors to monitor urban air quality, in part to get around the corruption and apathy in some city governments.

"I see a revolution in environmental data collection coming because of computing power, satellite mapping, remote sensing and other such information technologies," says Mr Esty. The arrival of hard data in this notoriously fuzzy area could cut down on environmental disputes by reducing uncertainty. One example is the long-running squabble between America's mid-western states, which rely heavily on coal, and the north-eastern states, which suffer from acid rain. Technology helped disprove claims by the mid-western states that New York's problems all resulted from home-grown pollution.

The arrival of good data would have other benefits as well, such as helping markets to work more robustly: witness America's pioneering scheme to trade emissions of sulphur dioxide, made possible by fancy equipment capable of monitoring emissions in real time. Mr Esty raises an even more intriguing possibility: "Like in the American West a hundred years ago, when barbed wire helped establish rights and prevent overgrazing, information technology can help establish 'virtual barbed wire' that secures property rights and so prevents overexploitation of the commons." He points to fishing in the waters between Australia and New Zealand, where tracking and monitoring devices have reduced over-exploitation.

Best of all, there are signs that the use of such fancy technology will not be confined to rich countries. Calestous Juma of Harvard University shares Mr Esty's excitement about the possibility of such a technology-driven revolution even in Africa: "In the past, the only environmental 'database' we had in Africa was our grandmothers. Now, with global information systems and such, the potential is enormous." Conservationists in Namibia, for example, already use satellite tracking to keep count of their elephants. Farmers in Mali receive satellite updates about impending storms on hand-wound radios. Mr Juma thinks the day is not far off when such technology, combined with ground-based monitoring, will help Africans measure trends in deforestation, soil erosion and climate change, and assess the effects on their local environment.

Make a Start

That is at once a sweeping vision and a modest one. Sweeping, because it will require heavy investment in both sophisticated hardware and nuts-and-bolts information infrastructure on the ground to make sense of all these new data. As the poor world clearly cannot afford to pay for all this, the rich world must help—partly for altruistic reasons, partly with the selfish aim of discovering in good time whether any global environmental calamities are in the making. A number of multilateral agencies now say they are willing to invest in this area as a "neglected global public good"—neglected especially by those agencies themselves. Even President Bush's administration has recently indicated that it will give environmental satellite data free to poor countries.

But that vision is also quite a modest one. Assuming that this data "revolution" does take place, all it will deliver is a reliable assessment of the health of the planet today. We will still not be able to answer the broader question of whether current trends are sustainable or not.

To do that, we need to look more closely at two very different sorts of environmental problems: global crises and local troubles. The global sort is hard to pin down, but can involve irreversible changes. The local kind is common and can have a big effect on the qualify of life, but is usually reversible. Data on both are predictably inadequate. We turn first to the most elusive environmental problem of all, global warming.

Blowing Hot and Cold

Climate change may be slow and uncertain, but that is no excuse for inaction.

What would Winston Churchill have done about climate change? Imagine that Britain's visionary wartime leader had been presented with a potential time bomb capable of wreaking global havoc, although not certain to do so. Warding it off would require concerted global action and economic sacrifice on the home front. Would he have done nothing?

Not if you put it that way. After all, Churchill did not dismiss the Nazi threat for lack of conclusive evidence of Hitler's evil intentions. But the answer might be less straightforward if the following provisos had been added: evidence of this problem would remain cloudy for decades; the worst effects might not be felt for a century; but the costs of tackling the problem would start biting immediately. That, in a nutshell, is the dilemma of climate change. It is asking a great deal of politicians to take action on behalf of voters who have not even been born yet.

One reason why uncertainty over climate looks to be with us for a long time is that the oceans, which absorb carbon from the atmosphere, act as a time-delay mechanism. Their massive thermal inertia means that the climate system responds only very slowly to changes in the composition of the atmosphere. Another complication arises from the relationship between carbon dioxide (CO_2), the principal greenhouse gas (GHG), and sulphur dioxide (SO_2), a common pollutant. Efforts to reduce man-made emissions of GHGs by cutting down on fossil-fuel use will reduce emissions of both gases. The reduction in CO_2 will cut warming, but the concurrent SO_2 cut may mask that effect by contributing to the warming.

There are so many such fuzzy factors—ranging from aerosol particles to clouds to cosmic radiation—that we are likely to see disruptions to familiar climate patterns for many years without knowing why they are happening or what to do about them. Tom Wigley, a leading climate scientist and member of the UN's Intergovernmental Panel on Climate Change (IPCC), goes further. He argues in an excellent book published by the Aspen Institute, "US Policies on Climate Change: What Next?", that whatever policy changes governments pursue, scientific uncertainties will "make it difficult to detect the effects of such changes, probably for many decades."

As evidence, he points to the negligible short- to medium-term difference in temperature resulting from an array of emissions "pathways" on which the world could choose to embark if it decided to tackle climate change. He plots various strategies for reducing GHGs (including the Kyoto one) that will lead in the next century to the stabilisation of atmospheric concentrations of CO_2 at 550 parts per million (ppm). That is roughly double the level which prevailed in pre-industrial times, and is often mooted by climate scientists as a reasonable target. But even by 2040, the temperature differences between the various options will still be tiny—and certainly within the magnitude of natural climatic variance. In short, in another four decades we will probably still not know if we have over- or undershot.

Ignorance Is Not Bliss

However, that does not mean we know nothing. We do know, for a start, that the "greenhouse effect" is real: without the heat-trapping effect of water vapour, CO_2, methane and other naturally occurring GHGs, our planet would be a lifeless $30°C$ or so colder. Some of these GHG emissions are captured and stored by "sinks", such as the oceans, forests and agricultural land, as part of nature's carbon cycle.

We also know that since the industrial revolution began, mankind's actions have contributed significantly to that greenhouse effect. Atmospheric concentrations of GHGs have risen from around 280ppm two centuries ago to around 370ppm today, thanks chiefly to mankind's use of fossil fuels and, to a lesser degree, to deforestation and other land-use changes. Both surface temperatures and sea levels have been rising for some time.

There are good reasons to think temperatures will continue rising. The IPCC has estimated a likely range for that increase of $1.4°C–5.8°C$ over the next century, although the lower end of that range is more likely. Since what matters is not just the absolute temperature level but the rate of change as well, it makes sense to try to slow down the increase.

The worry is that a rapid rise in temperatures would lead to climate changes that could be devastating for many (though not all) parts of the world. Central America, most of Africa, much of south Asia and northern China could all be hit by droughts, storms and floods and otherwise made miserable. Because they are poor and have the misfortune to live near the tropics, those most likely to be affected will be least able to adapt.

The colder parts of the world may benefit from warming, but they too face perils. One is the conceivable collapse of the Atlantic "conveyor belt", a system of currents that gives much of Europe its relatively mild climate; if temperatures climb too high, say scientists, the system may undergo radical changes that damage both Europe and America. That points to the biggest fear: warming may trigger irreversible changes that transform the earth into a largely uninhabitable environment.

Given that possibility, extremely remote though it is, it is no comfort to know that any attempts to stabilise atmospheric concentrations of GHGs at a particular level will take a very long time. Because of the oceans' thermal inertia, explains Mr Wigley, even once atmospheric concentrations of GHGs are stabilised, it will take decades or centuries for the climate to follow suit. And even then the sea level will continue to rise, perhaps for millennia.

This is a vast challenge, and it is worth bearing in mind that mankind's contribution to warming is the only factor that can be controlled. So the sooner we start drawing up a long-term strategy for climate change, the better.

What should such a grand plan look like? First and foremost, it must be global. Since CO_2 lingers in the atmosphere for a century or more, any plan must also extend across several generations.

The plan must recognise, too, that climate change is nothing new: the climate has fluctuated through history, and mankind has adapted to those changes—and must continue doing so. In the rich world, some of the more obvious measures will include building bigger dykes and flood defences. But since the most vulnerable people are those in poor countries, they too have to be helped to adapt to rising seas and unpredictable storms. Infrastructure improvements will be useful, but the best investment will probably be to help the developing world get wealthier.

It is essential to be clear about the plan's long-term objective. A growing chorus of scientists now argues that we need to keep temperatures from rising by much more than 2–3° C in all. That will require the stabilisation of atmospheric concentrations of GHGs. James Edmonds of the University of Maryland points out that because of the long life of CO_2, stabilisation of CO_2 concentrations is not at all the same thing as stabilisation of CO_2 emissions. That, says Mr Edmonds, points to an unavoidable conclusion: "In the very long term, global net CO_2 emissions must eventually peak and gradually decline toward zero, regardless of whether we go for a target of 350ppm or 1,000ppm."

A Low-Carbon World

That is why the long-term objective for climate policy must be a transition to a low-carbon energy system. Such a transition can be very gradual and need not necessarily lead to a world powered only by bicycles and windmills, for two reasons that are often overlooked.

One involves the precise form in which the carbon in the ground is distributed. According to Michael Grubb of the Carbon Trust, a British quasi-governmental body, the long-term problem is coal. In theory, we can burn all of the conventional oil and natural gas in the ground and still meet the most ambitious goals for tackling climate change. If we do that, we must ensure that the far greater amounts of carbon

trapped as coal (and unconventional resources like tar sands) never enter the atmosphere.

The snag is that poor countries are likely to continue burning cheap domestic reserves of coal for decades. That suggests the rich world should speed the development and diffusion of "low carbon" technologies using the energy content of coal without releasing its carbon into the atmosphere. This could be far off, so it still makes sense to keep a watchful eye on the soaring carbon emissions from oil and gas.

The other reason, as Mr Edmonds took care to point out, is that it is net emissions of CO_2 that need to peak and decline. That leaves scope for the continued use of fossil fuels as the main source of modern energy if only some magical way can be found to capture and dispose of the associated CO_2. Happily, scientists already have some magic in the works.

One option is the biological "sequestration" of carbon in forests and agricultural land. Another promising idea is capturing and storing CO_2—underground, as a solid or even at the bottom of the ocean. Planting "energy crops" such as switch-grass and using them in conjunction with sequestration techniques could even result in negative net CO_2 emissions, because such plants use carbon from the atmosphere. If sequestration is combined with techniques for stripping the hydrogen out of this hydrocarbon, then coal could even offer a way to sustainable hydrogen energy.

But is anyone going to pay attention to these long-term principles? After all, over the past couple of years all participants in the Kyoto debate have excelled at producing short-sighted, selfish and disingenuous arguments. And the political rift continues: the EU and Japan pushed ahead with ratification of the Kyoto treaty a month ago, whereas President Bush reaffirmed his opposition.

However, go back a decade and you will find precisely those principles enshrined in a treaty approved by the elder George Bush and since reaffirmed by his son: the UN Framework Convention on Climate Change (FCCC). This treaty was perhaps the most important outcome of the Rio summit, and it remains the basis for the international climate-policy regime, including Kyoto.

The treaty is global in nature and long-term in perspective. It commits signatories to pursuing "the stabilisation of GHG concentrations in the atmosphere at a level that would prevent dangerous interference with the climate system." Note that the agreement covers GHG concentrations, not merely emissions. In effect, this commits even gas-guzzling America to the goal of declining emissions.

Better than Kyoto

Crucially, the FCCC treaty not only lays down the ends but also specifies the means: any strategy to achieve stabilisation of GHG concentrations, it insists, "must not be disruptive of the global economy". That was the stumbling block for the Kyoto treaty, which is built upon the FCCC agreement: its targets and timetables proved unrealistic.

Any revised Kyoto treaty or follow-up accord (which must include the United States and the big developing countries) should rest on the three basic pillars. First, governments everywhere (but especially in Europe) must understand that a reduction in emissions has to start modestly. That is because the capital stock involved in the global energy system is vast and long-lived, so a dash to scrap fossil-fuel production would be hugely expensive. However, as Mr Grubb points out, that pragmatism must be flanked by policies that encourage a switch to low-carbon technologies when replacing existing plants.

Second, governments everywhere (but especially in America) must send a powerful signal that carbon is going out of fashion. The best way to do this is to levy a carbon tax. However, whether it is done through taxes, mandated restrictions on GHG emissions or market mechanisms is less important than that the signal is sent clearly, forcefully and unambiguously. This is where President Bush's mixed signals have done a lot of harm: America's industry, unlike Europe's, has little incentive to invest in low-carbon technology. The irony is that even some coal-fired utilities in America are now clamouring for CO_2 regulation so that they can invest in new plants with confidence.

The third pillar is to promote science and technology. That means encouraging basic climate and energy research, and giving incentives for spreading the results. Rich countries and aid agencies must also find ways to help the poor world adapt to climate change. This is especially important if the world starts off with small cuts in emissions, leaving deeper cuts for later. That, observes Mr Wigley, means that by mid-century "very large investments would have to have been made—and yet the 'return' on these investments would not be visible. Continued investment is going to require more faith in climate science than currently appears to be the case."

Even a visionary like Churchill might have lost heart in the face of all this uncertainty. Nevertheless, there is a glimmer of hope that today's peacetime politicians may rise to the occasion.

Miracles Sometimes Happen

Two decades ago, the world faced a similar dilemma: evidence of a hole in the ozone layer. Some inconclusive signs suggested that it was man-made, caused by the use of chlorofluorocarbons (CFCs). There was the distant threat of disaster, and the knowledge about a concerted global response was required. Industry was reluctant at first, yet with leadership from Britain and America the Montreal Protocol was signed in 1987. That deal has proved surprisingly successful. The manufacture of CFCs is nearly phased out, and there are already signs that the ozone layer is on the way to recovery.

This story holds several lessons for the admittedly far more complex climate problem. First, it is the rich world which has caused the problem and which must lead the way in solving it. Second, the poor world must agree to help, but is right to insist on being given time—as well as money and technology—to help it adjust. Third, industry holds the key: in the ozone-depletion story, it was only after DuPont and ICI broke ranks with the rest of the CFC manufacturers that a deal became possible. On the climate issue, BP and Shell have similarly broken ranks with Big Oil, but the American energy industry—especially the coal sector—remains hostile.

The final lesson is the most important: that the uncertainty surrounding a threat such as climate change is no excuse for inaction. New scientific evidence shows that the threat from ozone depletion had been much deadlier than was thought at the time when the world decided to act. Churchill would surely have approved.

Local Difficulties

Greenery is for the poor too, particularly on their own doorstep.

Why should we care about the environment? Ask a European, and he will probably point to global warming. Ask the two little boys playing outside a newsstand in Da Shilan, a shabby neighbourhood in the heart of Beijing, and they will tell you about the city's notoriously foul air: "It's bad—like a virus!"

Given all the media coverage in the rich world, people there might believe that global scares are the chief environmental problems facing humanity today. They would

be wrong. Partha Dasgupta, an economics professor at Cambridge University, thinks the current interest in global, future-oriented problems has "drawn attention away from the economic misery and ecological degradation endemic in large parts of the world today. Disaster is not something for which the poorest have to wait; it is a frequent occurrence."

Every year in developing countries, a million people die from urban air pollution and twice that number from exposure to stove smoke inside their homes. Another 3m unfortunates die prematurely every year from water-related diseases. All told, premature deaths and illnesses arising from environmental factors account for about a fifth of all diseases in poor countries, bigger than any other preventable factor, including malnutrition. The problem is so serious that Ian Johnson, the World Bank's vice-president for the environment, tells his colleagues, with a touch of irony, that he is really the bank's vice-president for health: "I say tackling the underlying environmental causes of health problems will do a lot more good than just more hospitals and drugs."

The link between environment and poverty is central to that great race for sustainability. It is a pity, then, that several powerful fallacies keep getting in the way of sensible debate. One popular myth is that trade and economic growth make poor countries' environmental problems worse. Growth, it is said, brings with it urbanisation, higher energy consumption and industrialisation—all factors that contribute to pollution and pose health risks.

In a static world, that would be true, because every new factory causes extra pollution. But in the real world, economic growth unleashes many dynamic forces that, in the longer run, more than offset that extra pollution. Traditional environmental risks (such as water-borne diseases) cause far more health problems in poor countries than modern environmental risks (such as industrial pollution).

Rigged Rules

However, this is not to say that trade and economic growth will solve all environmental problems. Among the reasons for doubt are the "perverse" conditions under which world trade is carried on, argues Oxfam. The British charity thinks the rules of trade are "unfairly rigged against the poor", and cites in evidence the enormous subsidies lavished by rich countries on industries such as agriculture, as well as trade protection offered to manufacturing industries such as textiles. These measurements hurt the environment because they force the world's poorest countries to rely heavily on commodities—a particularly energy-intensive and ungreen sector.

Mr Dasgupta argues that this distortion of trade amounts to a massive subsidy of rich-world consumption paid by the world's poorest people. The most persuasive critique of all goes as follows: "Economic growth is not sufficient for turning environmental degradation around. If economic incentives facing producers and consumers do not change with higher incomes, pollution will continue to grow unabated with the growing scale of economic activity." Those words come not from some anti-globalist green group, but from the World Trade Organisation.

Another common view is that poor countries, being unable to afford greenery, should pollute now and clean up later. Certainly poor countries should not be made to adopt American or European environmental standards. But there is evidence to suggest that poor countries can and should try to tackle some environmental problems now, rather than wait till they have become richer.

This so-called "smart growth" strategy contradicts conventional wisdom. For many years, economists have observed that as agrarian societies industrialised, pollution increased at first, but as the societies grew wealthier it declined again. The trouble is that this applies only to some pollutants, such as sulphur dioxide, but not to others, such as carbon dioxide. Even more troublesome, those smooth curves going up, then down, turn out to be misleading. They are what you get when you plot data for poor and rich countries together at a given moment in time, but actual levels of various pollutants in any individual country plotted over time wiggle around a lot more. This suggests that the familiar bell-shaped curve reflects no immutable law, and that intelligent government policies might well help to reduce pollution levels even while countries are still relatively poor.

Developing countries are getting the message. From Mexico to the Philippines, they are now trying to curb the worst of the air and water pollution that typically accompanies industrialisation. China, for example, was persuaded by outside experts that it was losing so much potential economic output through health troubles caused by pollution (according to one World Bank study, somewhere between 3.5% and 7.7% of GDP) that tackling it was cheaper than ignoring it.

One powerful—and until recently ignored—weapon in the fight for a better environment is local people. Old-fashioned paternalists in the capitals of developing countries used to argue that poor villagers could not be relied on to look after natural resources. In fact, much academic research has shown that the poor are more often victims than perpetrators of resource depletion: it tends to be rich locals or outsiders who are responsible for the worst exploitation.

Local people usually have a better knowledge of local ecological conditions than experts in faraway capitals, as well as a direct interest in improving the quality of life in their village. A good example of this comes from the bone-dry state of Rajasthan in India, where local activism and indigenous know-how about rainwater "harvesting" provided the people with reliable water supplies—something the government had failed to do. In Bangladesh, villages with active community groups or concerned mullahs proved greener than less active neighbouring villages.

Community-based forestry initiatives from Bolivia to Nepal have shown that local people can be good custodians of nature. Several hundred million of the world's poorest people live in and around forests. Giving those villagers an incentive to preserve forests by allowing sustainable levels of harvesting, it turns out, is a far better way to save those forests than erecting tall fences around them.

To harness local energies effectively, it is particularly important to give local people secure property rights, argues

24

Mr Dasgupta. In most parts of the developing world, control over resources at the village level is ill-defined. This often means that local elites usurp a disproportionate share of those resources, and that individuals have little incentive to maintain and upgrade forests or agricultural land. Authorities in Thailand tried to remedy this problem by distributing 5.5m land titles over a 20-year period. Agricultural output increased, access to credit improved and the value of the land shot up.

Name and Shame

Another powerful tool for improving the local environment is the free flow of information. As local democracy flourishes, ordinary people are pressing for greater environmental disclosure by companies. In some countries, such as Indonesia, governments have adopted a "sunshine" policy that involves naming and shaming companies that do not meet environmental regulations. It seems to achieve results.

Bringing greenery to the grass roots is good, but on its own it will not avert perceived threats to global "public goods" such as the climate or biodiversity. Paul Portney of Resources for the Future explains: "Brazilian villagers may think very carefully and unselfishly about their future descendants, but there's no reason for them to care about and protect species or habitats that no future generation of Brazilians will care about."

That is why rich countries must do more than make pious noises about global threats to the environment. If they believe that scientific evidence suggests a credible threat, they must be willing to pay poor countries to protect such things as their tropical forests. Rather than thinking of this as charity, they should see it as payment for environmental services (say, for carbon storage) or as a form of insurance.

In the case of biodiversity, such payments could even be seen as a trade in luxury goods: rich countries would pay poor countries to look after creatures that only the rich care about. Indeed, private green groups are already buying up biodiversity "hot spots" to protect them. One such initiative, led by Conservation International and the International Union for the Conservation of Nature (IUCN), put the cost of buying and preserving 25 hot spots exceptionally rich in species diversity at less than $30 billion. Sceptics say it will cost more, as hot spots will need buffer zones of "sustainable harvesting" around them. Whatever the right figure, such creative approaches are more likely to achieve results than bullying the poor into conservation.

It is not that the poor do not have green concerns, but that those concerns are very different from those of the rich. In Beijing's Da Shilan, for instance, the air is full of soot from the many tiny coal boilers. Unlike most of the neighbouring districts, which have recently converted from coal to natural gas, this area has been considered too poor to make the transition. Yet ask Liu Shihua, a shopkeeper who has lived in the same spot for over 20 years, and he insists he would readily pay a bit more for the cleaner air that would come from using natural gas. So would his neighbours.

To discover the best reason why poor countries should not ignore pollution, ask those two little boys outside Mr Liu's shop what colour the sky is. "Grey!" says one tyke, as if it were the most obvious thing in the world. "No, stupid, it's blue!" retorts the other. The children deserve blue skies and clean air. And now there is reason to think they will see them in their lifetime.

Working Miracles

Can technology save the planet?

"Nothing endures but change." That observation by Heraclitus often seems lost on modern environmental thinkers. Many invoke scary scenarios assuming that resources—both natural ones, like oil, and man-made ones, like knowledge—are fixed. Yet in real life man and nature are entwined in a dynamic dance of development, scarcity, degradation, innovation and substitution.

The nightmare about China turning into a resource-guzzling America raises two questions: will the world run out of resources? And even if it does not, could the growing affluence of developing nations lead to global environmental disaster?

The first fear is the easier to refute; indeed, history has done so time and again. Malthus, Ricardo and Mill all worried that scarcity of resources would snuff out growth. It did not. A few decades ago, the limits-to-growth camp raised worries that the world might soon run out of oil, and that it might not be able to feed the world's exploding population. Yet there are now more proven reserves of petroleum than three decades ago; there is more food produced than ever; and the past decade has seen history's greatest economic boom.

What made these miracles possible? Fears of oil scarcity prompted investment that led to better ways of producing oil, and to more efficient engines. In food production, technological advances have sharply reduced the amount of land required to feed a person in the past 50 years. Jesse Ausubel of Rockefeller University calculates that if in the next 60 to 70 years the world's average farmer reaches the yield of today's average (not best) American maize grower, then feeding 10 billion people will require just half of today's cropland. All farmers need to do is maintain the 2%-a-year productivity gain that has been the global norm since 1960.

"Scarcity and Growth", a book published by Resources for the Future, sums it up brilliantly: "Decades ago Vermont granite was only building and tombstone material; now it is a potential fuel, each ton of which has a usable energy content (uranium) equal to 150 tons of coal. The notion of an absolute limit to natural resource availability is untenable when the definition of resources changes drastically and unpredictably over time." Those words were written by Harold Barnett and Chandler Morse in 1963, long before the limits-to-growth bandwagon got rolling.

Giant Footprint

Not so fast, argue greens. Even if we are not going to run out of resources, guzzling ever more resources could still do irreversible damage to fragile ecosystems.

WWF, an environmental group, regularly calculates mankind's "ecological footprint", which it defines as the "biologically productive land and water areas required to produce the resources consumed and assimilate the wastes generated by a given population using prevailing technology." The group reckons the planet has around 11.4 billion "biologically productive" hectares of land available to meet continuing human needs. WWF thinks mankind has recently been using more than that. This is possible because a forest harvested at twice its regeneration rate, for example, appears in the footprint accounts at twice its area—an unsustainable practice which the group calls "ecological overshoot."

Any analysis of this sort must be viewed with scepticism. Everyone knows that environmental data are incomplete. What is more, the biggest factor by far is the land required to absorb CO_2 emissions of fossil fuels. If that problem could be managed some other way, then mankind's ecological footprint would look much more sustainable.

Even so, the WWF analysis makes an important point: if China's economy were transformed overnight into a clone of America's, an ecological nightmare could ensue. If a billion eager new consumers were suddenly to produce CO_2 emissions at American rates, they would be bound to accelerate global warming. And if the whole of the developing world were to adopt an American lifestyle tomorrow, local environmental crises such as desertification, aquifer depletion and topsoil loss could make humans miserable.

So is this cause for concern? Yes, but not for panic. The global ecological footprint is determined by three factors: population size, average consumption per person and technology. Fortunately, global population growth now appears to be moderating. Consumption per person in poor countries is rising as they become better off, but there are signs that the rich world is reducing the footprint of its consumption (as this survey's final section explains). The most powerful reason for hope—innovation—was foreshadowed by WWF's own definition. Today's "prevailing technologies" will, in time, be displaced by tomorrow's greener ones.

"The rest of the world will not live like America," insists Mr Ausubel. Of course poor people around the world covet the creature comforts that Americans enjoy, but they know full well that the economic growth needed to improve their lot will take time. Ask Wu Chengjian, an environmental official in booming Shanghai, what he thinks of the popular notion that his city might become as rich as today's Hong Kong by 2020: "Impossible—that's just not enough time." And that is Shanghai, not the impoverished countryside.

Leaps of Faith

This extra time will allow poor countries to embrace new technologies that are more efficient and less environmentally damaging. That still does not guarantee a smaller ecological footprint for China in a few decades' time than for America now, but it greatly improves the chances. To see why, consider the history of "dematerialisation" and "decarbonisation". Viewed across very long spans of time, productivity improvements allow economies to use ever fewer material inputs—and to emit ever fewer pollutants—per unit of economic output. Mr Ausubel concludes: "When China has today's American mobility, it will not have today's American cars," but the cleaner and more efficient cars of tomorrow.

The snag is that consumers in developing countries want to drive cars not tomorrow but today. The resulting emissions have led many to despair that technology (in the form of vehicles) is making matters worse, not better.

Can they really hope to "leapfrog" ahead to cleaner air? The evidence from Los Angeles—a pioneer in the fight against air pollution—suggests the answer is yes. "When I moved to Los Angeles in the 1960s, there was so much soot in the air that it felt like there was a man standing on your chest most of the time," says Ron Loveridge, the mayor of Riverside, a city to the east of LA that suffers the worst of the region's pollution. But, he says, "We have come an extraordinary distance in LA."

Four decades ago, the city had the worst air quality in America. The main problem was the city's infamous "smog" (an amalgam of "smoke" and "fog"). It took a while to figure out that this unhealthy ozone soup developed as a result of complex chemical reactions between nitrogen oxides and volatile organic compounds that need sunlight to trigger them off.

Arthur Winer, an atmospheric chemist at the University of California at Los Angeles, explains that tackling smog required tremendous perseverance and political will. Early regulatory efforts met stiff resistance from business interests, and began to falter when they failed to show dramatic results.

Clean-air advocates like Mr Loveridge began to despair: "We used to say that we needed a 'London fog' [a reference to an air-pollution episode in 1952 that may have killed 12,000 people in that city] here to force change." Even so, Californian officials forged ahead with an ambitious plan that combined regional regulation with stiff mandates for cleaner air. Despite uncertainties about the cause of the problem, the authorities introduced a sequence of controversial measures: unleaded and low-sulphur petrol, on-board diagnostics for cars to minimise emissions, three-way catalytic converters, vapour-recovery attachments for petrol nozzles and so on.

As a result, the city that two decades ago hardly ever met federal ozone standards has not had to issue a single alert in the past three years. Peak ozone levels are down by 50% since the 1960s. Though the population has shot up in recent years, and the vehicle-miles driven by car-crazy Angelenos have tripled, ozone levels have fallen by two-thirds. The city's air is much cleaner than it was two decades ago.

"California, in solving its air-quality problem, has solved it for the rest of the United States and the world—but it doesn't get credit for it," says Joe Norbeck of the University of California at Riverside. He is adamant that the poor world's cities can indeed leapfrog ahead by embracing some of the cleaner technologies developed specifically for the Californian market. He points to China's vehicle fleet as an example: "China's typical car has the emissions of a 1974 Ford Pinto, but the new Buicks sold there use 1990s emissions technology." The typical car sold today produces less than a tenth of the local pollution of a comparable model from the 1970s.

That suggests one lesson for poor cities such as Beijing that are keen to clean up: they can order polluters to meet high emissions standards. Indeed, from Beijing to Mexico city, regulators are now imposing rich-world rules, mandating new, cleaner technologies. In China's cities, where pollution from sooty coal fires in homes and industrial boilers had been a particular hazard, officials are keen to switch to natural-gas furnaces.

However, there are several reasons why such mandates—which worked wonders in LA—may be trickier to achieve in impoverished or politically weak cities. For a start, city officials must be willing to pay the political price of reforms that raise prices for voters. Besides, higher standards for new cars, useful though they are, cannot do the trick on their own. Often, clean technologies such as catalytic converters will require cleaner grades of petrol too. Introducing cleaner fuels, say experts, is an essential lesson from LA for poor countries. This will not come free either.

There is another reason why merely ordering cleaner new cars is inadequate: it does nothing about the vast stock of dirty old ones already on the streets. In most cities of the developing world, the oldest fifth of the vehicles on the road is likely to produce over half of the total pollution caused by all vehicles taken together. Policies that encourage a speedier turnover of the fleet therefore make more sense than "zero emissions" mandates.

Policy Matters

In sum, there is hope that the poor can leapfrog at least some environmental problems, but they need more than just technology. Luisa and Mario Molina of the Massachusetts Institute of Technology, who have studied such questions closely, reckon that technology is less important than the institutional capacity, legal safeguards and financial resources to back it up: "The most important underlying factor is political will." And even a techno-optimist such as Mr Ausubel accepts that: "There is nothing automatic about technological innovation and adoption; in fact, at the micro level, it's bloody."

Clearly innovation is a powerful force, but government policy still matters. That suggests two rules for policymakers. First, don't do stupid things that inhibit innovation. Second, do sensible things that reward the development and adoption of technologies that enhance, rather than degrade, the environment.

The greatest threat to sustainability may well be the rejection of science. Consider Britain's hysterical reaction to genetically modified crops, and the European Commission's recent embrace of a woolly "precautionary principle". Precaution applied case-by-case is undoubtedly a good thing, but applying any such principle across the board could prove disastrous.

Explaining how not to stifle innovation that could help the environment is a lot easier than finding ways to encourage it. Technological change often goes hand-in-hand with greenery by saving resources, as the long history of dematerialisation shows—but not always. Sports utility vehicles, for instance, are technologically innovative, but hardly green. Yet if those SUVs were to come with hydrogen-powered fuel cells that emit little pollution, the picture would be transformed.

The best way to encourage such green innovations is to send powerful signals to the market that the environment matters. And there is no more powerful signal than price, as the next section explains.

The Invisible Green Hand

Markets could be a potent force for greenery—if only greens could learn to love them.

"**M**andate, regulate and litigate." That has been the environmentalists' rallying cry for ages. Nowhere in the green manifesto has there been much mention of the market. And, oddly, it was market-minded America that led the dirigiste trend. Three decades ago, Congress passed a sequence of laws, including the Clean Air Act, which set lofty goals and generally set rigid technological standards. Much of the world followed America's lead.

This top-down approach to greenery has long been a point of pride for groups such as the Natural Resources Defence Council (NRDC), one of America's most influential environmental outfits. And with some reason, for it has had its successes: the air and water in the developed world is undoubtedly cleaner than it was three decades ago, even though the rich world's economies have grown by leaps and bounds. This has convinced such groups stoutly to defend the green status quo.

But times may be changing. Gus Speth, now head of Yale University's environment school and formerly head of the World Resources Institute and the UNDP, as well as one of the founders of the NRDC, recently explained how he was converted to market economics: "Thirty years ago, the economists at Resources for the Future were pushing the idea of pollution taxes. We lawyers at NRDC thought they were nuts, and feared that they would derail command-and-control measures like the Clean Air Act, so we opposed them. Looking back, I'd have to say this was the single biggest failure in environmental management—not getting the prices right."

A remarkable mea culpa; but in truth, the command-and-control approach was never as successful as its advocates claimed. For example, although it has cleaned up the air and water in rich countries, it has notably failed in dealing with waste management, hazardous emissions and fisheries depletion. Also, the gains achieved have come at a needlessly high price. That is because technology mandates and bureaucratic edicts stifle innovation and ignore local realities, such as varying costs of abatement. They also fail to use cost-benefit analysis to judge trade-offs.

Command-and-control methods will also be ill-suited to the problems of the future, which are getting trickier. One reason is that the obvious issues—like dirty air and water—have been tackled already. Another is increasing technological complexity: future problems are more likely to involve subtle linkages—like those involved in ozone depletion and global warming—that will require sophisticated responses. The most important factor may be society's ever-rising expectations; as countries grow wealthier, their people start clamouring for an ever-cleaner environment. But because the cheap and simple things have been done, that is proving increasingly expensive. Hence the greens' new interest in the market.

Carrots, Not Just Sticks

In recent years, market-based greenery has taken off in several ways. With emissions trading, officials decide on a pollution target and then allocate tradable credits to companies based on that target. Those that find it expensive to cut emissions can buy credits from those that find it cheaper, so the target is achieved at the minimum cost and disruption.

The greatest green success story of the past decade is probably America's innovative scheme to cut emissions of sulphur dioxide (SO_2). Dan Dudek of Environmental Defence, a most unusual green group, and his market-minded colleagues persuaded the elder George Bush to agree to an amendment to the sacred Clean Air Act that would introduce an emissions-trading system to achieve sharp cuts in SO_2. At the time, this was hugely controversial: America's power industry insisted the cuts were prohibitively costly, while nearly every other green group decried the measure as a sham. In the event, ED has been vindicated. America's scheme has surpassed its initial objectives, and at far lower cost than expected. So great is the interest worldwide in trading that ED is now advising groups ranging from hard-nosed oilmen at BP to bureaucrats in China and Russia.

Europe, meanwhile, is forging ahead with another sort of market-based instrument: pollution taxes. The idea is to levy charges on goods and services so that their price reflects their "externalities"—jargon for how much harm they do to the environment and human health. Sweden introduced a sulphur tax a decade ago, and found that the sulphur content of fuels dropped 50% below legal requirements.

Though "tax" still remains a dirty word in America, other parts of the world are beginning to embrace green tax reform by shifting taxes from employment to pollution. Robert Williams of Princeton University has looked at energy use (especially the terrible effects on health of particulate pollution) and concluded that such externalities are comparable in size to the direct economic costs of producing that energy.

Externalities are only half the battle in fixing market distortions. The other half involves scrapping environmentally harmful subsidies. These range from prices below market levels for electricity and water to shameless cash handouts for industries such as coal. The boffins at the OECD reckon that stripping away harmful subsidies, along with introducing taxes on carbon-based fuels and chemicals use, would result in dramatically lower emissions by 2020 than current policies would be able to achieve. If the revenues raised were then used to reduce other taxes, the cost of these virtuous policies would be less than 1% of the OECD's economic output in 2020.

Such subsidies are nothing short of perverse, in the words of Norman Myers of Oxford University. They do double damage, by distorting markets and by encouraging behaviour that harms the environment. Development banks say such subsidies add up to $700 billion a year, but Mr Myers reckons the true sum is closer to $2 trillion a year. Moreover, the numbers do not fully reflect the harm done. For example, EU countries subsidise their fishing fleets to the tune of $1 billion a year, but that has encouraged enough overfishing to drive many North Atlantic fishing grounds to near-collapse.

Fishing is an example of the "tragedy of the commons", which pops up frequently in the environmental debate. A resource such as the ocean is common to many, but an individual "free rider" can benefit from plundering that commons or dumping waste into it, knowing that the costs of his actions will probably be distributed among many neighbours. In the case of shared fishing grounds, the absence of individual ownership drives each fisherman to snatch as many fish as he can—to the detriment of all.

Of Rights and Wrongs

Assigning property rights can help, because providing secure rights (set at a sustainable level) aligns the interests of the individual with the wider good of preserving nature. This is what sceptical conservationists have observed in New Zealand and Iceland, where schemes for tradable quotas have helped revive fishing stocks. Similar rights-based approaches have led to revivals in stocks of African elephants in southern Africa, for example, where the authorities stress property rights and private conservation.

All this talk of property rights and markets makes many mainstream environmentalists nervous. Carl Pope, the boss of the Sierra Club, one of America's biggest green groups, does not reject market forces out of hand, but expresses deep scepticism about their scope. Pointing to the difficult problem of climate change, he asks: "Who has property rights over the commons?"

Even so, some greens have become converts. Achim Steiner of the IUCN reckons that the only way forward is rights-based conservation, allowing poor people "sustainable use" of their local environment. Paul Faeth of the World Resources Institute goes further. He says he is convinced that

market forces could deliver that holy grail of environmentalism, sustainability—"but only if we get prices right."

The Limits to Markets

Economic liberals argue that the market itself is the greatest price-discovery mechanism known to man. Allow it to function freely and without government meddling, goes the argument, and prices are discovered and internalised automatically. Jerry Taylor of the Cato Institute, a libertarian think-tank, insists that "The world today is already sustainable—except those parts where western capitalism doesn't exist." He notes that countries that have relied on central planning, such as the Soviet Union, China and India, have invariably misallocated investment, stifled innovation and fouled their environment far more than the prosperous market economies of the world have done.

All true. Even so, markets are currently not very good at valuing environmental goods. Noble attempts are under way to help them do better. For example, the Katoomba Group, a collection of financial and energy companies that have linked up with environmental outfits, is trying to speed the development of markets for some of forestry's ignored "co-benefits" such as carbon storage and watershed management, thereby producing new revenue flows for forest owners. This approach shows promise: water consumers ranging from officials in New York City to private hydro-electric operators in Costa Rica are now paying people upstream to manage their forests and agricultural land better. Paying for greenery upstream turns out to be cheaper than cleaning up water downstream after it has been fouled.

Economists too are getting into the game of helping capitalism "get prices right." The World Bank's Ian Johnson argues that conventional economic measures such as gross domestic product are not measuring wealth creation properly because they ignore the effects of environmental degradation. He points to the positive contribution to China's GDP from the logging industry, arguing that such a calculation completely ignores the billions of dollars-worth of damage from devastating floods caused by over-logging. He advocates a more comprehensive measure the Bank is working on, dubbed "genuine GDP", that tries (imperfectly, he accepts) to measure depletion of natural resources.

That could make a dramatic difference to how the welfare of the poor is assessed. Using conventional market measures, nearly the whole of the developing world save Africa has grown wealthier in the past couple of decades. But when the degradation of nature is properly accounted for, argues Mr Dasgupta at Cambridge, the countries of Africa and south Asia are actually much worse off today than they were a few decades ago—and even China, whose economic "miracle" has been much trumpeted, comes out barely ahead.

The explanation, he reckons, lies in a particularly perverse form of market distortion: "Countries that are exporting resource-based products (often among the poorest) may be subsidising the consumption of countries that are doing the importing (often among the richest)." As evidence, he points to the common practice in poor countries of encouraging resource extraction. Whether through licenses granted at below-market rates, heavily subsidised exports or corrupt officials tolerating illegal exploitation, he reckons the result is the same: "The cruel paradox we face may well be that contemporary economic development is unsustainable in poor countries because it is sustainable in rich countries."

One does not have to agree with Mr Dasgupta's conclusion to acknowledge that markets have their limits. That should not dissuade the world from attempting to get prices right—or at least to stop getting them so wrong. For grotesque subsidies, the direction of change should be obvious. In other areas, the market itself may not provide enough information to value nature adequately. This is true of threats to essential assets, such as nature's ability to absorb and "recycle" CO_2, that have no substitute at any price. That is when governments must step in, ensuring that an informed public debate takes place.

Robert Stavins of Harvard University argues that the thorny notion of sustainable development can be reduced to two simple ideas: efficiency and intergenerational equity. The first is about making the economic pie as large as possible; he reckons that economists are well equipped to handle it, and that market-based policies can be used to achieve it. On the second (the subject of the next section), he is convinced that markets must yield to public discourse and government policy: "Markets can be efficient, but nobody ever said they're fair. The question is, what do we owe the future?"

Insuring a Brighter Future

How to hedge against tomorrow's environmental risks.

So what do we owe the future? A precise definition for sustainable development is likely to remain elusive but, as this survey has argued, the hazy outline of a useful one is emerging from the experience of the past decade.

For a start, we cannot hope to turn back the clock and return nature to a pristine state. Nor must we freeze nature in the state it is today, for that gift to the future would impose an unacceptable burden on the poorest alive today. Besides, we cannot forecast the tastes, demands or concerns of future generations. Recall that the overwhelming pollution problem a century ago was horse manure clogging up city streets: a century hence, many of today's problems will surely seem equally irrelevant. We should therefore think of our debt to the future as including not just natural resources but also technology, institutions and especially the capacity to innovate. Robert Solow got it mostly right a decade ago: the most important thing to leave future generations, he said, is the capacity to live as well as we do today.

However, as the past decade has made clear, there is a limit to that argument. If we really care about the "sustainable" part of sustainable development, we must be much more watchful about environmental problems with critical thresholds. Most local problems are reversible and hence no cause for alarm. Not all, however: the depletion of aquifers and the loss of topsoil could trigger irreversible changes that would leave future generations worse off. And global or long-term threats, where victims are far removed in time and space, are easy to brush aside.

In areas such as biodiversity, where there is little evidence of a sustainability problem, a voluntary approach is best. Those in the rich world who wish to preserve pandas, or hunt for miracle drugs in the rainforest, should pay for their predilections. However, where there are strong scientific indications of unsustainability, we must act on behalf of the future—even at the price of today's development. That may be expensive, so it is prudent to try to minimise those risks in the first place.

A Riskier World

Human ingenuity and a bit of luck have helped mankind stay a few steps ahead of the forces degrading the environment this past century, the first full one in which the planet has been exposed to industrialisation. In the century ahead, the great race between development and degradation could well become a closer call.

On one hand, the demands of development seem sure to grow at a cracking pace in the next few decades as the Chinas, Indias and Brazils of this world grow wealthy enough to start enjoying not only the necessities but also some of the luxuries of life. On the other hand, we seem to be entering a period of huge technological advances in emerging fields such as biotechnology that could greatly increase resource productivity and more than offset the effect of growth on the environment. The trouble is, nobody knows for sure.

Since uncertainty will define the coming era, it makes sense to invest in ways that reduce that risk at relatively low cost.

Governments must think seriously about the future implications of today's policies. Their best bet is to encourage the three powerful forces for sustainability outlined in this survey: the empowerment of local people to manage local resources and adapt to environmental change; the encouragement of science and technology, especially innovations that reduce the ecological footprint of consumption; and the greening of markets to get prices right.

To advocate these interventions is not to call for a return to the hubris of yesteryear's central planners. These measures would merely give individuals the power to make greener choices if they care to. In practice, argues Chris Heady of the OECD, this may still not add up to sustainability "because we might still decide to be greedy, and leave less for our children."

Happily, there are signs of an emerging bottom-up push for greenery. Even such icons of western consumerism as Unilever and Procter & Gamble now sing the virtues of "sustainable consumption." Unilever has vowed that by 2005 it will be buying fish only from sustainable sources, and P&G is coming up with innovative products such as detergents that require less water, heat and packaging. It would be naive to label such actions as expressions of "corporate social responsibility": in the long run, firms will embrace greenery only if they see profit in it. And that, in turn, will depend on choices made by individuals.

Such interventions should really be thought of as a kind of insurance that tilts the odds of winning that great race just a little in humanity's favour. Indeed, even some of the world's most conservative insurance firms increasingly see things this way. As losses from weather-related disasters have risen of late, the industry is getting more involved in policy debates on long-term environmental issues such as climate change.

Bruno Porro, chief risk officer at Swiss Re, argues that: "The world is entering a future in which risks are more concentrated and more complex. That is why we are pressing for policies that reduce those risks through preparation, adaptation and mitigation. That will be cheaper than covering tomorrow's losses after disaster strikes."

Jeffrey Sachs of Columbia University agrees: "When you think about the scale of risk that the world faces, it is clear that we grossly underinvest in knowledge . . . we have enough income to live very comfortably in the developed world and to prevent dire need in the developing world. So we should have the confidence to invest in longer-term issues like the environment. Let's help insure the sustainability of this wonderful situation."

He is right. After all, we have only one planet, now and in the future. We need to think harder about how to use it wisely.

Acknowledgements

In addition to those cited in the text, the author would like to thank Robert Socolow, David Victor, Geoffrey Heal, and experts at Tsinghua University, Friends of the Earth, the European Commission, the World Business Council for Sustainable Development, the International Energy Agency, the OECD and the UN for sharing their ideas with him. A list of sources can be found on *The Economist's* website.

Five Meta-Trends Changing the World

Global, overarching forces such as modernization and widespread interconnectivity are converging to reshape our lives. But human adaptability—itself a "meta-trend"—will help keep our future from spinning out of control, assures THE FUTURIST's lifestyles editor.

DAVID PEARCE SNYDER

Last year, I received an e-mail from a long-time Australian client requesting a brief list of the "meta-trends" having the greatest impact on global human psychology. What the client wanted to know was, which global trends would most powerfully affect human consciousness and behavior around the world?

The Greek root *meta* denotes a transformational or transcendent phenomenon, not simply a big, pervasive one. A meta-trend implies multidimensional or catalytic change, as opposed to a linear or sequential change.

What follows are five meta-trends I believe are profoundly changing the world. They are evolutionary, system-wide developments arising from the simultaneous occurrence of a number of individual demographic, economic, and technological trends. Instead of each being individual freestanding global trends, they are composites of trends.

Trend 1—Cultural Modernization

Around the world over the past generation, the basic tenets of modern cultures—including equality, personal freedom, and self-fulfillment—have been eroding the domains of traditional cultures that value authority, filial obedience, and self-discipline. The children of traditional societies are growing up wearing Western clothes, eating Western food, listening to Western music, and (most importantly of all) thinking Western thoughts. Most Westerners—certainly most Americans—have been unaware of the personal intensities of this culture war because they are so far away from the "battle lines." Moreover, people in the West regard the basic institutions of modernization, including universal education, meritocracy, and civil law, as benchmarks of social progress, while the defenders of traditional cultures see them as threats to social order.

Demographers have identified several leading social indicators as key measures of the extent to which a nation's culture is modern. They cite the average level of education for men and for women, the percentage of the salaried workforce that is female,

and the percentage of population that lives in urban areas. Other indicators include the percentage of the workforce that is salaried (as opposed to self-employed) and the percentage of GDP spent on institutionalized socioeconomic support services, including insurance, pensions, social security, civil law courts, worker's compensation, unemployment benefits, and welfare.

As each of these indicators rises in a society, the birthrate in that society goes down. The principal measurable consequence of cultural modernization is declining fertility. As the world's developing nations have become better educated, more urbanized, and more institutionalized during the past 20 years, their birthrates have fallen dramatically. In 1988, the United Nations forecast that the world's population would double to 12 billion by 2100. In 1992, their estimate dropped to 10 billion, and they currently expect global population to peak at 9.1 billion in 2100. After that, demographers expect the world's population will begin to slowly decline, as has already begun to happen in Europe and Japan.

Three signs that a culture is modern: its citizens' average level of education, the number of working women, and the percentage of the population that is urban. As these numbers increase, the birthrate in a society goes down, writes author David Pearce Snyder.

The effects of cultural modernization on fertility are so powerful that they are reflected clearly in local vital statistics. In India, urban birthrates are similar to those in the United States, while rural birthrates remain unmanageably high. Cultural modernization is the linchpin of human sustainability on the planet.

The forces of cultural modernization, accelerated by economic globalization and the rapidly spreading wireless telecommunications info-structure, are likely to marginalize the world's traditional cultures well before the century is over. And because the wellsprings of modernization—secular industrial economies—are so unassailably powerful, terrorism is the only means by which the defenders of traditional culture can fight to preserve their values and way of life. In the near-term future, most observers believe that ongoing cultural conflict is likely to produce at least a few further extreme acts of terrorism, security measures not withstanding. But the eventual intensity and duration of the overt, violent phases of the ongoing global culture war are largely matters of conjecture. So, too, are the expert pronouncements of the probable long-term impacts of September 11, 2001, and terrorism on American priorities and behavior.

After the 2001 attacks, social commentators speculated extensively that those events would change America. Pundits posited that we would become more motivated by things of intrinsic value—children, family, friends, nature, personal self-fulfillment—and that we would see a sharp increase in people pursuing *pro bono* causes and public-service careers. A number of media critics predicted that popular entertainment such as television, movies, and games would feature much less gratuitous violence after September 11. None of that has happened. Nor have Americans become more attentive to international news coverage. Media surveys show that the average American reads less international news now than before September 11. Event-inspired changes in behavior are generally transitory. Even if current conflicts produce further extreme acts of terrorist violence, these seem unlikely to alter the way we live or make daily decisions. Studies in Israel reveal that its citizens have become habituated to terrorist attacks. The daily routine of life remains the norm, and random acts of terrorism remain just that: random events for which no precautions or mind-set can prepare us or significantly reduce our risk.

In summary, cultural modernization will continue to assault the world's traditional cultures, provoking widespread political unrest, psychological stress, and social tension. In developed nations, where the great majority embrace the tenets of modernization and where the threats from cultural conflict are manifested in occasional random acts of violence, the ongoing confrontation between tradition and modernization seems likely to produce security measures that are inconvenient, but will do little to alter our basic personal decision making, values, or day-to-day life. Developed nations are unlikely to make any serious attempts to restrain the spread of cultural modernization or its driving force, economic globalization.

Trend 2—Economic Globalization

On paper, globalization poses the long-term potential to raise living standards and reduce the costs of goods and services for people everywhere. But the short-term marketplace consequences of free trade threaten many people and enterprises in both developed and developing nations with potentially insurmountable competition. For most people around the world, the threat from foreign competitors is regarded as much greater than the threat from foreign terrorists. Of course, risk

and uncertainty in daily life is characteristically high in developing countries. In developed economies, however, where formal institutions sustain order and predictability, trade liberalization poses unfamiliar risks and uncertainties for many enterprises. It also appears to be affecting the collective psychology of both blue-collar and white-collar workers—especially males—who are increasingly unwilling to commit themselves to careers in fields that are likely to be subject to low-cost foreign competition.

Strikingly, surveys of young Americans show little sign of xenophobia in response to the millions of new immigrant workers with whom they are competing in the domestic job market. However, they feel hostile and helpless at the prospect of competing with Chinese factory workers and Indian programmers overseas. And, of course, economic history tells us that they are justifiably concerned. In those job markets that supply untariffed international industries, a "comparable global wage" for comparable types of work can be expected to emerge worldwide. This will raise workers' wages for freely traded goods and services in developing nations, while depressing wages for comparable work in mature industrial economies. To earn more than the comparable global wage, labor in developed nations will have to perform *incomparable* work, either in terms of their productivity or the superior characteristics of the goods and services that they produce. The assimilation of mature information technology throughout all production and education levels should make this possible, but developed economies have not yet begun to mass-produce a new generation of high-value-adding, middle-income jobs.

Meanwhile, in spite of the undeniable short-term economic discomfort that it causes, the trend toward continuing globalization has immense force behind it. Since World War II, imports have risen from 6% of world GDP to more than 22%, growing steadily throughout the Cold War, and even faster since 1990. The global dispersion of goods production and the uneven distribution of oil, gas, and critical minerals worldwide have combined to make international interdependence a fundamental economic reality, and corporate enterprises are building upon that reality. Delays in globalization, like the September 2003 World Trade Organization contretemps in Cancun, Mexico, will arise as remaining politically sensitive issues are resolved, including trade in farm products, professional and financial services, and the need for corporate social responsibility. While there will be enormous long-term economic benefits from globalization in both developed and developing nations, the short-term disruptions in local domestic employment will make free trade an ongoing political issue that will be manageable only so long as domestic economies continue to grow.

Trend 3—Universal Connectivity

While information technology (IT) continues to inundate us with miraculous capabilities, it has given us, so far, only one new power that appears to have had a significant impact on our collective behavior: our improved ability to communicate with each other, anywhere, anytime. Behavioral researchers have found that cell phones have blurred or changed the boundaries between work and social life and between personal and public life. Cell phones have also increased users' propensity to "micromanage their lives, to be more spontaneous, and, therefore, to be late for everything," according to Leysia Palen, computer science professor at the University of Colorado at Boulder.

Cell phones have blurred the lines between the public and the private. Nearly everyone is available anywhere, anytime—and in a decade cyberspace will be a town square, writes Snyder.

Most recently, instant messaging—via both cell phones and online computers—has begun to have an even more powerful social impact than cell phones themselves. Instant messaging initially tells you whether the person you wish to call is "present" in cyberspace—that is, whether he or she is actually online at the moment. Those who are present can be messaged immediately, in much the same way as you might look out the window and call to a friend you see in the neighbor's yard. Instant messaging gives a physical reality to cyberspace. It adds a new dimension to life: A person can now be "near," "distant," or "in cyberspace." With video instant messaging—available now, and widely available in three years—the illusion will be complete. We will have achieved what Frances Cairncross, senior editor of *The Economist,* has called "the death of distance."

Universal connectivity will be accelerated by the integration of the telephone, cell phone, and other wireless telecom media with the Internet. By 2010, all long-distance phone calls, plus a third of all local calls, will be made via the Internet, while 80% to 90% of all Internet access will be made from Web-enabled phones, PDAs, and wireless laptops. Most important of all, in less than a decade, one-third of the world's population—2 billion people—will have access to the Internet, largely via Web-enabled telephones. In a very real sense, the Internet will be the "Information Highway"—the infrastructure, or infostructure, for the computer age. The infostructure is already speeding the adoption of flexplace employment and reducing the volume of business travel, while making possible increased "distant collaboration," outsourcing, and offshoring.

Corporate integrity and openness will grow steadily under pressure from watchdog groups and ordinary citizens demanding business transparency. The leader of tomorrow must adapt to this new openness or risk business disaster.

As the first marketing medium with a truly global reach, the Internet will also be the crucible from which a global consumer culture will be forged, led by the first global youth peer culture. By 2010, we will truly be living in a global village, and cyberspace will be the town square.

Trend 4—Transactional Transparency

Long before the massive corporate malfeasance at Enron, Tyco, and WorldCom, there was a rising global movement toward greater transparency in all private and public enterprises. Originally aimed at kleptocratic regimes in Africa and the former Soviet states, the movement has now become universal, with the establishment of more stringent international accounting standards and more comprehensive rules for corporate oversight and record keeping, plus a new UN treaty on curbing public-sector corruption. Because secrecy breeds corruption and incompetence, there is a growing worldwide consensus to expose the principal transactions and decisions of *all* enterprises to public scrutiny.

But in a world where most management schools have dropped all ethics courses and business professors routinely preach that government regulation thwarts the efficiency of the marketplace, corporate and government leaders around the world are lobbying hard against transparency mandates for the private sector. Their argument: Transparency would "tie their hands," "reveal secrets to their competition," and "keep them from making a fair return for their stockholders."

Most corporate management is resolutely committed to the notion that secrecy is a necessary concomitant of leadership. But pervasive, ubiquitous computing and comprehensive electronic documentation will ultimately make all things transparent, and this may leave many leaders and decision makers feeling uncomfortably exposed, especially if they were not provided a moral compass prior to adolescence. Hill and Knowlton, an international public-relations firm, recently surveyed 257 CEOs in the United States, Europe, and Asia regarding the impact of the Sarbanes-Oxley Act's reforms on corporate accountability and governance. While more than 80% of respondents felt that the reforms would significantly improve corporate integrity, 80% said they also believed the reforms would not increase ethical behavior by corporate leaders.

While most consumer and public-interest watchdog groups are demanding even more stringent regulation of big business, some corporate reformers argue that regulations are often counterproductive and always circumventable. They believe that only 100% transparency can assure both the integrity and competency of institutional actions. In the world's law courts—and in the court of public opinion—the case for transparency will increasingly be promoted by nongovernmental organizations (NGOs) who will take advantage of the global infostructure to document and publicize environmentally and socially abusive behaviors by both private and public enterprises. The ongoing battle between institutional and socioecological imperatives will become a central theme of Web newscasts, Netpress publications, and

Weblogs that have already begun to supplant traditional media networks and newspaper chains among young adults worldwide. Many of these young people will sign up with NGOs to wage undercover war on perceived corporate criminals.

In a global marketplace where corporate reputation and brand integrity will be worth billions of dollars, businesses' response to this guerrilla scrutiny will be understandably hostile. In their recently released *Study of Corporate Citizenship,* Cone/Roper, a corporate consultant on social issues, found that a majority of consumers "are willing to use their individual power to punish those companies that do not share their values." Above all, our improving comprehension of humankind's innumerable interactions with the environment will make it increasingly clear that total transparency will be crucial to the security and sustainability of a modern global economy. But there will be skullduggery, bloodshed, and heroics before total transparency finally becomes international law—15 to 20 years from now.

Trend 5—Social Adaptation

The forces of cultural modernization—education, urbanization, and institutional order—are producing social change in the developed world as well as in developing nations. During the twentieth century, it became increasingly apparent to the citizens of a growing number of modern industrial societies that neither the church nor the state was omnipotent and that their leaders were more or less ordinary people. This realization has led citizens of modern societies to assign less weight to the guidance of their institutions and their leaders and to become more self-regulating. U.S. voters increasingly describe themselves as independents, and the fastest-growing Christian congregations in America are nondenominational.

Since the dawn of recorded history, societies have adapted to their changing circumstances. Moreover, cultural modernization has freed the societies of mature industrial nations from many strictures of church and state, giving people much more freedom to be individually adaptive. And we can be reasonably certain that modern societies will be confronted with a variety of fundamental changes in circumstance during the next five, 10, or 15 years that will, in turn, provoke continuous widespread adaptive behavior, especially in America.

Reaching retirement age no longer always means playing golf and spoiling the grandchildren. Seniors in good health who enjoy working probably won't retire, slowing the prophesied workforce drain, according to author David Pearce Snyder

During the decade ahead, *infomation*—the automated collection, storage, and application of electronic data—will dramatically reduce paperwork. As outsourcing and off-shoring eliminate millions of U.S. middle-income jobs, couples are likely to work two lower-pay/lower-skill jobs to replace lost income. If our employers ask us to work from home to reduce the company's office rental costs, we will do so, especially if the arrangement permits us to avoid two hours of daily commuting or to care for our offspring or an aging parent. If a wife is able to earn more money than her spouse, U.S. males are increasingly likely to become househusbands and take care of the kids. If we are in good health at age 65, and still enjoy our work, we probably won't retire, even if that's what we've been planning to do all our adult lives. If adult children must move back home after graduating from college in order to pay down their tuition debts, most families adapt accordingly.

Each such lifestyle change reflects a personal choice in response to an individual set of circumstances. And, of course, much adaptive behavior is initially undertaken as a temporary measure, to be abandoned when circumstances return to normal. During World War II, millions of women voluntarily entered the industrial workplace in the United States and the United Kingdom, for example, but returned to the domestic sector as soon as the war ended and a prosperous normalcy was restored. But the Information Revolution and the aging of mature industrial societies are scarcely temporary phenomena, suggesting that at least some recent widespread innovations in lifestyle—including delayed retirements and "sandwich households"—are precursors of long-term or even permanent changes in society.

The current propensity to delay retirement in the United States began in the mid-1980s and accelerated in the mid-1990s. Multiple surveys confirm that delayed retirement is much more a result of increased longevity and reduced morbidity than it is the result of financial necessity. A recent AARP survey, for example, found that more than 75% of baby boomers plan to work into their 70s or 80s, regardless of their economic circumstances. If the baby boomers choose to age on the job, the widely prophesied mass exodus of retirees will not drain the workforce during the coming decade, and Social Security may be actuarially sound for the foreseeable future.

The Industrial Revolution in production technology certainly produced dramatic changes in society. Before the steam engine and electric power, 70% of us lived in rural areas; today 70% of us live in cities and suburbs. Before industrialization, most economic production was home- or family-based; today, economic production takes place in factories and offices. In preindustrial Europe and America, most households included two or three adult generations (plus children), while the great majority of households today are nuclear families with one adult generation and their children.

Current trends in the United States, however, suggest that the three great cultural consequences of industrialization—the urbanization of society, the institutionalization of work, and the atomization of the family—may all be reversing, as people adapt to their changing circumstances. The U.S. Census Bureau reports that, during the 1990s, Americans began to migrate out of cities and suburbs into exurban and rural areas for the first time in the twentieth century. Simultaneously, information work has begun to migrate out of offices and into households. Given the recent accelerated growth of telecommuting, self-employment, and contingent work, one-fourth to one-third of all gainful employment is likely to take place at home within 10 years. Meanwhile, growing numbers of baby boomers find themselves living with both their debt-burdened, underemployed adult children and their own increasingly dependent aging parents. The recent emergence of the "sandwich household" in America resonates powerfully with the multigenerational, extended families that commonly served as society's safety nets in preindustrial times.

Leadership in Changing Times

The foregoing meta-trends are not the only watershed developments that will predictably reshape daily life in the decades ahead. An untold number of inertial realities inherent in the common human enterprise will inexorably change our collective circumstances—the options and imperatives that confront society and its institutions. Society's adaptation to these new realities will, in turn, create further changes in the institutional operating environment, among customers, competitors, and constituents. There is no reason to believe that the Information Revolution will change us any less than did the Industrial Revolution.

In times like these, the best advice comes from ancient truths that have withstood the test of time. The Greek philosopher-historian Heraclitus observed 2,500 years ago that "nothing about the future is inevitable except change." Two hundred years later, the mythic Chinese general Sun Tzu advised that "the wise leader exploits the inevitable." Their combined message is clear: "The wise leader exploits change."

DAVID PEARCE SNYDER is the lifestyles editor of *THE FUTURIST* and principal of The Snyder Family Enterprise, a futures consultancy located at 8628 Garfield Street, Bethesda, Maryland 20817. Telephone 301-530-5807; e-mail davidpearcesnyder@earthlink.net; Web site www.the-futurist.com.

Originally published in the July/August 2004 issue of *The Futurist*, pp. 22–27. Copyright © 2004 by World Future Society, 7910 Woodmont Avenue, Suite 450, Bethesda, MD 20814. Telephone: 301/656-8274; Fax: 301/951-0394; http://www.wfs.org. Used with permission from the World Future Society.

Globalization's Effects on the Environment

Jo Kwong

In recent years, globalization has become a remarkably polarizing issue. In particular, discussions about globalization and its environmental impacts generate ferocious debate among policy analysts, environmental activists, economists and other opinion leaders. Is globalization a solution to serious economic and social problems of the world? Or is it a profit-motivated process that leads to oppression and exploitation of the world's less fortunate?

This article examines alternative perspectives about globalization and the environment. It offers an explanation for the conflicting visions that are frequently expressed and suggests elements of an institutional framework that can align the benefits of globalization with the objective of enhanced environmental protection.

Globalization, free of the emotional rhetoric, is simply about removing barriers so goods, services, people, and ideas, can freely move from place to place. At its most rudimentary level, globalization describes a process whereby people can make their own decisions about who their trading partners are and what opportunities they wish to pursue.

While this may seem fairly innocuous, globalization certainly raises many concerns. In developed nations, some people worry about globalization's impacts on culture, traditional ways of living, and indigenous control in less developed parts of the world. They wonder, "What's to stop profit-motivated companies from developing some of the pristine environments and fragile natural resources found in the developing world?" These critics of open trade fear that residents of developing nations will be the losers in more ways than one—stripped of their land's natural resources and hopelessly in debt to exploitative developed countries. This group takes a rather paternalistic view of the problems facing the world's poor.

Others—free marketers—believe that the developed world can produce positive benefits by exporting knowledge and technology to the developing world. By avoiding mistakes made in the developed world, it is argued that developing countries can advance in manners that sidestep some of the errors that occurred in others' development processes. Third-world poverty is cited as an important reason to foster greater economic growth in the developing world. To proponents of globalization, trade is seen as a way to lift the third world from poverty and enable local people to help themselves.

Moreover, there are divided views within the developing world. Some argue against so-called "eco-imperialism." "Why are others dictating whether or not we can develop our own resources? Who are these environmental activists that say billions of people in China shouldn't have cars because this will greatly accelerate global warming?" they ask. But others question, "Who are these corporations that come in and buy huge tracts of land in third-world interiors and develop large-scale forestry or oil developments, seemingly without concern about the impact on the local environment?"

In many ways, these alternative perspectives can be viewed as a "conflicts of visions" to steal a phrase from Thomas Sowell. Some people simply view the world fundamentally differently. In the globalization context, for example, one view values the protection of indigenous ways of life, even if that means living with greater poverty and fewer individual choices. Others believe economic efficiency is key—getting the most from our resources to provide the greatest amount of financial wealth and opportunity. Most likely, however, most people fall somewhere in between.

This discussion will offer an additional factor other than a "conflict of visions" that can help us understand the broad disparities in perspectives and understandings about the question, "Is globalization good for the environment?" In particular, it raises the possibility that perhaps we are not asking the right questions to address the set of concerns at hand.

In the 1990s, a number of economists sought to empirically answer the question of whether globalization helps or harms the environment. Some of the most often-cited findings are those from economists Gene Grossman and Alan Krueger. Grossman and Krueger investigated the relationship between the scale of economic activity and environmental quality for a broad set of environmental indicators. They found that environmental degradation and income have an inverted U-shaped relationship, with pollution increasing with income at low levels of income and decreasing with income at high levels of income. The turning point at which economic growth and pollution emissions switch from a positive to a negative relationship depends on the particular emissions and air quality measure tracked. For NOx, SOx and biological oxygen demand (BOD), the turning point appears to be around $5,000 per capita gross domestic product (GDP). This observation supports the view that countries can grow out of pollution problems with wealth.

These findings were followed by further studies that examined this "Environmental Kuznets Curve", as this inverted U-shaped curve was labeled, generating a new set of policy implications that supported the idea that trade can be good for the environment. If economic growth is good for the environment, policies that stimulate growth (trade liberalization, economic restructuring, and free markets) should also be good for the environment.

The most basic description of how this inverted curve can occur is to think about the types of activities that countries experience as they develop. At the most rudimentary level, people are burning cow dung and other readily available materials for heat and cooking sources. No controls are in place; the pollutants are released directly into the air. As economic activity increases and the economy reaches a point at which it can begin making investments, catalytic converters, furnaces, etc., pollution levels are reduced, and hence the inverted curve.

In "Poverty, Wealth and Waste," Barun Mitra compares patterns of waste distribution in India to those of the developed world. He addresses the myth that poor countries have lower levels of pollution:

> The painstaking efforts to recycle materials do not mean that a poor country like India is pollution-free. Indeed, the low quantity of waste generated in an economy with little capital and technological backwardness keeps the waste industry from graduating above small-scale local initiatives. And higher pollution occurs because there isn't the technology to capture highly dispersed waste such as sulfur dioxide from smokestacks or heavy metals that flow into wastewater.

A number of possible explanations for this observed relationship between pollution and income were advanced:

- As local economies grow and develop, they will inevitably change the way they use resources, creating different types of impacts upon the environment. A simple example is the pollution tradeoffs involved from our transition in transportation modes from horses to cars. Horses generated plenty of pollution in terms of manure, carcass disposal, etc. Cars, of course, generate an entirely different brand of pollution concerns. In other words, some environmental degradation along a country's development path is inevitable, especially during the take-off process of industrialization.
- Growth is associated with an increasing share of services and high-technology production, both of which tend to be more environment-friendly than production processes in earlier stages of industrialization.
- Knowledge and technology from the developed world can help ease this transition and lessen its duration, moving countries more quickly to the levels at which pollution will be decreasing. Free trade can promote a quicker diffusion of environment-friendly technologies and lead to a more efficient allocation of resources.
- The prosperity generated from economic activity will lead to more investments and higher standards of living that enable still greater investments in cleaner and newer technologies and processes. When a certain level of per capita income is reached, economic growth helps to undo the damage done in earlier years. As free trade expands, each 1 percent increase in per capita income tends to drive pollution concentrations down by 1.25 to 1.5 percent because of the movement to cleaner techniques of production.
- As individuals become richer they are willing to spend more on non-material goods, such as a cleaner environment. This point is made by Indur M. Goklany, in his description of earlier stages of development, "Society [initially] places a much higher priority on acquiring basic public health and other services such as sewage treatment, water supply, and electricity than on environmental quality, which initially worsens. But as the original priorities are met, environmental problems become higher priorities. More resources are devoted to solving those problems. Environmental degradation is arrested and then reversed."

These findings and explanations, unsurprisingly, generated an outpouring of negative response from environmental activists and anti-globalization proponents. "How can these economists be serious?" they, in effect, asked. "Do they really think it is wise to advocate policies that predictably increase pollution? Are we supposed to believe pollution will eventually decrease if we continue with the polluting activities? How absolutely ludicrous!"

Typical responses to the "growth is good" thesis include:

- Globalization will result in a "race to the bottom" as polluting companies relocate to countries with lax environmental standards.
- Trading with countries that do not have suitable environmental laws will lower environmental standards for all countries.
- Multinationals will exploit pristine environments in the developing world, reaping the resources for short-term growth, and then pulling out to repeat the process elsewhere—growth ruins the environment.
- Free trade provides a license to pollute—it is bad for the environment. Stronger environmental regulations at national and international levels are needed.

The Sierra Club summarized the widespread critiques to the Grossman and Krueger studies, drawing from research studies produced by the World Wide Fund for Nature and others. It argued that the findings were sufficiently over-generalized to dispense with any notion that they justify complacency about trade and the environment, pointing out several facts.

The empirical estimates of where "turning points" occur for different pollutants vary so widely as to cast doubt on the validity of any one set of results. For instance, where Grossman and Krueger found turning points for certain air pollutants at less than $5,000 per capita, others found turning points above $8,000 per capita.

For some air pollutants, Grossman and Krueger found that emissions levels don't follow an inverted U-curve, but following an S-curve that starts to rise again as incomes rise. For instance, they found that sulphur dioxide emissions start to rise when income increases above $14,000 per capita. The implication is that efficiency gains from improved technology at medium levels of per capita income are eventually overwhelmed by the growing size of the economy.

Since most of the world's population earns per capita incomes well below estimated turning points, global air pollution levels will continue to rise for nearly another century. By that time, emissions of some pollutants will be anywhere from two to four times higher than current levels.

Even for the limited number of pollutants that Grossman and Krueger study, they only demonstrate a correlation between changing per capita income and changing levels of environmental quality. They do not demonstrate a causal connection. The positive relationships they describe could actually be caused by noneconomic factors, such as the adoption of environmental legislation.

Both camps seem to have reasonable grounds for their views. Clearly there is a conflict of visions that is rooted in very different value systems. Can these two opposing perspectives be reconciled sufficiently to reach some type of consensus?

As noted earlier, many studies have re-examined the Environmental Kuznets curve since the publication of the Grossman and Krueger analysis in 1991, each attempting to prove or disprove the relationship between economic growth and environmental quality, or to isolate variables that may explain the observed relationships. In that same year, a fascinating monograph was published in London, called The Wealth of Nations and the Environment. Author Mikhail Bernstam set out to analyze the contention that economic growth negatively impacts the environment by examining how institutional structure impacts this relationship.

Bernstam examined and contrasted the impact of economic growth upon the environment in both capitalist and socialist countries. Interestingly, he found that the environmental Kuznets curve does in fact exist, but it does not apply to countries across the board. The Kuznets curve, he found, applies to market economies, but not to socialist ones. The difference, according to Bernstam, has its roots in the different structures of incentives and property rights of these two economic systems.

Under market economies with secure property rights and open trade, the pursuit of profits leads to the husbanding of resources. These capitalist economies use fewer resources to produce the equivalent level of output and hence do less damage to the environment. In contrast, in socialist countries, the managers of state enterprises operate under incentives that encourage them to maximize inputs, with little regard towards economic waste or damage to the environment.

More recently, a 2001 study by economists Werner Antweiler, Brian R. Copeland, and M. Scott Taylor asked, "Is Free Trade Good for the Environment?" They analyzed data on sulfur dioxide over the period 1971 to 1996, a time when trade barriers were coming down and international trade was expanding. They found that countries that opened up to trade generated faster economic growth. Although economic growth produced more pollution, the greater wealth and higher incomes also generated a demand for a cleaner environment.

To separate these effects, the Antweiler model looked at the negative environmental consequences of increases in economic activity (the scale effect), the positive environmental consequences of increases in income that lead to cleaner production methods (the technique effect), and the impact of trade-induced changes in the composition of output upon pollution concentrations (the composition effect). When the scale, technique and composition effects estimates were combined, the Antweiler et al.model yielded the conclusion that free trade is good for the environment. For example, when analyzing sulfur dioxide, the authors estimate that for each 1 percent increase in per capita income in a nation, pollution *falls* by 1 percent.

The critical explanatory factor is that wealthier countries value environmental amenities more highly and enhance their production by employing environmentally friendly technologies. However, like Bernstam, these authors specified that it is important to distinguish between communist and non-communist countries. Communist countries provided the exception to their rule about globalization's positive impacts upon the environment.

The studies, which consider the impact of institutional structures, make an important contribution to our understanding of the "economics vs. environment" debates. They suggest we consider other factors in our analysis of the effects of globalization. It is true that we often do find examples of disastrous environmental conditions, particularly when we look at socialist countries. But it is misleading to attribute the disasters to globalization. Instead, we need to examine the institutional arrangements in a particular country to see what role they play in economic development and environmental protection.

Positive Globalization

As described earlier, at its most rudimentary level, globalization simply embodies a process of free and open trade, whereby people can make decisions about who their trading partners are and what opportunities they will choose to pursue.

But the cautions of the environmental activists are worthy of consideration. Free trade, in and of itself, will not guarantee positive outcomes. We also need guiding rules that essentially create the terms for fair and civil interaction.

In *Property Rights: A Practical Guide to Freedom & Prosperity,* Terry Anderson and Laura Huggins describe the importance of institutional rules. They use the example of children playing together and inventing games. In essence, the children work together to form rules that are fair. When they cannot agree on rules, chaos typically results and their play breaks down. The same is true for civil society. Institutional rules, in the form of constitutions, common law, and so on, provide the structure for human activity.

The critical role of institutions in shaping human behavior gained international attention in 1993 when Douglass C. North received the Nobel Prize in economics. North's groundbreaking research in economic history integrated economics, sociology, statistics and history to explain the role that institutions play in economic growth.

For several decades, North looked at the question, "Why do some countries become rich, while others remain poor?" In seeking answers to this query, he came to understand that institutions establish the formal and informal sets of rules that govern the behavior of human beings in a society. His research showed that, depending on their structure and enforcement, institutional arrangements can either foster or restrain economic development.

For the past nine years, the *Index of Economic Freedom,* jointly published by the Wall Street Journal and the Heritage Foundation (Washington, DC), has provided fascinating empirical evidence of the relationship of various institutions to economic prosperity. The study analyzes and ranks the economic freedom of 161 countries according to 10 institutional factors (trade policy, property rights, regulation, and black market, for example) in an effort to trace the path to economic prosperity.

The key finding of the research, supported year after year, is that countries with the most economic freedom enjoy higher rates of long-term economic growth and prosperity than those with less economic freedom. But, more relevant to this discussion, is the finding that economic freedom, which enables people to choose who and where their trading partners are, ultimately leads to more efficient resource use.

In another comparative index, *Economic Freedom of the World 2002,* published by the Fraser Institute in conjunction with public policy institutes around the world, Nobel laureate Milton Friedman describes the importance of private property and the rule of law as a basis for economic freedom. He spells out the three key ingredients key to establishing economic freedom as follows: "First of all, and most important, the rule of law, which extends to the protection of property. Second, widespread private ownership of the means of production. Third, freedom to enter or to leave industries, freedom of competition, freedom of trade. Those are essentially the basic requirements." These same factors also provide a framework for positive environmental development.

In the 1980s, a team of economists affiliated with the Property and Environmental Research Center (PERC) in Bozeman, MT, began developing a new paradigm for environmental policy. Their model, which eventually was coined "Free Market Environmentalism" described how incentives are the key to environmental stewardship. Not surprisingly, people who face little or no consequences for environmentally destructive actions face no incentive to protect the environment. Alternatively, people who are rewarded for good stewardship are much more likely to invest in environmental protection. The key, according to economists John Baden, Richard Stroup and Terry Anderson, are the very same three elements that Milton Friedman mentioned for economic prosperity: free and open markets, clearly established property rights, and rule of law.

Free and open markets. One of the most important benefits produced by a market economy is information, conveyed in the form of prices. Prices of natural and environmental resources provide clear signals about their availability. As a resource becomes scarcer, its price increases. And of course, the reverse is also true: When a resource becomes more abundant, the price decreases.

Many people fear that the profit motive leads to the depletion or degradation of environmental resources. As counterintuitive as it may sound, the profit motive actually works to the benefit of the environment.

Businesses face incentives to carefully consider the prices of the various natural resources that they use in their production processes. If a particular resource is in short supply, its price will be higher than others that are more readily available. It makes little sense for a producer to over utilize, or "waste," a high-priced resource.

High prices also encourage the search for, and development of, appropriate substitutes or alternatives. As companies search for ways to reduce costs, they naturally tend toward utilizing lower-priced, more abundant resources. Thus, the pursuit of profits is actually a driving force to conserve resources. In essence, under free market systems, entrepreneurs compete in developing low cost, efficient means to solve contemporary resource problems.

Property rights. Clearly established property rights generate another incentive for environmental stewardship. It makes no sense for private landowners, for example, to exploit and destroy their own property. Ownership creates a long-term perspective that leads to preserving and protecting property.

Careless destruction, however, does make sense for those who are only loosely held accountable for their actions. Politicians, bureaucrats, or others, who may be short-term managers, face the incentive to maximize immediate returns, even if this means long-term environmental damage. Even managers with longer tenures realize they can simply turn to the federal government for more funds to address the problems that short-sighted decision making may have created.

Rule of law. In many ways, the "rule of law" is the glue that holds market transactions and property rights together. Freedom to exchange is meaningless if individuals do not have secure rights to property, including the fruits of their labors. Failure of a country's legal system to provide for the security of property rights, enforcement of contracts, and the mutually agreeable settlement of disputes will undermine the operation of a market-exchange system. If individuals and businesses lack confidence that contracts will be enforced and the returns from their productive activity protected, their incentive to engage in innovative activities will be eroded.

With these elements in place, the economists' explanations prevail—globalization will enable local cultures to pick and choose the development and environmental paths that they wish to traverse. But without these institutional arrangements, the likelihood of negative consequences increases.

In countries that lack property rights and rule of law and that promote barriers to trade, an institutional structure develops that fosters destruction of the environment. For example, in Liberia, former President Charles Taylor rapidly sold off many of the

In the time it takes to read this introduction, another 5,000 people will inhabit the Earth. Thousands or even millions of people cannot do much environmental damage to planet Earth, but with a population in excess of 6.6 billion people, resources are bound to be consumed without being replenished, and long-term environmental damage is almost unavoidable. For generations, demographers (those who study population trends) have predicted that the Earth was nearing its carrying capacity, and every time they were proven wrong. Population has doubled since 1960. In 1800, it was less than 1 billion compared to today's 6 billion. A historic event occurred in the mid 1960s, where, for the first time, the *rate* of population growth slowed. The World's population is still increasing, but just at a slower rate than before. Even so, the World's population is expected to approach 10 billion by the year 2050 (U.S. Census Bureau). One thing is clear: population growth is undoubtedly limited by the availability of resources and these resources are not infinite. Sustainable development is necessary for our population to continue with a lifestyle that is not seriously compromised.

To meet the needs of the growing population, more marginal lands are being cultivated and more effective methods are being used to maximize crop production. Unfortunately, these efforts, such as increased fertilization and disruption of soil, have long-term deleterious effects. The nutrients in fertilizers create algal blooms in rivers and the oceans, adversely affecting the natural ecosystems. Soil erosion is an irreversible process that can and has led to expansive desertification of marginal farmlands. This, in turn, leads to even greater mass migration. Monocrops crowd out the natural biological community, and genetic engineering, even with all of its potential benefits, is an unknown commodity. The potential disruption to the natural ecosystem could have irreconcilable consequences for the finely tuned and delicate ecological balance that exists in the Earth's biomes.

Population growth puts pressure on our water resources as well. The greatest concern for the coming century is not a lack of petroleum, but a lack of water. Aquifers are being pumped with little thought for the future. Water supplies are often polluted, or have dried up. Crop failures, starvation, and mass migration can be the tragic outcome.

Ironically, population growth is highest in regions that can least afford the additional people. A number of so-called 'first world' nations are experiencing negative population growth rates. Contrast this with several poor equatorial African nations where annual growth rates exceed 3% and even 4%. Only with proper education, economic security, and better health care, can we hope to stabilize population growth in these regions.

In "Population and Consumption: What We Know, What We Need to Know" MacArthur Fellow Robert Kates summarizes the

Flat Earth Images

problem of population growth and consumption and proposes action that would minimize the future hazards, namely reducing our need for ever-greater levels of consumption. Gregory Foster follows with "A New Security Paradigm" and a recognition that nations need to band together in order to combat the negative effects of global warming. Only if climate change is viewed as a national and global threat far more serious than terrorism, will the necessary action be taken to combat it.

The environmental damage incurred by explosive population growth is exacerbated by climate change caused by global warming. As the planet warms, droughts are expected to become more extreme. Rising sea level and shoreline erosion displaces low-lying communities. Desertification expands, displacing entire populations. As we become aware of the problem, numerous solutions have been proposed. One idea to reduce carbon emissions is 'carbon offsets,' in which companies or individuals buy into green technology. Alan Zarembo examines this trend in "Can You Buy a Greener Conscience," and finds that it offers mostly hype and little benefits. The remaining articles reaffirm the strong commitments that must be made at all levels to limit the predicted damage due to anthropogenic factors. Strong leadership and international agreements are necessary to reduce the levels of greenhouse gases. And at the local level, a change in our attitudes towards 'green technology' must occur. All is not lost, however. An educated public, concerned about the environment, is necessary to insure a change in our ways.

Population and Consumption
What We Know, What We Need to Know

ROBERT W. KATES

Thirty years ago, as Earth Day dawned, three wise men recognized three proximate causes of environmental degradation yet spent half a decade or more arguing their relative importance. In this classic environmentalist feud between Barry Commoner on one side and Paul Ehrlich and John Holdren on the other, all three recognized that growth in population, affluence, and technology were jointly responsible for environmental problems, but they strongly differed about their relative importance. Commoner asserted that technology and the economic system that produced it were primarily responsible.[1] Ehrlich and Holdren asserted the importance of all three drivers: population, affluence, and technology. But given Ehrlich's writings on population,[2] the differences were often, albeit incorrectly, described as an argument over whether population or technology was responsible for the environmental crisis.

Now, 30 years later, a general consensus among scientists posits that growth in population, affluence, and technology are jointly responsible for environmental problems. This has become enshrined in a useful, albeit overly simplified, identity known as IPAT, first published by Ehrlich and Holdren in *Environment* in 1972[3] in response to the more limited version by Commoner that had appeared earlier in *Environment* and in his famous book *The Closing Circle*.[4] In this identity, various forms of environmental or resource impacts (I) equals population (P) times affluence (A) (usually income per capita) times the impacts per unit of income as determined by technology (T) and the institutions that use it. Academic debate has now shifted from the greater or lesser importance of each of these driving forces of environmental degradation or resource depletion to debate about their interaction and the ultimate forces that drive them.

However, in the wider global realm, the debate about who or what is responsible for environmental degradation lives on. Today, many Earth Days later, international debates over such major concerns as biodiversity, climate change, or sustainable development address the population and the affluence terms of Holdrens' and Ehrlich's identity, specifically focusing on the character of consumption that affluence permits. The concern with technology is more complicated because it is now widely recognized that while technology can be a problem, it can be a solution as well. The development and use of more environmentally benign and friendly technologies in industrialized countries have slowed the growth of many of the most pernicious forms of pollution that originally drew Commoner's attention and still dominate Earth Day concerns.

A recent report from the National Research Council captures one view of the current public debate, and it begins as follows:

> *For over two decades, the same frustrating exchange has been repeated countless times in international policy circles. A government official or scientist from a wealthy country would make the following argument: The world is threatened with environmental disaster because of the depletion of natural resources (or climate change or the loss of biodiversity), and it cannot continue for long to support its rapidly growing population. To preserve the environment for future generations, we need to move quickly to control global population growth, and we must concentrate the effort on the world's poorer countries, where the vast majority of population growth is occurring.*

Government officials and scientists from low-income countries would typically respond:

> *If the world is facing environmental disaster, it is not the fault of the poor, who use few resources. The fault must lie with the world's wealthy countries, where people consume the great bulk of the world's natural resources and energy and cause the great bulk of its environmental degradation. We need to curtail overconsumption in the rich countries which use far more than their fair share, both to preserve the environment and to allow the poorest people on earth to achieve an acceptable standard of living.*[5]

It would be helpful, as in all such classic disputes, to begin by laying out what is known about the relative responsibilities of both population and consumption for the environmental crisis, and what might need to be known to address them. However, there is a profound asymmetry that must fuel the frustration of the developing countries' politicians and scientists: namely, how much people know about population and how little they know about consumption. Thus, this article begins by examining these differences in knowledge and action and concludes with the alternative actions needed to go from more to enough in both population and consumption.[6]

Population

What population is and how it grows is well understood even if all the forces driving it are not. Population begins with people and their key events of birth, death, and location. At the margins, there is some debate over when life begins and ends or whether residence is temporary or permanent, but little debate in between. Thus, change in the world's population or any place is the simple arithmetic of adding births, subtracting deaths, adding immigrants, and subtracting outmigrants. While whole subfields of demography are devoted to the arcane details of these additions and subtractions, the error in estimates of population for almost all places is probably within 20 percent and for countries with modern statistical services, under 3 percent—better estimates than for any other living things and for most other environmental concerns.

Current world population is more than six billion people, growing at a rate of 1.3 percent per year. The peak annual growth rate in all history—about 2.1 percent—occurred in the early 1960s, and the peak population increase of around 87 million per year occurred in the late 1980s. About 80 percent or 4.8 billion people live in the less developed areas of the world, with 1.2 billion living in industrialized countries. Population is now projected by the United Nations (UN) to be 8.9 billion in 2050, according to its medium fertility assumption, the one usually considered most likely, or as high as 10.6 billion or as low as 7.3 billion.[7]

A general description of how birth rates and death rates are changing over time is a process called the demographic transition.[8] It was first studied in the context of Europe, where in the space of two centuries, societies went from a condition of high births and high deaths to the current situation of low births and low deaths. In such a transition, deaths decline more rapidly than births, and in that gap, population grows rapidly but eventually stabilizes as the birth decline matches or even exceeds the death decline. Although the general description of the transition is widely accepted, much is debated about its cause and details.

The world is now in the midst of a global transition that, unlike the European transition, is much more rapid. Both births and deaths have dropped faster than experts expected and history foreshadowed. It took 100 years for deaths to drop in Europe compared to the drop in 30 years in the developing world. Three is the current global average births per woman of reproductive age. This number is more than halfway between the average of five children born to each woman at the post World War II peak of population growth and the average of 2.1 births required to achieve eventual zero population growth.[9] The death transition is more advanced, with life expectancy currently at 64 years. This represents three-quarters of the transition between a life expectancy of 40 years to one of 75 years. The current rates of

decline in births outpace the estimates of the demographers, the UN having reduced its latest medium expectation of global population in 2050 to 8.9 billion, a reduction of almost 10 percent from its projection in 1994.

Demographers debate the causes of this rapid birth decline. But even with such differences, it is possible to break down the projected growth of the next century and to identify policies that would reduce projected populations even further. John Bongaarts of the Population Council has decomposed the projected developing country growth into three parts and, with his colleague Judith Bruce, has envisioned policies that would encourage further and more rapid decline.[10] The first part is unwanted fertility, making available the methods and materials for contraception to the 120 million married women (and the many more unmarried women) in developing countries who in survey research say they either want fewer children or want to space them better. A basic strategy for doing so links voluntary family planning with other reproductive and child health services.

Yet in many parts of the world, the desired number of children is too high for a stabilized population. Bongaarts would reduce this desire for large families by changing the costs and benefits of childrearing so that more parents would recognize the value of smaller families while simultaneously increasing their investment in children. A basic strategy for doing so accelerates three trends that have been shown to lead to lower desired family size: the survival of children, their education, and improvement in the economic, social, and legal status for girls and women.

However, even if fertility could immediately be brought down to the replacement level of two surviving children per woman, population growth would continue for many years in most developing countries because so many more young people of reproductive age exist. So Bongaarts would slow this momentum of population growth by increasing the age of childbearing, primarily by improving secondary education opportunity for girls and by addressing such neglected issues as adolescent sexuality and reproductive behavior.

How much further could population be reduced? Bongaarts provides the outer limits. The population of the developing world (using older projections) was expected to reach 10.2 billion by 2100. In theory, Bongaarts found that meeting the unmet need for contraception could reduce this total by about 2 billion. Bringing down desired family size to replacement fertility would reduce the population a billion more, with the remaining growth—from 4.5 billion today to 7.3 billion in 2100—due to population momentum. In practice, however, a recent U.S. National Academy of Sciences report concluded that a 10 percent reduction is both realistic and attainable and could lead to a lessening in projected population numbers by 2050 of upwards of a billion fewer people.[11]

Consumption

In contrast to population, where people and their births and deaths are relatively well-defined biological events, there is no consensus as to what consumption includes. Paul Stern of the

National Research Council has described the different ways physics, economics, ecology, and sociology view consumption.[12] For physicists, matter and energy cannot be consumed, so consumption is conceived as transformations of matter and energy with increased entropy. For economists, consumption is spending on consumer goods and services and thus distinguished from their production and distribution. For ecologists, consumption is obtaining energy and nutrients by eating something else, mostly green plants or other consumers of green plants. And for some sociologists, consumption is a status symbol—keeping up with the Joneses—when individuals and households use their incomes to increase their social status through certain kinds of purchases. These differences are summarized in the box below.

In 1977, the councils of the Royal Society of London and the U.S. National Academy of Sciences issued a joint statement on consumption, having previously done so on population. They chose a variant of the physicist's definition:

> *Consumption is the human transformation of materials and energy. Consumption is of concern to the extent that it makes the transformed materials or energy less available for future use, or negatively impacts biophysical systems in such a way as to threaten human health, welfare, or other things people value.*[13]

On the one hand, this society/academy view is more holistic and fundamental than the other definitions; on the other hand, it is more focused, turning attention to the environmentally damaging. This article uses it as a working definition with one modification, the addition of information to energy and matter, thus completing the triad of the biophysical and ecological basics that support life.

In contrast to population, only limited data and concepts on the transformation of energy, materials, and information exist.[14] There is relatively good global knowledge of energy transformations due in part to the common units of conversion between different technologies. Between 1950 and today, global energy production and use increased more than fourfold.[15] For material transformations, there are no aggregate data in common units on a global basis, only for some specific classes of materials including materials for energy production, construction, industrial minerals and metals, agricultural crops, and water.[16] Calculations of material use by volume, mass, or value lead to different trends.

Trend data for per capita use of physical structure materials (construction and industrial minerals, metals, and forestry products) in the United States are relatively complete. They show an inverted S shaped (logistic) growth pattern: modest doubling between 1900 and the depression of the 1930s (from two to four metric tons), followed by a steep quintupling with economic recovery until the early 1970s (from two to eleven tons), followed by a leveling off since then with fluctuations related to economic downturns (see Figure 1).[17] An aggregate analysis of all current material production and consumption in the United States averages more than 60 kilos per person per day (excluding water). Most of this material flow is split between energy and related products (38 percent) and minerals

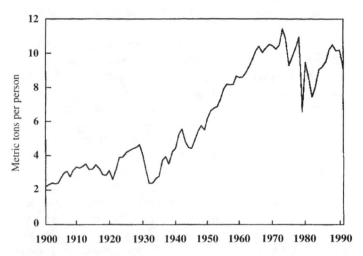

Figure 1 Consumption of Physical Structure Materials in the United States, 1900–1991.

Source: I. Wernick, "Consuming Materials: The American Way," *Technological Forecasting and Social Change*, 53 (1996): 114.

for construction (37 percent), with the remainder as industrial minerals (5 percent), metals (2 percent), products of fields (12 percent), and forest (5 percent).[18]

A massive effort is under way to catalog biological (genetic) information and to sequence the genomes of microbes, worms, plants, mice, and people. In contrast to the molecular detail, the number and diversity of organisms is unknown, but a conservative estimate places the number of species on the order of 10 million, of which only one-tenth have been described.[19] Although there is much interest and many anecdotes, neither concepts nor data are available on most cultural information. For example, the number of languages in the world continues to decline while the number of messages expands exponentially.

Trends and projections in agriculture, energy, and economy can serve as surrogates for more detailed data on energy and material transformation.[20] From 1950 to the early 1990s, world population more than doubled (2.2 times), food as measured by grain production almost tripled (2.7 times), energy more than quadrupled (4.4 times), and the economy quintupled (5.1 times). This 43-year record is similar to a current 55-year projection (1995–2050) that assumes the continuation of current trends or, as some would note, "business as usual." In this 55-year projection, growth in half again of population (1.6 times) finds almost a doubling of agriculture (1.8 times), more

---

What Is Consumption?

Physicist: "What happens when you transform matter/energy"
Ecologist: "What big fish do to little fish"
Economist: "What consumers do with their money"
Sociologist: "What you do to keep up with the Joneses"

than twice as much energy used (2.4 times), and a quadrupling of the economy (4.3 times).[21]

Thus, both history and future scenarios predict growth rates of consumption well beyond population. An attractive similarity exists between a demographic transition that moves over time from high births and high deaths to low births and low deaths with an energy, materials, and information transition. In this transition, societies will use increasing amounts of energy and materials as consumption increases, but over time the energy and materials input per unit of consumption decrease and information substitutes for more material and energy inputs.

Some encouraging signs surface for such a transition in both energy and materials, and these have been variously labeled as decarbonization and dematerialization.[22] For more than a century, the amount of carbon per unit of energy produced has been decreasing. Over a shorter period, the amount of energy used to produce a unit of production has also steadily declined. There is also evidence for dematerialization, using fewer materials for a unit of production, but only for industrialized countries and for some specific materials. Overall, improvements in technology and substitution of information for energy and materials will continue to increase energy efficiency (including decarbonization) and dematerialization per unit of product or service. Thus, over time, less energy and materials will be needed to make specific things. At the same time, the demand for products and services continues to increase, and the overall consumption of energy and most materials more than offsets these efficiency and productivity gains.

What to Do about Consumption

While quantitative analysis of consumption is just beginning, three questions suggest a direction for reducing environmentally damaging and resource-depleting consumption. The first asks: *When is more too much for the life-support systems of the natural world and the social infrastructure of human society?* Not all the projected growth in consumption may be resource-depleting—"less available for future use"—or environmentally damaging in a way that "negatively impacts biophysical systems to threaten human health, welfare, or other things people value."[23] Yet almost any human-induced transformations turn out to be either or both resource-depleting or damaging to some valued environmental component. For example, a few years ago, a series of eight energy controversies in Maine were related to coal, nuclear, natural gas, hydroelectric, biomass, and wind generating sources, as well as to various energy policies. In all the controversies, competing sides, often more than two, emphasized environmental benefits to support their choice and attributed environmental damage to the other alternatives.

Despite this complexity, it is possible to rank energy sources by the varied and multiple risks they pose and, for those concerned, to choose which risks they wish to minimize and which they are more willing to accept. There is now almost 30 years of experience with the theory and methods of risk assessment and 10 years of experience with the identification and setting of environmental priorities. While there is still no readily accepted methodology for separating resource-depleting or environmentally damaging consumption from general consumption or for identifying harmful transformations from those that are benign, one can separate consumption into more or less damaging and depleting classes and *shift* consumption to the less harmful class. It is possible to *substitute* less damaging and depleting energy and materials for more damaging ones. There is growing experience with encouraging substitution and its difficulties: renewables for nonrenewables, toxics with fewer toxics, ozone-depleting chemicals for more benign substitutes, natural gas for coal, and so forth.

The second question, *Can we do more with less?,* addresses the supply side of consumption. Beyond substitution, shrinking the energy and material transformations required per unit of consumption is probably the most effective current means for reducing environmentally damaging consumption. In the 1997 book, *Stuff: The Secret Lives of Everyday Things,* John Ryan and Alan Durning of Northwest Environment Watch trace the complex origins, materials, production, and transport of such everyday things as coffee, newspapers, cars, and computers and highlight the complexity of reengineering such products and reorganizing their production and distribution.[24]

Yet there is growing experience with the three Rs of consumption shrinkage: reduce, recycle, reuse. These have now been strengthened by a growing science, technology, and practice of industrial ecology that seeks to learn from nature's ecology to reuse everything. These efforts will only increase the existing favorable trends in the efficiency of energy and material usage. Such a potential led the Intergovernmental Panel on Climate Change to conclude that it was possible, using current best practice technology, to reduce energy use by 30 percent in the short run and 50–60 percent in the long run.[25] Perhaps most important in the long run, but possibly least studied, is the potential for and value of substituting information for energy and materials. Energy and materials per unit of consumption are going down, in part because more and more consumption consists of information.

The third question addresses the demand side of consumption —*When is more enough?*[26] Is it possible to reduce consumption by more satisfaction with what people already have, by *satiation,* no more needing more because there is enough, and by *sublimation,* having more satisfaction with less to achieve some greater good? This is the least explored area of consumption and the most difficult. There are, of course, many signs of *satiation* for some goods. For example, people in the industrialized world no longer buy additional refrigerators (except in newly formed households) but only replace them. Moreover, the quality of refrigerators has so improved that a 20-year or more life span is commonplace. The financial pages include frequent stories of the plight of this industry or corporation whose markets are saturated and whose products no longer show the annual growth equated with profits and progress. Such enterprises are frequently viewed as failures of marketing or entrepreneurship rather than successes in meeting human needs sufficiently and efficiently. Is it possible to reverse such views, to create a standard of satiation, a satisfaction in a need well met?

Can people have more satisfaction with what they already have by using it more intensely and having the time to do so? Economist Juliet Schor tells of some overworked Americans who would willingly exchange time for money, time to spend with family and using what they already have, but who are constrained by an uncooperative employment structure.[27] Proposed U.S. legislation would permit the trading of overtime for such compensatory time off, a step in this direction. *Sublimation,* according to the dictionary, is the diversion of energy from an immediate goal to a higher social, moral, or aesthetic purpose. Can people be more satisfied with less satisfaction derived from the diversion of immediate consumption for the satisfaction of a smaller ecological footprint?[28] An emergent research field grapples with how to encourage consumer behavior that will lead to change in environmentally damaging consumption.[29]

A small but growing "simplicity" movement tries to fashion new images of "living the good life."[30] Such movements may never much reduce the burdens of consumption, but they facilitate by example and experiment other less-demanding alternatives. Peter Menzel's remarkable photo essay of the material goods of some 30 households from around the world is powerful testimony to the great variety and inequality of possessions amidst the existence of alternative life styles.[31] Can a standard of "more is enough" be linked to an ethic of "enough for all"? One of the great discoveries of childhood is that eating lunch does not feed the starving children of some far-off place. But increasingly, in sharing the global commons, people flirt with mechanisms that hint at such—a rationing system for the remaining chlorofluorocarbons, trading systems for reducing emissions, rewards for preserving species, or allowances for using available resources.

A recent compilation of essays, *Consuming Desires: Consumption, Culture, and the Pursuit of Happiness,*[32] explores many of these essential issues. These elegant essays by 14 well-known writers and academics ask the fundamental question of why more never seems to be enough and why satiation and sublimation are so difficult in a culture of consumption. Indeed, how is the culture of consumption different for mainstream America, women, inner-city children, South Asian immigrants, or newly industrializing countries?

Why We Know and Don't Know

In an imagined dialog between rich and poor countries, with each side listening carefully to the other, they might ask themselves just what they actually know about population and consumption. Struck with the asymmetry described above, they might then ask: "Why do we know so much more about population than consumption?"

The answer would be that population is simpler, easier to study, and a consensus exists about terms, trends, even policies. Consumption is harder, with no consensus as to what it is, and with few studies except in the fields of marketing and advertising. But the consensus that exists about population comes from substantial research and study, much of it funded by governments and groups in rich countries, whose asymmetric concern readily identifies the troubling fertility behavior of others and only reluctantly considers their own consumption behavior. So while consumption is harder, it is surely studied less (see Table 1).

The asymmetry of concern is not very flattering to people in developing countries. Anglo-Saxon tradition has a long history of dominant thought holding the poor responsible for their condition—they have too many children—and an even longer tradition of urban civilization feeling besieged by the barbarians at their gates. But whatever the origins of the asymmetry, its persistence does no one a service. Indeed, the stylized debate of population versus consumption reflects neither popular understanding nor scientific insight. Yet lurking somewhere beneath the surface concerns lies a deeper fear.

Consumption is more threatening, and despite the North–South rhetoric, it is threatening to all. In both rich and poor countries alike, making and selling things to each other, including unnecessary things, is the essence of the economic system. No longer challenged by socialism, global capitalism seems inherently based on growth—growth of both consumers and their consumption. To study consumption in this light is to risk concluding that a transition to sustainability might require profound changes in the making and selling of things and in the opportunities that this provides. To draw such conclusions, in the absence of convincing alternative visions, is fearful and to be avoided.

What We Need to Know and Do

In conclusion, returning to the 30-year-old IPAT identity—a variant of which might be called the Population/Consumption (PC) version—and restating that identity in terms of population and consumption, it would be: $I = P^*C/P^*I/C$, where I equals environmental degradation and/or resource depletion; P equals the number of people or households; and C equals the transformation of energy, materials, and information (see Figure 2).

With such an identity as a template, and with the goal of reducing environmentally degrading and resource-depleting influences, there are at least seven major directions for research and policy. To reduce the level of impacts per unit of consumption, it is necessary to separate out more damaging consumption and shift to less harmful forms, *shrink* the amounts of environmentally damaging energy and materials per unit of consumption, and *substitute* information for energy and materials. To reduce

Table 1   A Comparison of Population and Consumption

Population	Consumption
Simpler, easier to study	More complex
Well-funded research	Unfunded, except marketing
Consensus terms, trends	Uncertain terms, trends
Consensus policies	Threatening policies

Source: Robert W. Kates.

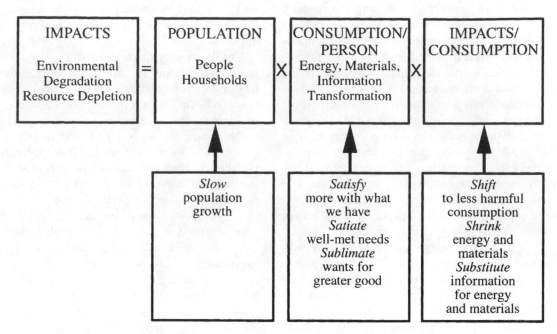

Figure 2 IPAT (Population/Consumption Version): A Template for Action.
Source: Robert W. Kates.

consumption per person or household, it is necessary to *satisfy* more with what is already had, *satiate* well-met consumption needs, and *sublimate* wants for a greater good. Finally, it is possible to *slow* population growth and then to *stabilize* population numbers as indicated above.

However, as with all versions of the IPAT identity, population and consumption in the PC version are only proximate driving forces, and the ultimate forces that drive consumption, the consuming desires, are poorly understood, as are many of the major interventions needed to reduce these proximate driving forces. People know most about slowing population growth, more about shrinking and substituting environmentally damaging consumption, much about shifting to less damaging consumption, and least about satisfaction, satiation, and sublimation. Thus the determinants of consumption and its alternative patterns have been identified as a key understudied topic for an emerging sustainability science by the recent U.S. National Academy of Science study.[33]

But people and society do not need to know more in order to act. They can readily begin to separate out the most serious problems of consumption, shrink its energy and material throughputs, substitute information for energy and materials, create a standard for satiation, sublimate the possession of things for that of the global commons, as well as slow and stabilize population. To go from more to enough is more than enough to do for 30 more Earth Days.

Notes

1. B. Commoner, M. Corr, and P. Stamler, "The Causes of Pollution," *Environment,* April 1971, 2–19.

2. P. Ehrlich, *The Population Bomb* (New York: Ballantine, 1966).

3. P. Ehrlich and J. Holdren, "Review of The Closing Circle," *Environment,* April 1972, 24–39.

4. B. Commoner, *The Closing Circle* (New York: Knopf, 1971).

5. P. Stern, T. Dietz, V. Ruttan, R. H. Socolow, and J. L. Sweeney, eds., *Environmentally Significant Consumption: Research Direction* (Washington, D.C.: National Academy Press, 1997), 1.

6. This article draws in part upon a presentation for the 1997 De Lange-Woodlands Conference, an expanded version of which will appear as: R. W. Kates, "Population and Consumption: From More to Enough," in *In Sustainable Development: The Challenge of Transition,* J. Schmandt and C. H. Wards, eds. (Cambridge, U.K.: Cambridge University Press, forthcoming), 79–99.

7. United Nations, Population Division, *World Population Prospects: The 1998 Revision* (New York: United Nations, 1999).

8. K. Davis, "Population and Resources: Fact and Interpretation," K. Davis and M. S. Bernstam, eds., in *Resources, Environment and Population: Present Knowledge, Future Options,* supplement to *Population and Development Review,* 1990: 1–21.

9. Population Reference Bureau, *1997 World Population Data Sheet of the Population Reference Bureau* (Washington, D.C.: Population Reference Bureau, 1997).

10. J. Bongaarts, "Population Policy Options in the Developing World," *Science,* 263: (1994), 771–776; and J. Bongaarts and J. Bruce, "What Can Be Done to Address Population Growth?" (unpublished background paper for The Rockefeller Foundation, 1997).

11. National Research Council, Board on Sustainable Development, *Our Common Journey: A Transition Toward Sustainability* (Washington, D.C.: National Academy Press, 1999).

12. 12. See Stern, et al., note 5 above.

13. Royal Society of London and the U.S. National Academy of Sciences, "Towards Sustainable Consumption," reprinted in *Population and Development Review,* 1977, 23 (3): 683–686.

14. For the available data and concepts, I have drawn heavily from J. H. Ausubel and H. D. Langford, eds., *Technological Trajectories and the Human Environment.* (Washington, D.C.: National Academy Press, 1997).

15. L. R. Brown, H. Kane, and D. M. Roodman, *Vital Signs 1994: The Trends That Are Shaping Our Future* (New York: W. W. Norton and Co., 1994).

16. World Resources Institute, United Nations Environment Programme, United Nations Development Programme, World Bank, *World Resources, 1996–97* (New York: Oxford University Press, 1996); and A. Gruebler, *Technology and Global Change* (Cambridge, Mass.: Cambridge University Press, 1998).

17. I. Wernick, "Consuming Materials: The American Way," *Technological Forecasting and Social Change,* 53 (1996): 111–122.

18. I. Wernick and J. H. Ausubel, "National Materials Flow and the Environment," *Annual Review of Energy and Environment,* 20 (1995): 463–492.

19. S. Pimm, G. Russell, J. Gittelman, and T. Brooks, "The Future of Biodiversity," *Science,* 269 (1995): 347–350.

20. Historic data from L. R. Brown, H. Kane, and D. M. Roodman, note 15 above.

21. One of several projections from P. Raskin, G. Gallopin, P. Gutman,
A. Hammond, and R. Swart, *Bending the Curve: Toward Global Sustainability,* a report of the Global Scenario Group, Polestar Series, report no. 8 (Boston: Stockholm Environmental Institute, 1995).

22. N. Nakićénovíc, "Freeing Energy from Carbon," in *Technological Trajectories and the Human Environment,* eds., J. H. Ausubel and H. D. Langford. (Washington, D.C.: National Academy Press, 1997); I. Wernick, R. Herman, S. Govind, and J. H. Ausubel, "Materialization and Dematerialization: Measures and Trends," in J. H. Ausubel and H. D. Langford, eds., *Technological Trajectories and the Human Environment* (Washington, D.C.: National Academy Press, 1997), 135–156; and see A. Gruebler, note 16 above.

23. Royal Society of London and the U.S. National Academy of Science, note 13 above.

24. 24. J. Ryan and A. Durning, *Stuff: The Secret Lives of Everyday Things* (Seattle, Wash.: Northwest Environment Watch, 1997).

25. R. T. Watson, M. C. Zinyowera, and R. H. Moss, eds., *Climate Change 1995: Impacts, Adaptations, and Mitigation of Climate Change—Scientific-Technical Analyses* (Cambridge, U.K.: Cambridge University Press, 1996).

26. A sampling of similar queries includes: A. Durning, *How Much Is Enough?* (New York: W. W. Norton and Co., 1992); Center for a New American Dream, *Enough!: A Quarterly Report on Consumption, Quality of Life and the Environment* (Burlington, Vt.: The Center for a New American Dream, 1997); and N. Myers, "Consumption in Relation to Population, Environment, and Development," *The Environmentalist,* 17 (1997): 33–44.

27. J. Schor, *The Overworked American* (New York: Basic Books, 1991).

28. A. Durning, *How Much Is Enough?: The Consumer Society and the Future of the Earth* (New York: W. W. Norton and Co., 1992); Center for a New American Dream, note 26 above; and M. Wackernagel and W. Ress, *Our Ecological Footprint: Reducing Human Impact on the Earth* (Philadelphia. Pa.: New Society Publishers, 1996).

29. W. Jager, M. van Asselt, J. Rotmans, C. Vlek, and P. Costerman Boodt, *Consumer Behavior: A Modeling Perspective in the Contest of Integrated Assessment of Global Change,* RIVM report no. 461502017 (Bilthoven, the Netherlands: National Institute for Public Health and the Environment, 1997); and P. Vellinga, S. de Bryn, R. Heintz, and P. Molder, eds., *Industrial Transformation: An Inventory of Research.* IHDP-IT no. 8 (Amsterdam, the Netherlands: Institute for Environmental Studies, 1997).

30. H. Nearing and S. Nearing. *The Good Life: Helen and Scott Nearing's Sixty Years of Self-Sufficient Living* (New York: Schocken, 1990); and D. Elgin, *Voluntary Simplicity: Toward a Way of Life That Is Outwardly Simple, Inwardly Rich* (New York: William Morrow, 1993).

31. P. Menzel, *Material World: A Global Family Portrait* (San Francisco: Sierra Club Books, 1994).

32. R. Rosenblatt, ed., *Consuming Desires: Consumption, Culture, and the Pursuit of Happiness* (Washington, D.C.: Island Press, 1999).

33. National Research Council, Board on Sustainable Development, *Our Common Journey: A Transition Toward Sustainability* (Washington, D.C.: National Academy Press, 1999).

ROBERT W. KATES is an independent scholar in Trenton, Maine; a geographer; university professor emeritus at Brown University; and an executive editor of *Environment.* The research for "Population and Consumption: What We Know, What We Need to Know" was undertaken as a contribution to the recent National Academies/National Research Council report, *Our Common Journey: A Transition Toward Sustainability.* The author retains the copyright to this article. Kates can be reached at RR1, Box 169B, Trenton, ME 04605.

Can You Buy a Greener Conscience?

A budding industry sells 'offsets' of carbon emissions, investing in environmental projects. But there are doubts about whether it works.

ALAN ZAREMBO

The Oscar-winning film "An Inconvenient Truth" touted itself as the world's first carbon-neutral documentary.

The producers said that every ounce of carbon emitted during production—from jet travel, electricity for filming and gasoline for cars and trucks—was counterbalanced by reducing emissions somewhere else in the world. It only made sense that a film about the perils of global warming wouldn't contribute to the problem.

Co-producer Lesley Chilcott used an online calculator to estimate that shooting the film used 41.4 tons of carbon dioxide and paid a middleman, a company called Native Energy, $12 a ton, or $496.80, to broker a deal to cut greenhouse gases elsewhere. The film's distributors later made a similar payment to neutralize carbon dioxide from the marketing of the movie.

It was a ridiculously good deal with one problem: So far, it has not led to any additional emissions reductions.

Beneath the feel-good simplicity of buying your way to carbon neutrality is a growing concern that the idea is more hype than solution.

According to Native Energy, money from "An Inconvenient Truth," along with payments from others trying to neutralize their emissions, went to the developers of a methane collector on a Pennsylvanian farm and three wind turbines in an Alaskan village.

As it turned out, both projects had already been designed and financed, and the contributions from Native Energy covered only a minor fraction of their costs. "If you really believe you're carbon neutral, you're kidding yourself," said Gregg Marland, a fossil-fuel pollution expert at Oak Ridge National Laboratory in Tennessee who has been watching the evolution of the new carbon markets. "You can't get out of it that easily."

The race to save the planet from global warming has spawned a budding industry of middlemen selling environmental salvation at bargain prices.

The companies take millions of dollars collected from their customers and funnel them into carbon-cutting projects, such as tree farms in Ecuador, windmills in Minnesota and no-till fields in Iowa.

In return, customers get to claim the reductions, known as voluntary carbon offsets, as their own. For less than $100 a year, even a Hummer can be pollution-free—at least on paper.

Driven by guilt, public relations or genuine concern over global warming, tens of thousands of people have purchased offsets to zero out their carbon impact on the planet.

"It made me feel better about driving my car," said Nicky Tenpas, a 29-year-old occupational therapist from Hermosa Beach, who bought offsets to neutralize emissions from the Jeep she always wanted.

The star of "An Inconvenient Truth," former Vice President Al Gore, says he and his family are carbon neutral, as are Dave Matthews Band concerts and Coldplay albums. The travel websites Expedia and Travelocity now offer passengers the option of counteracting their flights, and Rupert Murdoch promises that his entire News Corp. will be carbon neutral by 2010, largely through the purchase of offsets.

Offset companies stress that they are not a cure-all for the world's greenhouse gas emissions, which are equivalent to 54 billion tons of carbon dioxide each year.

Tom Boucher, chief executive of Native Energy, said people should first reduce their energy consumption and waste, and then buy offsets—"the only way to really get to zero unless you stop driving, stop traveling."

But the industry is clouded by an approach to carbon accounting that makes it easy to claim reductions that didn't occur. Many projects that have received money from offset companies would have reduced emissions by the same amount anyway.

The growing popularity of offsets has now prompted the Federal Trade Commission to begin looking into the $55-million-a-year industry.

"Everybody would like to find happy-face, win-win solutions that don't cost anything," said Robert Stavins, an environmental economist at Harvard University. "Unfortunately, they don't exist."

Selling Clean Air

In the rolling hills of southwestern Pennsylvania, outside the town of Berlin, Dave Van Gilder's family has been raising cows for four decades. He and his twin sons, Jason and Justin, tend to their 400 Holsteins while his wife, Connie, keeps the books.

The smell of manure has long been the sweet exhaust of a dairy farm running full tilt.

Millions of pounds of cow excrement over the decades were funneled from the barns to a 3.3-million-gallon lagoon, where it decayed, burping invisible clouds of the potent greenhouse gas methane.

In the days of Van Gilder's father, nobody cared about the greenhouse gases.

But things began to change a few years ago. Van Gilder didn't know it, but his lagoon had become an economic opportunity.

A local congressman urged him to apply for a state alternative energy grant to build a system that would capture methane from cow manure and burn it to generate electricity.

The whole project, known as a methane digester, would cost about $750,000—$631,000 of it coming from the state and the U.S. Department of Agriculture.

Van Gilder had to make up the difference, but he figured he could earn that back—and in several years start making money—by supplying electricity for the farm and selling the excess to the local utility.

A year before construction began, Van Gilder was contacted by Native Energy, which wanted to buy his emissions reduction, along with the reductions of others who had won state energy grants.

Van Gilder had never heard of the company or the idea of selling clean air.

Nothing came of his discussions with the company until construction started on the massive tank for heating manure. He gladly signed a contract to sell Native Energy 29,000 tons in carbon dioxide reductions—the company's estimate of how much greenhouse gas the digester will keep out of the atmosphere over the next 20 years.

"There wasn't a lot of negotiation," said Van Gilder, who was happy to accept whatever the company was offering.

Family members said the contract prohibited them from disclosing the payment, but based on a contract with another dairy farmer, signed with Native Energy, it was about $70,000, or $2.40 a ton.

Justin Van Gilder said the money had nothing to do with the family's decision to build its methane plant. "It was a free bonus," he said.

"We still don't understand it all," Connie Van Gilder said. "It's hard for us to fathom, to see what it is doing."

The situation was similar for the Alaska Village Electric Cooperative, a power utility for dozens of remote communities.

In early 2006, account manager Brent Petrie was at an Anchorage environmental conference talking about a windmill project that the cooperative was building in the Yup'ik Eskimo village of Kasigluk, a soggy patch of tundra on the remote Yukon-Kuskokwim Delta in western Alaska.

Rising 100 feet over the landscape, the three 100-kilowatt turbines were intended to reduce the area's dependence on diesel generators, whose fuel must be shipped in on barges. Federal grants were covering $2.8 million of the project's $3.1-million cost.

Petrie had barely finished his presentation when he was cornered by representatives from two brokers—Native Energy and the Bonneville Environmental Foundation—eager to buy the project's offsets.

The cooperative sold 25 years of carbon dioxide reductions to Native Energy for $36,000—roughly $4 a ton.

Native Energy had contributed just over 1% of the total cost of the project yet claimed 100% of its carbon reductions.

"If you look at the costs of these projects, it's a tiny, tiny fraction," said cooperative president Meera Kohler. The payment did "not determine whether those blades turn or not."

At best, Kohler said, the money could cover some maintenance costs.

An Untapped Market

Despite its relatively small role in the project, Native Energy counts the windmills as a success, demonstrating the power of carbon offsets to encourage clean energy.

"Every kilowatt-hour they produce means one fewer kilowatt-hour is generated by the diesel generators that otherwise provide power for this village," the company's website says.

Wind power has long been a fascination for Boucher, Native Energy's co-founder. As an electrical engineering major in the 1970s at the University of Vermont, he built a 25-foot-high wind turbine in his parents' backyard, carving the blades from a piece of redwood.

He later worked at a Vermont utility and helped develop one of the first wind farms in the northeast. Boucher started his own company in the 1990s to sell alternative energy but soon came upon a simpler and possibly more lucrative product: voluntary carbon offsets.

It was a new twist on an old idea.

About 30 years ago, the U.S. government began fostering emissions markets that allowed industrial polluters to buy offsets for such gases as nitrogen oxides and sulfur oxides. One of their successes has been in reducing acid rain in the Northeast.

The Europeans have recently adopted a similar model to regulate carbon dioxide emissions, allowing the continent's dirtiest industries to buy and sell rights to spew greenhouse gases.

The key to these regulated markets is a gradually falling cap on total emissions, forcing factories to either reduce their own emissions or buy someone else's reductions at increasing prices.

Boucher and other environmental entrepreneurs, however, believed there was an untapped market for carbon reductions: people and companies who would buy them voluntarily.

"What was coming was a way for folks to take actions against global warming," said Boucher, a bearded 52-year-old, who has long believed that alternative energy can be competitive with cheaper power from fossil fuels.

Native Energy, based in Charlotte, Vt., was one of the first offset companies in the United States and has become one of the most respected in environmental circles.

It has sold offsets to thousands of individuals and companies, including Levi Strauss & Co. and Ben & Jerry's. It has been big with political campaigns, providing offsets to Democratic presidential candidates Hillary Rodham Clinton and John Edwards. The producers of the 2005 film "Syriana," which claims to be the first carbon-neutral major motion picture, bought 2,040 tons of carbon dioxide reductions from the company.

After Native Energy's name was mentioned in the final credits for "An Inconvenient Truth," visits to the company's website jumped 1,100%, marketing director Billy Connelly said.

As a private company, it doesn't report its revenue, but Connelly said he expects it will double its sales this year, reaching a total of about 1 million tons of carbon dioxide neutralized since its founding.

"Things have really taken off in the last year or two," said Boucher. The company has about 20 employees.

Native Energy now finds itself facing competition from nearly three dozen other offset firms worldwide. Some are nonprofit, but most of the biggest are in business to make money.

In 2006, offset companies sold greenhouse gas reductions equivalent to at least 14.8 million tons of carbon dioxide, more than double the previous year, said Katherine Hamilton, carbon project manager for Ecosystem Marketplace, which tracks the industry.

Sales are expected to double again this year.

Requirements are Vague

For all the money spent, nobody can say if the offsets have done much to alleviate global warming.

The problem is whether the voluntary reductions really exist. The buzzword in the industry is "additionality"—the idea that offset purchases actually lead to additional greenhouse gas reductions.

The concept should be simple: Pay for a project, monitor its actual reductions, then claim your share.

Instead, offset companies often have vague requirements to determine if their potential investments would actually lead to additional reductions.

Native Energy says it looks for projects that need offset revenue to survive—a difficult standard, since the projects are expensive and the offset payments are relatively small. But even if a project can stand on its own, it can still qualify for the money if it is novel or simply "not business as usual," according to the company's website.

That definition has allowed Native Energy and other offset companies to claim the carbon reductions from projects in which they have played minor roles. Still, Native Energy's contract requires projects to certify that whatever offset money they receive "is a necessary component of the project's economic viability."

The company has struggled with whether its funding matters. Boucher said the windmills in Alaska were debated for weeks inside Native Energy since the project already had been funded by the government. "This is a case of one of the more difficult determinations," he said.

Native Energy, he said, eventually concluded that its contribution, if used as a reserve fund for emergency repairs, was meaningful. It helps "to make sure these turbines will run as well as they can, and to further the chances that other wind farms will be built," he said.

In the case of the methane digester, Boucher said the reductions were additional since the offset payments helped cover a significant portion of Van Gilder's out-of-pocket expenses.

The best way to ensure additionality, according to Native Energy, is to pay a project for a decade or more of offsets while the developers are still arranging the financing.

The downside is that the carbon reductions might not occur for a decade or more.

One of Native Energy's expected future projects is a windmill development being planned for a South Dakota Sioux reservation. Part of the offset money from "An Inconvenient Truth" has been earmarked for the $45-million project, known as the Owl Feather War Bonnet Wind Farm.

Native Energy and the developers are still negotiating, but the payment for the next 25 years of carbon reductions could be a few million dollars.

Even in this ideal case, the developer is hesitant to say that the money he will get from Native Energy is essential.

"Could we do it without it?" said Dale Osborn, head of Distributed Generation Systems Inc. in Lakewood, Colo. "Maybe. But that would require charging more for the energy or the investors accepting a lower rate of return."

Several environmental and clean energy groups have also raised concerns about verifying projects, monitoring their actual carbon reductions and ensuring that each carbon offset is not sold more than once.

"People are trying to do the right thing," said Peter Knight, a partner with Al Gore in Generation Investment Management, which invests in environmentally responsible companies. "It's a new field . . . and it's going through some growing pains."

Price of Feeling Good

Without government regulation and mandatory caps on emissions, all that is left to drive offset sales is guilt and marketing. Offset companies charge what the market will bear.

"How much are you willing to spend to feel good or to impress your neighbors?" asked Marland, of Oak Ridge National Laboratory.

For Katy Hansen, a 29-year-old law school student, the answer was $429.99.

She wanted to offset the travel of the guests to her May wedding in Madison, Wis., and paid San Francisco-based offset company TerraPass Inc. to neutralize 40 tons of carbon dioxide.

"The wedding felt out of character in a way," she said. "It was more overdone than anything in our lives."

It was a well-intended gesture, but one that some scientists say does carry a price. Offsets are so convenient that they may foster a false sense that global warming can be easily solved when the hard and expensive work remains undone.

A United Nations panel recently concluded that actually reducing emissions to avoid the worst perils of global warming would cost trillions of dollars a year over the next several decades.

"These offsets are not addressing the problem that must be addressed now," said James Hansen, NASA's top climate researcher. "If we just fool around with marginal things, we will be up a creek without a paddle in the rather near future."

Despite the imperfections of the offset market, some customers say, it still helps people think about their own responsibility for global warming.

"It's a powerful first step," said Davis Guggenheim, director of "An Inconvenient Truth." "The choice of doing this rather than nothing is not a choice."

He acknowledged the skepticism about voluntary offsets, but he added, "All of us knew when you're doing offsets that the theoretical and symbolic quality to doing this is as important as the practical quality."

Ethan Prochnik, a 43-year-old freelance television producer in Los Angeles, said the enormous threat of global warming means that everyone has to help in reducing emissions.

He foresees a time when a carbon tax will be factored into the price of every tomato, every pair of jeans, every computer.

Until then, offset payments are a sort of self-imposed pollution tax.

For months, Prochnik and his fiancee saved to offset a year of their lives. He finally went to the website for an Oregon company called the Climate Trust. He entered their state of residence, the size of their Montecito Heights house, their cars, their annual mileage and the number of flights they took in a year.

"It's like a poor man's Prius," he said, explaining that he couldn't afford to trade his Isuzu Rodeo SUV for a hybrid.

"It does make you feel a little better," said his fiancee, Amber Vovola, 29.

Prochnik entered his credit card number and pressed the "enter" key. The total came to 21.22 metric tons of carbon dioxide, or $200.28.

Figuring that the website calculator didn't cover all their pollution, he decided to buy an additional $300 in offsets, just to be safe.

A New Security Paradigm

It's easy to equate "national security" or "global security" with military defense against rogue states and terrorism, but a leading U.S. military expert says that view is far too narrow—and could lead to catastrophe if not changed.

GREGORY D. FOSTER

Whatever else the year 2004 might be noted for by future historians—the U.S. political wars, the genocide in Darfur, the strategic debacle in Iraq—it may well turn out to have been a seminal year for the field of environmental security—the intellectual, operational, and policy space where environmental conditions and security concerns converge.

So too, one hopes, might it have been the year when U.S. policymakers and the American public began to awaken, however belatedly, to the need for an entirely new approach to security and for the fundamental strategic transformation necessary to achieve such security.

The highlight of the year in this regard was, for two essentially countervailing reasons, the award of the Nobel Peace Prize to Kenyan environmental activist Wangari Maathai. On one hand, by broadening the definitional bounds of peace, the award gave new legitimacy to those who would embrace unconventional conceptions of security, especially involving the environment. "This is the first time environment sets the agenda for the Nobel Peace Prize, and we have added a new dimension to peace," said committee chairman Ole Danbolt Mjoes in announcing the award. "Peace on earth depends on our ability to secure our living environment."

On the other hand, no less noteworthy were critics of the award, whose expressions of disparagement typified and reaffirmed the stultifying hold of traditionalist thinking on the conduct of international relations. Espen Barth Eide, former Norwegian deputy foreign minister, observed: "The one thing the Nobel Committee does is define the topic of this epoch in the field of peace and security. If they widen it too much, they risk undermining the core function of the Peace Prize; you end up saying everything that is good is peace." Traditionalists everywhere, including most of the U.S. policy establishment, no doubt took succor from such self-righteous indignation and resolved to perpetuate the received truths of the past that have made real peace so elusive and illusory to date.

Beyond the Peace Prize, two other events ten months apart served as defining bookends for what could turn out to have been the undeclared Year of Environmental Security. The first was an attention-grabbing article, "The Pentagon's Weather Nightmare," that appeared in the February 9, 2004 issue of *Fortune* magazine. Describing a report two futurists—Peter Schwartz and Doug Randall of Global Business Network—had recently prepared for the Defense Department on the national security implications of abrupt climate change, the article generated a flurry of intense but short-lived excitement and speculation on whether, why, and to what extent the Pentagon was finally taking climate change seriously.

The second bookend event came at the end of the year with the issuance of the final report of the internationally distinguished, 16-member High-Level Panel on Threats, Challenges and Change that UN Secretary-General Kofi Annan had appointed in November 2003 to examine the major threats and challenges the world faces in the broad field of peace and security.

These two particular events, potentially significant enough in their own right, should be viewed in the larger context of several other magnifying events that occurred over the course of the year.

For starters, Sir David King, chief science adviser to British prime minister Tony Blair, raised eyebrows and hackles with a controversial article in the January 9, 2004 issue of *Science* magazine. King cited climate change as "the most severe problem that we are facing today—more serious even than the threat of terrorism," and accused the U.S. government of "failing to take up the challenge of global warming." In a subsequent speech to the American Association for the Advancement of Science, he added: "Climate change is real. Millions will increasingly be exposed to hunger, drought, flooding and debilitating diseases such as malaria. Inaction due to questions over the science [a thinly veiled reference to Bush administration foot-dragging] is no longer defensible."

In March, former UN chief weapons inspector Hans Blix added further fuel to the fire in a BBC television interview with David Frost: "I think we still overestimate the danger of terror. There are other things that are of equal, if not greater, magnitude, like the environmental global risks." This statement reinforced an equally pointed one Blix had made a year earlier: "To me the question of the environment is more ominous than that of peace and war. . . . I'm more worried about global warming than I am of any major military conflict."

In May, the blockbuster 20th Century Fox disaster movie *The Day After Tomorrow,* portraying the cataclysmic global consequences of accelerated climate change, was released to theaters nationwide (with European release scheduled for October). Some, such as Sir David King and former vice president Al Gore, promoted or endorsed the movie, clearly not because of its admittedly unrealistic compression of time and exaggeration of catastrophic effects, but because of its potential for awakening and sensitizing the public to the plausibility and seriousness of abrupt climate change. Others fiercely criticized the movie for trivializing such a vital issue. Anti-doomsayer Gregg Easterbrook, senior editor of *The New Republic,* assailed the "cheapo, third-rate disaster movie" for its "imbecile-caliber" science: "By presenting global warming in a laughably unrealistic way, the movie will only succeed in making audiences think that climate change is a big joke, when in fact the real science case for greenhouse-gas reform gets stronger all the time."

In a major September address in London, Tony Blair, faced with continuing criticism from his opposition, called climate change "the world's greatest environmental challenge . . . a challenge so far-reaching in its impact and irreversible in its destructive power, that it alters radically human existence." "Apart from a diminishing handful of skeptics," he said, "there is a virtual worldwide scientific consensus on the scope of the problem."

Then in October, the United Nations Environment Programme's Division of Early Warning and Assessment issued a thought-provoking new report, *Understanding Environment, Conflict, and Cooperation,* that resulted from the deliberations of participants in a new initiative to leverage environmental activities, policies, and actions for promoting international conflict prevention, peace, and cooperation. The subject matter of the report is not new, but the question it implies is: whether new life can be breathed into what was, throughout most of the 1990s, a lively debate over whether and how the environment and security are related and interact. Since the Kyoto negotiations of 1997, that debate has been largely moribund, to the detriment of both U.S. policy and strategic discourse more generally.

Revivifying Environmental Security

Even if the events recounted above had not occurred this past year, the findings and recommendations of the UN High-Level Threat Panel and the introduction into the public imagination of abrupt climate change as a matter of prospective national security concern would stand as forceful stimuli for policy practitioners, scholars, and the general public to accord environmental security more serious and immediate attention.

This article goes to press before the actual release of the High-Level Panel's final report; but publicly available preliminary work by the United Nations Foundation's United Nations & Global Security Initiative, in cooperation with the Environmental Change & Security Project of the Woodrow Wilson Center, prefigures how the Panel's thinking is likely to be guided on environmental matters. This introductory passage from a discussion summary presented to the Panel is indicative of that thrust:

> *Environmental changes can threaten global, national, and human security. Environmental issues include land degradation, climate change, water quality and quantity, and the management and distribution of natural-resource assets (such as oil, forests, and minerals). These factors can contribute directly to conflict, or can be linked to conflict, by exacerbating other causes such as poverty, migration, small arms, and infectious diseases. For example, experts predict that climate change will trigger enormous physical and social changes like water shortages, natural disasters, decreased agricultural productivity, increased rates and scope of infectious diseases, and shifts in human migration; these changes could significantly impact international security by leading to competition for natural resources, destabilizing weak states, and increasing humanitarian crises. However, managing environmental issues and natural resources can also build confidence and contribute to peace through cooperation across lines of tension.*

Add to this Secretary-General Annan's own words in announcing the High-Level Panel to the UN General Assembly in September 2003, and it seems clear that the Panel will endorse the environment-security linkage and acknowledge that environmental degradation, resource scarcity, and climate change are threats or challenges that face the world and demand collective response:

> *All of us know there are new threats that must be faced—or, perhaps, old threats in new and dangerous combinations: new forms of terrorism, and the proliferation of weapons of mass destruction. But, while some consider these threats as self-evidently the main challenge to world peace and security, others feel more immediately menaced by small arms employed in civil conflict, or by so-called "soft threats" such as the persistence of extreme poverty, the disparity of income between and within societies, and the spread of infectious diseases, or climate change and environmental degradation.*

The February 2004 *Fortune* article was a dispassionate but revealing summary of a Pentagon-commissioned study that, though unclassified, ordinarily wouldn't have received much—if any—public exposure. Substantively, the article did two things. First, judging from the volume and intensity of follow-up commentary it generated, it clearly raised expectations—positive and negative—about the content and ramifications of the Pentagon report. Was the military actually interested in climate change? Why? Enough to do something about it? To what end and with what effect (especially on the military's principal mission)?

Second, the article—and the report it reported—upped the ante in the continuing debate over climate change. In addressing *abrupt* climate change, it accentuated an emerging thesis that gives urgency to what otherwise is considered (by some, perhaps many) to be a long-term, gradual phenomenon that, if real, can be passed off, without present political or economic regret, for future generations to deal with. And in tying abrupt climate change to national security, the article and report give added—ultimate—importance to the subject. National security is, after all, the public-policy holy of holies—the iconic totem that takes precedence over all else. National security is about endangerment and survival, the thinking goes. So if something can be shown to have national security implications (however defined), then perhaps it too is about such things; perhaps it too, therefore, warrants serious attention and the commitment of resources.

For people familiar with the U.S. military's normal modes of communication, the release of the Schwartz-Randall report to *Fortune* was unusual enough to cause speculation about whether the man who commissioned it, Andrew W. Marshall—the Pentagon's director of net assessment for the past 30 years—may have been signaling concerns that went well beyond the report's scientific message: first, that the institution he works for is intractably parochial and resistant to change; second, that the Pentagon is particularly inbred and close-minded about matters as esoteric and ideologically encumbered as the environment; third, that since imaginative futurists had prepared the report, it could more easily be dismissed as speculative fantasy by bureaucratic pragmatists who prefer to think they are grounded in reality; fourth, that however long he (Marshall) may have served in the Pentagon, he has little clout in influencing the military to actually take action based on his office's analytical products; fifth, that going public therefore offers more hope for forcing internal Pentagon awakening (if not change) in response to external pressure from arguably less parochial outside parties such as Congress and the media; and sixth, that perhaps the most potent force for movement on this particular front is the business community, which has the most to both gain and lose from climate change—especially when the political regime in power opts for dogmatic inaction in deference to the cosmic invisible hand of the marketplace.

The importance of this episode, as well as its relevance for the future, lies in both the message and the method of the Schwartz-Randall report itself. The implicit message is that even worse than climate change is the not unrealistic possibility of *abrupt* climate change. For those who had not heard of it, the article made clear that abrupt climate change is not just global warming speeded up, but a wholly different kind of event triggered by the baseline climate change we already know. In brief, the global warming now taking place could conceivably lead to a halting of the ocean currents that now keep Europe temperate—global *warming* thus ironically leading to regionally much colder conditions and widespread accelerations of the catastrophic effects already commonly associated with "normal" climate change: floods, droughts, windstorms, wave events, wildfires, disease epidemics, species loss, famine, and more. The explicit message is that the concatenation of such effects could then lead to additional, national security consequences—most notably military confrontations between states over access to scarce food, water,

and energy supplies, or what the authors describe as a "world of warring states."

Paradoxically, portraying what is relevant to national security as essentially that which invites or involves military force is perhaps necessary to grab the attention of purported experts on the subject, but it thereby also betrays the shallowness and narrowness of the canonical security paradigm most of us have unthinkingly bought into. This state of affairs is reinforced by the methodology of the Schwartz-Randall report, which seeks not to predict whether, when, or how abrupt climate change and its attendant effects would occur, but merely to present a plausible scenario of what might happen if and when it does. In the authors' words, "The duration of this event could be decades, centuries, or millennia and it could begin this year or many years in the future." Despite this caveat, the theme of abrupt climate change as a national security concern may be sufficiently eye-opening and provocative that, in conjunction with the other motivating forces of the year just past, it could take public consciousness of environmental security to a new level.

Rethinking Security, Reassessing the Threat

However many people there may now be who recognize that environmental conditions precipitate or contribute to other conditions—violent conflict, civil unrest, instability, regime or state failure—regularly associated with security as usually defined, they are vastly outnumbered by those who either openly oppose the environment-security linkage or ignore it as irrelevant or inconsequential.

These oppositionists come from two different but overlapping camps: ideological conservatives and the mainstream traditionalists who dominate the national security community. This distinction is crucial because the latter—the technocratic mandarins inside and outside government—have the final exegetical say about what security is and what therefore is allowed to be a legitimate part of the security dialogue.

Oppositionists treat the environment as a purely *ideological* issue, and climate change as the most ideological of all—accordingly as dismissible as feminism, the homosexual agenda, or any other reflection of "political correctness." This despite the fact that, in a purer ontological sense, the environment is an inherently strategic matter, and climate change the most strategic of all. The environment is everywhere. It respects no borders, physical or otherwise. In its reach, its effects, and its consequences, it is truly global. And, fully understood, it brings into question all of our prevailing notions of sovereignty, territorial integrity, and even aggression and intervention. Nonetheless, just as to a hammer everything looks like a nail, to an ideologue everything looks ideological—to be accepted or rejected on the basis not of reason but of internalized dogma.

One of the major issues that has most divided those who debate the environment-security relationship is how broadly or narrowly to define security. Oppositionists invariably take the narrow road—basically equating security with defense, just as they similarly equate power with force. To them, security has axiomatic meaning that derives from its historical roots.

Ironically, on this particular count the oppositionists are abetted by a shadow contingent of like-minded liberal environmentalists, who believe that linking the environment to security is dangerous because it will inevitably militarize the former and relinquish vital resources needed for environmental protection to an already bloated, profligate military establishment. In their fear of militarizing the environment, they risk getting into bed with betes noir who are fully committed to militarizing our entire strategic posture.

The counterpoise to this narrow construction of security begins with the recognition that security is, at root, a psychological and sociological phenomenon that starts—and ends—with the individual. To be secure is, literally, to be free—from harm and danger, threat and intimidation, doubt and fear, need and want. In the hierarchy of human needs, security is one of the most basic impulses—exceeded in its primacy only by the even more basic physiological needs for food, water, shelter, and the like, each of which is dependent on environmental well-being. Such primal needs translate into the natural rights that all human beings deserve to enjoy and that governments, as we have learned from America's founders, are instituted to secure.

Individual or human security, then, is the necessary precondition for national security, not merely its residual by-product. Accordingly, *assured security* stands as the primary overarching strategic aim a democratic society such as ours must seek to attain. In this supernal sense, security is something much more robust than defense. It encompasses the totality of conditions enumerated in America's security credo, the Preamble to the Constitution—not just the common defense, but no less importantly, national unity (a more perfect union), justice, domestic tranquility, the general welfare, and liberty. Only where all of these conditions exist in adequate measure is there true security. Where even one—liberty, say, or the general welfare—is sacrificed or compromised for another—say the common defense—the result is some degree of insecurity. Thus, in the final analysis, everything is related to security; everything is related to *national* security.

However broadly or narrowly security is defined, whatever endangers it or places it at risk is a threat; and whatever constitutes or qualifies as a threat is crucial because, in the idealized protocol of traditional national security planning, threats are the ostensible starting point for determining the requirements that produce capabilities and programs for countering these threats. (In reality, of course, capabilities and programs acquire their own bureaucratic life and thus are more likely to determine than to derive from threats.)

Oppositionists generally accept as legitimate threats only those parties or phenomena that, beyond their perceived potential for harm, are considered capable of or the product of malevolent intent. *Intentionality* is the key legitimizing element. Terrorism fills this bill, just as state-based adversaries traditionally have. Weapons of mass destruction seem to qualify because, though inert entities in themselves, in human hands they can be ominous instruments of harm. Climate change and assorted forms of environmental degradation, though, typically don't pass muster

as credible threats, no matter how much death and destruction they can wreak. Instead, they are implicitly written off as pure acts of nature, assuming metaphysical proportions that place them beyond human control and therefore outside the bounds of either preventive or retributive concern.

Such blinkered threat assessment is entirely characteristic of the policy establishment. To cite just a few notable examples:

- The 2002 White House national security strategy, in 34 pages of text, mentions the word environment in only one short paragraph about U.S. trade negotiations.

- In his February 2004 "Worldwide Threat Briefing" to Congress, Director of Central Intelligence George Tenet devoted five pages of testimony each to terrorism, Iraq, and proliferation, three paragraphs to global narcotics, a paragraph each to population trends, infectious disease, and humanitarian food insecurity, but nothing at all to environmental matters.

- The much bally-hooed, future-oriented Hart-Rudman Commission, whose members extolled their own prescience for adumbrating 9/11-type terrorist attacks on the United States, gave only the most cursory treatment to 21st-century environmental challenges in its initial September 1999 report. Arguing innocuously that pollution can be—and implicitly will be—counteracted by economic growth and the spread of remediation technologies, the commission essentially dismissed the subject with this (dare we say, ideological) statement: "There is fierce disagreement over several major environmental issues. Many are certain that global warming will produce major social traumas within 25 years, but the scientific evidence does not yet support such a conclusion. Nor is it clear that recent weather patterns result from anthropogenic activity as opposed to natural fluctuations."

- Somewhat in contrast, the National Intelligence Council's *Global Trends 2015* report, issued in December 2000 (before the following year's 9/11 attacks), identified natural resources and the environment as one of the most important "drivers and trends that will shape the world of 2015." Focusing principally on food, water, and energy security developments, the experts who collaborated on the report acknowledged the persistence and growth of global environmental problems in the years ahead, a growing consensus on the need to deal with such problems, and the prospect that "global warming will challenge the international community."

Typifying the thinking of policymakers and other members of the national security mandarinate, such assessments also seem more representative than not of general public sentiment. A particularly revealing indication of this is the most recent Chicago Council on Foreign Relations study of U.S. public opinion on international issues, *Global Views 2004*. Asked to identify the most critical threats to U.S. vital interests, the public ranks global warming a distant seventh (37% of respondents), behind the likes of international terrorism (75%), chemical and biological weapons (66%), unfriendly countries becoming nuclear powers (64%), immigration into the United States (58%), and other developments. Another recent (February 2004) poll by Gallup found that environmental concerns don't even make the public's top-eleven list of possible threats to U.S. vital interests—international terrorism and the spread of weapons of mass destruction far outpacing all other prospective threats.

That environmental matters should be of such little overall public concern is a reflection of how limited and unstrategic our thinking about security actually is. Perhaps if we were to pay more attention to the documented effects of particular conditions and events, rather than to the nebulous, abstract notion of intentionality implicitly embedded in our prevailing standards of threat-worthiness, we could see the world differently—and more accurately.

Look, for example, at comparative fatalities from the highly credible threat of terrorism and the highly dubious threat of natural disasters. Since 1968, there have been 19,114 incidents of terrorism world wide, resulting in a total of 23,961 deaths and 62,502 associated injuries. However disturbing these figures may be, they pale in comparison to those resulting from natural disasters.

The average *annual* death toll over the past century due to drought, famine, floods, windstorms, temperature extremes, wave surges, and wildfires has been 243,577. Thus, even if we ignore earthquakes, volcanic eruptions, and disease epidemics, and don't count injuries or other harmful effects (such as homelessness), three times as many people die each year on average in natural disasters that could be linked to—and exacerbated by—climate change as have been killed and injured together in 37 years of terrorist incidents. And lest the use of a century-long average seem skewed, consider that just since 1990, there have been more than 207,000 fatalities from the foregoing types of disasters in South Asia alone, more than 23,000 in Central America and Mexico, and tens of thousands more in other parts of the world.

These figures are startling in their empirical exactitude, more so if one accepts estimates that average annual economic losses to such disasters were on the order of $660 billion in the 1990s. They lead us to consider a final argument that ideological conservatives invoke to discredit environmental and climate concerns—the need for sounder, more defensible science—and the associated argument national security mandarins use to deny or ignore the environment-security linkage—the lack of unequivocal evidence that environmental conditions actually cause diminished security in the form of violence.

Both arguments are excuses for denial and inaction; and both are suffused with hypocrisy. Those who demand conclusive *proof* that environmental conditions *cause* violence set a disingenuously unattainable legitimizing standard that permits them to perpetuate their own established preference for dealing with visible, immediate, politically remunerative symptoms. Terrorism is a cardinal example of this—singularly symptomatic,

never causative, except at some advanced, derivative level, where violence produces further violence.

Those who call for science as the only proper basis for public policy—at least climate policy—pretend to be motivated by a rigorous quest for objective (nonpolitical, non-ideological) truth. Yet they shamelessly accept or reject truth claims, labeling them "scientific" or not, based on whether those claims support or contradict their pre-established ideological beliefs. President Bush, for example, has repeatedly stated that climate policy must be based on better science (that is not yet available). But when asked about embryonic stem-cell research in this past year's second presidential debate, he stated that science is important, but it must be balanced by ethics. So, when the issue is stem-cell research—or perhaps abortion or homosexuality or capital punishment—ethics can take precedence over non-cooperative science; but when the issue is climate change or the environment more generally, not so. Maybe the earth really is flat.

Of more immediate relevance to this discussion is the practice common to many who call for better climate science. Paradoxically, they are perfectly content to unquestioningly accept and espouse demonstrably unscientific *assertions* from the military—especially concerning the degradation of military readiness that allegedly results from the so-called "encroachment" of environmental restrictions (e.g., species protection) on military installations. This despite the fact that the General Accounting Office has strongly criticized the military for failing to document whether and how much encroachment has actually degraded readiness.

Senator James Inhofe (R-OK) and the Senate Republican Policy Committee both exemplify this particular hypocrisy. Inhofe has said that "catastrophic global warming is a hoax"—"alarmism not based on objective science" even as he has said that "readiness problems . . . are caused by an ever-growing maze of environmental procedures and regulations in which we are losing the ability to prepare our patriot children, our war fighters, for war." Similarly the Senate Republican Policy Committee claims that "what scientists do agree on [with regard to climate change] is not policy-relevant, and on policy-relevant issues, there is little scientific agreement," while also asserting: "Among the most burdensome [examples of encroachment] are environmental laws and lawsuits that hinder or even ban military training and testing—thereby impairing readiness. . . . The evidence of detrimental impact is ample."

Searching for a Strategic Response

What the foregoing contradictions suggest, among other things, is that the prevailing paradigm of security, according primacy as it does to the military and the use of force, long ago hijacked us intellectually and continues to hold us hostage; and, moreover, that in the absence of countervailing strategic thought of any consequence, ideology inevitably rushes in to fill the intellectual void, as it has in the case of environmental security, thereby forcing out rationality and blinding us to the future. The only remedy for this state of affairs, the only hope that the environment, climate change in particular, and, for that matter, other unconventional threats and challenges might be taken seriously as matters of serious security concern, is for fundamental strategic transformation to take place.

It seems insultingly obvious that strategic threats demand strategic response. But let us grasp the magnitude of this statement, for in the media age in which we live, there is virtually nothing—however obscure, however remote that is without almost instantaneous strategic consequence: Let us further understand why being strategic is therefore so intrinsically important. First, it is a moral obligation of government—to take the long view, to grasp the big picture, to anticipate and prevent, to appreciate the hidden, residual consequences of action or inaction, to recognize and capitalize on the interrelatedness of all things otherwise seemingly discrete and unrelated.

Second, being strategic inoculates us against crisis. Where crisis occurs, be it a terrorist incident or a natural disaster, strategic thinking has failed—with the unwanted result that decisionmaking must be artificially compressed and forced, and resources diverted from their intended purposes. Thus does *crisis prevention* stand alongside assured security as an overarching strategic aim of democratic society.

Third, being strategic provides the intellectual basis for both the strategic leadership expected of a superpower and the enduring, broad-based consensus necessary to galvanize a diverse, pluralistic society in common cause in the face of uncertainty, complexity, and ambiguity.

Four strategic imperatives should guide our future. The first let us call *targeted causation management*—focusing our thinking and our actions on identifying and eradicating the underlying causes of insecurity, thereby curing the disease rather than treating the symptoms. Environmental degradation and climate change take us much farther along the path to ultimate causes than terrorism ever could, especially if we acknowledge that the social, political, economic, and military conditions we prefer to deal with and attribute violence to may mask disaffection and unrest more deeply attributable to an environmentally degraded quality of life.

A second strategic imperative, *institutionalized anticipatory response,* calls for institutionalizing—giving permanence and legitimacy to the capacity and inclination for preventive action. This would enhance the prospects that conditions and events can be dealt with when they are manageable, before they mutate out of control and demand forceful response. Examples could range from a Manhattan Project-like effort to develop alternative energy sources and technologies, to greater inter-jurisdictional intelligence sharing, to massive disaster-resistant infrastructure development in the developing world.

A third strategic imperative is *appropriate situational tailoring*—dealing with conditions and events on their own geographic, cultural, and political terms rather than, as we are wont to do, inviting failure by imposing our preferred capabilities and approaches on the situations at hand. In a purely institutional sense, such tailoring might take the form, for example, of new multilateral collective security regimes in each region of the world, with major environmental preparedness and enforcement arms.

The fourth strategic imperative is *comprehensive operational integration*—achieving fuller organizational, doctrinal, procedural, and technological integration across military-nonmilitary, governmental-non-governmental, and national-international lines. In a conceptual policy sense, this might assume the form of an overarching strategic architecture for unifying the activities of five organizational and cultural pillars—sustainable development, sustainable energy, sustainable business, sustainable consumption, and sustainable security. In a purely structural sense, the recognition that reorganization may be required to give birth and life to needed rethinking might produce such measures as the addition of a new Cabinet-level secretary of energy and environmental affairs to formal National Security Council membership, the creation of a UN under secretary-general for environmental affairs (or environmental security), or the expansion of the United Nations Environment Programme into an organization with operational capabilities and enforcement authority. In any event, all such measures would have to be underwritten by a firm commitment to more thoroughgoing transparency and multilateralism.

Finally, let us turn to the military. On the one hand, military action represents the least strategic option available for addressing environmental security (or virtually anything else for that matter). At least this is true so long as the military continues to be configured and oriented as it is and always has been—that is, for warfighting. On the other hand, the military is so central to our governing conception of security that true strategic transformation can take place only if it includes, or perhaps is preceded by, far-reaching military transformation—making real what until now has been only tiresome rhetoric from the Pentagon.

If the military has shown itself serious to date about environmental matters—even to the extent of crediting itself with being an excellent steward of nature—it is entirely a reflection of a distinctly engineering and management orientation dedicated principally to installation cleanup and remediation. Environmental security—the stuff of operations and intelligence, rather than of engineering and logistics—has been largely alien to the military ethos and identity. One need only consider the military's efforts under President Clinton to seek and gain selected exemptions from the Kyoto Protocol, or its tireless (if not entirely successful) attempts under President Bush to seek exemption from an array of environmental laws alleged to degrade readiness.

Two overriding considerations must guide military transformation. The first is the realization that what we ought to want is a military that is not just militarily effective—an instrument of force that serves the state—but that is *strategically effective*—an instrument of power that serves the larger aims of society and even humanity. The second overriding consideration is the concomitant realization that the military must be, and be seen to be, not a warfighting machine so much as a self-contained, self-sufficient enterprise that is capable of being projected over long distances for sustained periods of time to effectively manage all stages of a full range of complex emergencies.

Such considerations, taken to heart, ideally would produce a completely revamped military organized, manned, equipped, and trained primarily for nation-building, peacekeeping, humanitarian assistance, and disaster response, and only residually for warfighting. Such a military not only would possess the requisite capabilities for fulfilling the strategic imperatives enumerated above; it also would project the all-important imagery of a force truly committed to the pursuit of peace rather than to the enduringly illogical proposition that peace can be purchased by practicing war.

If we are to think and act strategically, which we must, we do well to recall the declaration from the Gayanashagowa, the Great Law of Peace of the Six Nations Iroquois Confederacy: "In our every deliberation we must consider the impact of our decisions on the next seven generations." And in applying this strategic precept to the matter at hand, which we must, we do no less well to take up the challenge issued recently by former Soviet President Mikhail Gorbachev. Interviewed some months ago, he was asked what he thought of the American doctrine of preemption. To which he responded:

Those who talk about leadership of the world all the time ought to exercise it. Rather than develop strategic doctrines of military preemption—as we've seen in Iraq, where no weapons of mass destruction have yet been found—let's act where the intelligence is clear: on climate change and other issues such as water, where today 2 billion people in the world don't have access to clean water. Let's talk instead about preempting global warming and the looming water crisis.

Indeed. Words for the self-proclaimed world's only superpower to act on.

GREGORY D. FOSTER is a professor at the Industrial College of the Armed Forces, National Defense University, Washington, D.C., where he previously has served as George C. Marshall Professor and J. Carlton Ward Distinguished Professor and Director of Research. The views presented here are strictly his own.

Where Oil and Water Do Mix

Environmental Scarcity and Future Conflict in the Middle East and North Africa

Jason J. Morrissette and Douglas A. Borer

"Many of the wars of the 20th century were about oil, but wars of the 21st century will be over water."

—Isamil Serageldin
World Bank Vice President

I n the eyes of a future observer, what will characterize the political landscape of the Middle East and North Africa? Will the future mirror the past or, as suggested by the quote above, are significant changes on the horizon? In the past, struggles over territory, ideology, colonialism, nationalism, religion, and oil have defined the region. While it is clear that many of those sources of conflict remain salient today, future war in the Middle East and North Africa also will be increasingly influenced by economic and demographic trends that do not bode well for the region. By 2025, world population is projected to reach eight billion.[1] As a global figure, this number is troubling enough; however, over 90 percent of the projected growth will take place in developing countries in which the vast majority of the population is dependent on local renewable resources. For instance, World Bank estimates place the present annual growth rate in the Middle East and North Africa at 1.9 percent versus a worldwide average of 1.4 percent.[2] In most of these countries, these precious renewable resources are controlled by small segments of the domestic political elite, leaving less and less to the majority of the population. As a result, if present population and economic trends continue, we project that many future conflicts throughout the region will be directly linked to what academic researchers term "environmental scarcity"[3]— the scarcity of renewable resources such as arable land, forests, and fresh water.

The purpose of this article is twofold. In the first section, we conceptualize how environmental scarcity is linked to domestic political unrest and the subsequent crisis of domestic political legitimacy that may ultimately result in conflict. We review the academic literature which suggests that competition over water is the key environmental variable that will play an increasing role in future domestic challenges to governments throughout

the region. We then describe how these crises of domestic political legitimacy may result in both intrastate and interstate conflict. Even though the Middle East can generally be characterized as an arid climate, two great river systems, the Nile and the Tigris/Euphrates, serve to anchor the major population centers in the region. Conflict over the water of the Nile may someday come to pass between Egypt, Sudan, and Ethiopia; while Turkey, Syria, and Iraq all are located along the Tigris/Euphrates watershed and compete for its resources. Further conflict over water may embroil Israel, Syria, and the Palestinians.

Despite many existing predictions of war over water, we investigate the intriguing question: How have governments in the Middle East thus far avoided conflict over dwindling water supplies? In the second section of the article, we discuss the concept of "virtual water" and use this concept to illustrate the important linkages between water usage and the global economy, showing how existing tangible water shortages have been ameliorated by a combination of economic factors, which may or may not be sustainable into the future.

Environmental Scarcity and Conflict: An Overview

Mostafa Dolatyar and Tim Gray identify water resources as "the principal challenge for humanity from the early days of civilization."[4] The 1998 United Nations Development Report estimates that almost a third of the 4.4 billion people currently living in the developing world have no access to clean water. The report goes on to note that the world's population tripled in the 20th century, resulting in a corresponding sixfold increase in the use of water resources. Moreover, infrastructure problems related to water supply abound in much of the developing world; the United Nations estimates that between 30 and 50 percent of the water presently diverted for irrigation purposes is lost through leaking pipes alone. In turn, roughly 20 countries in the developing world presently suffer from water stress (defined as having less than 1,000 cubic meters of available freshwater per capita), and 25 more are expected to join that list by 2050.[5] In response to these trends, the United Nations resolved in 2002 to reduce by half the proportion of people in the developing world who are unable to reach—or afford—safe drinking water.

In turn, numerous scholars in recent years have conceptualized water in security terms as a key strategic resource in many regions of the world. Thomas Naff maintains that water scarcity holds significant potential for conflict in large part because it is fundamentally essential to life. Naff identifies six basic characteristics that distinguish water as a vital and potentially contentious resource. (1) Water is necessary for sustaining life and has no substitute for human or animal use. (2) Both in terms of domestic and international policy, water issues are typically addressed by policymakers in a piecemeal fashion rather than comprehensively. (3) Since countries typically feel compelled by security concerns to control the ground on or under which water flows, by its nature, water is also a terrain security issue. (4) Water issues are frequently perceived as zero-sum, as actors compete for the same limited water resources. (5) As a result of the competition for these limited resources, water presents a constant potential for conflict. (6) International law concerning water resources remains relatively "rudimentary" and "ineffectual."[6] As these factors suggest, water is a particularly volatile strategic issue, especially when it is in severe shortage.

Arguing that environmental concerns have gained prominence in the post-Cold War era, Alwyn R. Rouyer establishes a basic paradigm of contemporary environmental conflict. Rouyer argues that "rapid population growth, particularly in the developing world, is putting severe stress on the earth's physical environment and thus creating a growing scarcity of renewable resources, including water, which in turn is precipitating violent civil and international conflict that will escalate in severity as scarcity increases."[7] Rouyer goes on to assert that this potential conflict over scarce resources will likely be most disruptive in states with rapidly expanding populations in which policymakers lack the political and economic capability to minimize environmental damage.

"Almost a third of the 4.4 billion people currently living in the developing world have no access to clean water."

Security concerns linked fundamentally to environmental scarcity are far from a contrivance of the post-Cold War era, however. Ulrich Küffner asserts that conflicts over water "have occurred between many countries in all climatic regions, but between countries in arid regions they appear to be unavoidable. Claims over water have led to serious tensions, to threats and counter threats, to hostilities, border clashes, and invasions."[8] Moreover, as Miriam Lowi notes, "Well before the emergence of the nation-state, the arbitrary political division of a unitary river basin . . . led to problems regarding the interests of the states and/or communities located within the basin and the manner in which conflicting interests should be resolved."[9] Lowi fundamentally frames the issue of water scarcity in terms

of a dilemma of collective action and failed cooperation—the archetypal "Tragedy of the Common"—in which communal resources are abused by the greediness of individuals. In many regions of the world, the international agreements and coordinating institutions necessary to lower the likelihood of conflict over water are either inadequate or altogether nonexistent.[10]

Thomas Homer-Dixon argues that the environmental resource scarcity that potentially results in conflict, including water scarcity, fundamentally derives from one of three sources. The first, *supply-induced scarcity*, is caused when a resource is either degraded (for example, when cropland becomes unproductive due to overuse) or depleted (for example, when cropland is converted into suburban housing). Throughout most of the Middle East and North Africa countries, both environmental and resource degradation and depletion are of relevant concern. For instance, many of these countries face significant decreases in the agricultural productivity of their arable soil as a result of ongoing trends of desertification, soil erosion, and pollution. This problem is coupled with the continued loss of croplands to urbanization, as rural dwellers move to cities in search of employment and opportunity. The second source of environmental scarcity, *demand-induced scarcity*, is caused by either an increase in per-capita consumption or by simple population growth. If the supply remains constant, and demand increases by existing users consuming more, or more users each consuming the same amount, eventually scarcity will result as demand overtakes supply. The third type of environmental scarcity is known as *structural scarcity*, a phenomenon that results when resource supplies are unequally distributed. In this case the "haves" in any given society generally control and consume an inordinate amount of the existing supply, which results in the more numerous "have-nots" experiencing the scarcity.[11]

These three sources of scarcity routinely overlap and interact in two common patterns: "resource capture" and ecological marginalization. Resource capture occurs when both demand-induced and supply-induced scarcities interact to produce structural scarcity. In this pattern, powerful groups within society foresee future shortages and act to ensure the protection of their vested interests by using their control of state structures to capture control of a valuable resource. An example of this pattern occurred in Mauritania (one of Algeria's neighbors) in the 1970s and 1980s when the countries bordering the Senegal River built a series of dams to boost agricultural production. As a result of the new dams, the value of land adjacent to the river rapidly increased—an economic development that motivated Mauritanian Moors to abandon their traditional vocation as cattle grazers located in the arid land in the north and, instead, to migrate south onto lands next to the river. However, black Mauritanians already occupied the land on the river's edge. As a result, the Moorish political elite that controlled the Mauritanian government rewrote the legislation on citizenship and land rights to effectively block black Mauritanians from land ownership. By declaring blacks as non-citizens, the Islamic Moors managed to capture the land through nominally legal (structural) means. As a result, high levels of violence later arose between Mauritania and Senegal, where hundreds of thousands of the black Mauritanians had become refugees after being driven from their land.[12]

The second pattern, ecological marginalization, occurs when demand-induced and structural scarcities interact in a way that results in supply-induced scarcity. An example of this pattern comes from the Philippines, a country whose agricultural lands traditionally have been controlled by a small group of dominant landowners who, prior to the election of former President Estrada, have controlled Filipino politics since colonial times. Population growth in the 1960s and 1970s forced many poor peasants to settle in the marginal soils of the upland interior. This more mountainous land could not sustain the lowland slash-and-burn farming practices that they brought with them. As a result, the Philippines suffered serious ecological damage in the form of water pollution, soil erosion, landslides, and changes in the hydrological cycle that led to further hardship for the peasantry as the land's capacity shrunk. As a result of their economic marginalization, many upland peasants became increasingly susceptible to the revolutionary rhetoric promoted by the communist-led New People's Army, or they supported the "People Power" movement that ousted US-backed Ferdinand Marcos from power in 1986.[13]

Thus, as shown in the Philippines, social pressures created by environmental scarcity can have a direct influence on the ruling legitimacy of the state, and may cause state power to crumble. Indeed, reductions in agricultural and economic production can produce objective socio-economic hardship; however, deprivation does not necessarily produce grievances against the government that result in serious domestic unrest or rebellion. One can look at the relative stability in famine-stricken North Korea as a poignant example of a polity whose citizens have suffered widespread physical deprivation under policies of the existing regime, but who are unwilling or unable to risk their lives to challenge the state.

This phenomenon is partly explained by conflict theorists who argue that individuals and groups have feelings of "relative deprivation" when they perceive a gap between what they believe they deserve and what in reality they actually have achieved.[14] In other words, can a government meet the expectations of the masses enough to avoid conflict? For example, in North Korea—a regime that tightly controls the information that its people receive—many people understand that they are suffering, but they may not know precisely how much they are suffering relative to others, such as their brethren in the South. The North Korean government indoctrinates its people to expect little other than hardship, which in turn it blames on outside enemies of the state. Thus, the people of North Korea have very low expectations, which their government has been able to meet. More important, then, is the question of whom do the people perceive as being responsible for their plight? If the answer is the people's own government—whether as a result of supply-induced, demand-induced, or structural resource scarcity—then social discord and rebellion are more likely to result in intrastate conflict, as citizens challenge the ruling legitimacy of the state itself. If the answer is someone else's government, then interstate conflict may result.

On numerous occasions, history has shown that governments whose people are suffering can remain in power for long periods of time by pointing to external sources for the people's hardship.[15] As noted above regarding political legitimacy, perception is politically more important than any standard of objective truth.[16] When faced with a crisis of legitimacy derived from environmental resource scarcity, any political regime essentially has a choice of two options in dealing with the situation. The regime may choose temporarily not to respond to looming challenges to its authority because water-induced stress may in fact pass when sufficient heavy rainfall occurs. However, most regimes in the Middle East and North Africa have sought more proactive ways to ensure their survival. Indeed, a people might forgive its government for one drought, but if governmental action is not taken, a subsequent drought-induced crisis of legitimacy could result in significant social upheaval by an unforgiving public. Furthermore, if the government itself is perceived to be the direct source of the scarcity—through structural arrangements, resource capture, or other means—these trends of social unrest are likely to be exacerbated. Thus, in order to survive, most states have developed policies to increase their water supplies and to address issues of environmental scarcity. The problem with doing so throughout most of the Middle East and North Africa, however, is that increasing supply in one state often creates environmental scarcity problems in another. If Turkey builds dams, Iraq and Syria are vulnerable; if Ethiopia or the Sudan builds dams, Egypt feels threatened. Thus far, interstate water problems leading to war have been avoided due to the economic interplay between oil wealth and the importation of "virtual water," which will be discussed at greater length below.

As noted above, resource scarcity issues centered on water are particularly prominent in the Middle East and North Africa. Ewan Anderson notes that resource geopolitics in the Middle East "has long been dominated by one liquid—oil. However, another liquid, water, is now recognized as the fundamental political weapon in the region."[17] Ecologically speaking, water scarcity in the Middle East and North Africa results from four primary causes: fundamentally dry climatic conditions, drought, desiccation (the degradation of land due to the drying up of the soil), and water stress (the low availability of water resulting from a growing population).[18] These resource scarcity problems are exacerbated in the Middle East by such factors as poor water quality and inadequate—and, at times, purposefully discriminatory—resource planning. As a result of these ecological and political trends, Nurit Kliot states, "water, not oil, threatens the renewal of military conflicts and social and economic disruptions" in the Middle East.[19] In the case of the Arab-Israeli conflict, Alwyn Rouyer suggests that "water has become inseparable from land, ideology, and religious prophecy."[20] Martin Sherman echoes these sentiments in the following passage, describing specifically the Arab-Israeli conflict:

In recent years, particularly since the late 1980s, water has become increasingly dominant as a bone of contention between the two sides. More than one Arab leader, including those considered to be among the most moderate, such as King Hussein of Jordan and former UN Secretary General, Boutros Boutros-Ghali of Egypt, have warned explicitly that water is the issue most likely to become the cause of a future Israeli-Arab war.[21]

"Water is a particularly volatile strategic issue, especially when it is in severe shortage."

While Jochen Renger contends that a conflict waged explicitly over water may not lie on the immediate horizon, he notes that "it is likely that water might be used as leverage during a conflict."[22] As a result of such geopolitical trends, managing these water resources in the Middle East and North Africa—and, in turn, managing the conflict over these resources—should be considered a primary concern of both scholars and policymakers.

Keeping the Peace: The Importance of Virtual Water

The warning signals that war over water may replace war over oil and other traditional sources of conflict are very real in recent history. Yet, for more than 25 years, despite increasing demand, water has not been the primary cause of war in the Middle East and North Africa. The scenarios outlined in the preceding section? have yet to fully address the fundamental questions of why and how governments in the region have thus far avoided major interstate conflict over water. In order to understand the likelihood of war, we must address the foundation of the past peace, testing whether or not this foundation remains strong for the foreseeable future. How have the governments of the region been able to avoid the apparently inevitable consequences of conflict that derive from the interlinked problems of water deficits, population growth, and weak economic performance? In this section of the article, we turn our attention to the important linkages between water usage and the global economy, showing how existing water shortages have been ameliorated by a combination of economic factors.

To understand the politics of water in the Middle East and North Africa, one must first look at the region's most fungible resource: oil. For much of the post-World War II era, the growing need for oil to fuel economic growth has served as the dominant motivating factor in US security policy in the Middle East. Conventional wisdom in the United States holds that US dependency on Middle Eastern oil is a strategic weakness. Indeed, the specter of a regional hegemonic power that controls the oil and that is also hostile to the United States strikes fear into the hearts of policymakers in Washington. Thus, for roughly the past 50 years, the United States has sought to prop up "moderate" (meaning pro-US) regimes while denying hegemony to "radical" (meaning anti-US) regimes.[23] However, we contend that both policymakers and the public at large in the United States generally misunderstand the politics of oil as they relate to water in the Middle East.

Country	Total Water Resources per Capita (cubic meters) in 2000	Percent of Population with Access to Adequate, Improved Water Source, 2000	GDP per Capita, 2000 Estimate
Algeria	477	94%	$5,500
Egypt	930	95%	$3,600
Iran	2,040	95%	$6,300
Iraq	1,544	85%	$2,500
Israel	180	99%	$18,900
Jordan	148	96%	$3,500
Kuwait	0	100%	$15,000
Lebanon	1,124	100%	$5,000
Libya	148	72%	$8,900
Morocco	1,062	82%	$3,500
Oman	426	39%	$7,700
Qatar	—	100%	$20,300
Saudi Arabia	119	95%	$10,500
Sudan	5,312	—	$1,000
Syria	2,845	80%	$3,100
Tunisia	434	—	$6,500
Turkey	3,162	83%	$6,800
Yemen	241	69%	$820

Figure 1 Water Resources and Economics in the Middle East and North Africa.

Sources: World Bank Development Indicators, Country-at-a-Glance Tables, Freshwater Resources, and *CIA World Factbook*, at http://www.worldbank.org and http://www.cia.gov.

In absolute terms, problems arising from US vulnerability to foreign oil are basically true—it would be better to be free of dependency on oil from any foreign source than to be dependent. However, the other side of the equation is often forgotten: oil-producing states are dependent on the United States and other major oil importers for their economic livelihood. More bluntly put, oil-exporting states are dependent on the influx of dollars, euros, and yen to purchase goods, services, and commodities that they lack. Thus, oil-producing countries in the Middle East and North Africa, few of whom have managed to successfully diversify their economies beyond the petroleum sector, exist in an interdependent world economy. The world depends on their oil, and they depend on the world's goods and services—including that most valuable life-sustaining resource, water.

On the surface, this perhaps seems to be a contentious claim. Outgoing oil tankers do not return with freshwater used to grow crops, and Middle East countries do not rely on the importation of bottled water for their daily consumption needs. However, according to hydrologists, each individual needs approximately 100 cubic meters of water each year for personal needs, and an additional 1,000 cubic meters are required to grow the food that person consumes. Thus, every person alive requires approximately 1,100 cubic meters of water every year. In 1970, the water needs of most Middle Eastern and North African countries could be met from sources within the region. During the colonial and early post-colonial eras, regional governments and their engineers had effectively managed supply to deliver new water to meet the requirements of the growing urban populations, industrial requirements, agricultural needs, and other demand-induced factors. What is clear is that in the past 30 years, the status of the region's water resources has significantly worsened as populations have increased (an example of demand-induced scarcity). Since the mid-1970s, most countries have been able to supply daily consumption and industrial needs; however, as indicated in Figure 1, the approximate 1,000 cubic meters of water per capita that is required for self-sufficient agricultural production represents a seemingly impossible challenge for some Middle Eastern and North African economies.

Simply put, many countries of the region cannot presently meet the irrigation requirements needed to feed their own growing populations.[24] Furthermore, for those countries that have sufficient resources to meet this need in aggregate (such as Syria), resource capture and structural distribution problems keep water out of the hands of many citizens. If this situation has been deteriorating for nearly three decades, the question remains: Why has there been no war over water? The answer, according to Tony Allen, lies in an extremely important hidden source of water, which he describes as "virtual water."[25] Virtual water is the water contained in the food that the region imports—from the United States, Australia, Argentina, New Zealand, the countries of the European Union, and other major food-exporting countries. If each person of the world consumes food that requires 1,000 cubic meters of water to grow, plus 100 additional cubic meters for drinking, hygiene, and industrial production, it is still possible that any country that cannot supply the water to produce food may have sufficient water to meet its needs—if it has the economic capacity to buy, or the political capacity to beg, the remaining virtual water in the form of imported food.

According to Allen, more water flows into the countries of the Middle East and North Africa as virtual water each year than flows down the Nile for Egypt's agriculture. Virtual water obtained in the food available on the global market has enabled the governments of the region's countries to augment their inadequate and declining water resources. For instance, despite its meager freshwater resources of 180 cubic meters per capita, Israel—otherwise self-sufficient in terms of food production—manages its problems of water scarcity in part by importing large supplies of grain each year. As noted in Figure 1, this pattern is replicated by eight other countries in the region that have less than 1,100 cubic meters of water per person. Thus, the global cereal grain commodity markets have proven to be a very accessible and effective system for importing virtual water needs. In the Middle East and North Africa, politicians and resource managers have thus far found this option a better choice than resorting to war over water with their neighbors. As a result, the strategic imperative for maintaining peace has been met through access to virtual water in the form of food imports from the global market.[26]

The global trade in food commodities has been increasingly accessible, even to poor economies, for the past 50 years. During the Cold War, food that could not be purchased was often provided in the form of grants by either the United States or the Soviet Union, and in times of famine, international relief efforts in various parts of the globe have fed the starving. Over time, competition by the generators of the global grain surplus—the United States, Australia, Argentina, and the European Community—brought down the global price of grain. As a result, the past quarter-century, the period during which water conflicts in the Middle East and North Africa have been most insistently predicted, was also a period of global commodity markets awash with surplus grain. This situation allowed the region's states to replace domestic water supply shortages with subsidized virtual water in the form of purchases from the global commodities market. For example, during the 1980s, grain was being traded at about $100 (US) a ton, despite costing about $200 a ton to produce.[27] Thus, US and European taxpayers were largely responsible for funding the cost of virtual water (in the form of significant agricultural subsidies they paid their own farmers) which significantly benefited the countries of the Middle East and North Africa.

For the most part we concur with Allen's evaluation that countries have not gone to war primarily over water, and that they have not done so because they have been able to purchase virtual water on the international market. However, the key question for the future is, Will this situation continue? If the answer is yes, and grain will remain affordable to the countries of the region, then it is relatively safe to conclude that conflict derived from environmental scarcity (in the form of water deficits) will not be a significant problem in the foreseeable future. However, if the answer is no, and grain will not be as affordable as it has been in the past, then future conflict scenarios based on environmental scarcity must be seriously considered.

Global Economic Restructuring: The World Trade Organization's Impact on Subsidies

Regrettably, a trend toward the answer "no" appears to be gaining some momentum due to ongoing structural changes in the global economy. The year 1995 witnessed a dramatic change in the world grain market, when wheat prices rose rapidly, eventually reaching $250 a ton by the spring of 1996. With the laws of supply and demand kicking in, this increased price resulted in greater production; by 1998, world wheat prices had fallen back to $140 a ton, but had risen again to over $270 by June 2001.[28] These rapid wheat price fluctuations reemphasize the strategic importance and volatility of virtual water. If the global price of food staples remains affordable, many countries in the Middle East and North Africa may struggle to meet the demand-induced scarcity resulting from their growing populations, but they most likely will succeed. However, if basic food staple prices rise significantly in the coming decades and the existing economic growth patterns that have characterized the region's economies over the past 30 years remain constant, an outbreak of war is more likely.

It is clear that recent structural changes in the world economy do not favor the continuation of affordable food prices for the region's countries in the future. As noted above, wheat that costs $200 a ton to produce has often been sold for $100 a ton on world markets. This situation is possible only when the supplier is compensated for the lost $100 per ton in the form of a subsidy. Historically, these subsidies have been paid by the governments of major cereal grain-producing countries, primarily the United States and members of the European Union. Indeed, for the last 100 years, farm subsidies have been a bedrock public policy throughout the food-exporting countries of the first world. However, with the steady embrace of global free-trade economics and the establishment of the World Trade Organization (WTO), agricultural subsidies have come under pressure in most major grain-producing countries. According to a recent US Department of Agriculture (USDA) study, "The elimination of agriculture trade and domestic policy distortions could raise world agriculture prices about 12 percent.[29]

> **"Many countries of the region cannot presently meet the irrigation requirements needed to feed their own growing populations."**

Thus, as the WTO gains systematic credibility over the coming decades, its free-trade policies will further erode the practice of farm price supports, and it is highly unlikely that the aggregate farm subsidies of the past will continue at historic levels in the future. Under the new WTO regime, global food production will be increasingly based on the real cost of production plus whatever profit is required to keep farmers in business. Therefore, as global food prices rise in the future, and American and European governments are restricted by the new global trading regime from subsidizing their farmers, the price of virtual water in the Middle East and North Africa and throughout the food-importing world will also rise. According to the USDA report mentioned above, both developed and developing countries will gain from WTO liberalization. Developed countries that are major food exporters will gain immediately from the projected $31 billion in increased global food prices, of which they will share $28.5 billion ($13.3 billion to the United States), with $2.6 billion going to food exporters in the developing world. However, the report also claims food-importing countries will gain because global food price increases will spur more efficient production in their own economies, thus enabling them a "potential benefit" of $21 billion.[30] Even if accepted at face value, it is clear that such benefits will occur mostly in those developing countries with an abundance of water resources. Indeed, developing countries that produce fruits, vegetables, and other high-value crops for export to first-world markets may indeed benefit from the reduction of farm subsidies, which today undercut their competitive advantage. But when it comes to basic foodstuffs—wheat, corn, and rice—the cereal grains that sustain life for most people, the developing world cannot compete with the highly efficient mechanized corporate farms of the first world.

In future research, basic intelligence is needed on two fronts. First, we must obtain a clearer understanding of the capacity of global commodity markets to meet future virtual water needs in the form of food. Second, we must identify which Middle East and North Africa governments will most likely have the economic capacity to meet their virtual water needs though food purchases—or, perhaps more important, which ones will not. In short, is there food available in the global market, and can countries afford to buy it? Countries that cannot afford virtual water may choose instead to pursue war as a means of achieving their national interest goals. Clearly the strongest countries, or those least susceptible to intrastate or interstate conflict arising from environmental scarcity, are those that have significant water resources or the economic capacity to purchase virtual water. However, it is also clear that the relative condition of peace that has existed in the Middle East and North Africa has been maintained historically through deeply buried linkages between American and European taxpayers, their massive farm subsidies programs, and world food prices. In the future, it appears that these hidden links may be radically altered if not broken by the World Trade Organization, and, as a result, the likelihood of conflict will increase.

Conclusion: Why War Will Come

Having moved away from the conventional understanding of water strictly as a zero-sum environmental resource by reconceptualizing it in more fungible economic terms, we nevertheless believe two incompatible social trends will collide to make war in the Middle East and North Africa virtually inevitable in the future. The first trend is economic globalization. As capitalism becomes ever more embraced as the global economic philosophy, and the world increasingly embraces free-trade economics, economic growth is both required and is inevitable. The WTO will facilitate this aggregate global growth, which, on the plus side, will undoubtedly increase the basic standard of living for the average world citizen. However, the global economy will be required to meet the needs of an estimated eight billion citizens in the year 2025. Achieving growth will demand an ever-greater share of the world's existing natural resources, including water. Thus, if present regional economic and demographic trends continue, resource shortfalls will occur, with water being the most highly stressed resource in the Middle East and North Africa.

Globalization is both a cause and a consequence of the rapid spread of information technology. Thus, in the globalized world, the figurative distance between cultures, philosophies of rule, and, perhaps more important, a basic understanding of what is possible in life, becomes much shorter. Personal computers, the internet, cellular phones, fax machines, and satellite television are all working in partnership to rewire the psychological infrastructure of the citizens of the Middle East and North Africa, and the world at large. As a result, by making visible what is possible in the outside world, this cognitive liberation will bring heightened material expectations of a better life, both economically and politically. Consequently, citizens will demand more from their governments. This emerging reality will collide head-on with the second trend—political authoritarianism—that characterizes most Middle East and North Africa governments.

Throughout the region there are few governments that allow for public expression of dissent. Although Turkey, Algeria, Tunisia, and Egypt are democracies in name, these states have exhibited a propensity to revert to authoritarian tactics when deemed necessary to limit political activity among their respective populaces.[31] Likewise, while Israel is institutionally a democracy, ethnic minorities are all but excluded from the democratic process. The remainder of the Middle East and North Africa states can be described only as authoritarian regimes. In retrospect, the most fundamental common denominator of all authoritarian regimes throughout history is their fierce resistance to change. Change is seen as a threat to the regime because most authoritarian regimes base their right to rule in some form of infallibility: the infallibility of the sultan, the king, or the ruling party and its ideology. Any admission that change is needed strikes at the foundation of this inflexible infallibility. Historically, most change has occurred in the Middle East and North Africa during times of intrastate unrest and interstate war. In the coming decades, globalization will bring change that will be resisted by governments of the region. As a result, to the distant observer the future will resemble the past: periods of wholesale peace will be a rare occurrence, intense competition and low-intensity conflict will be the norm, and major wars will occur at sporadic intervals.

The wild card in this equation may be post-2004 Iraq. Operation Iraqi Freedom and the ouster of Saddam Hussein have altered the strategic political landscape. If a sustainable democracy indeed emerges in Iraq, the country may turn away from future conflicts with its neighbors. Potential conflict between Turkey and Iraq over water may now be averted due to the fact that both countries may choose nonviolent solutions to their disputes. If President Bush's vision of a democratic Middle East comes to fruition, war may be averted. After all, there is a rich body of scholarly research regarding the "democratic peace" that suggests liberal democracies are significantly less likely to resort to war to resolve interstate disputes, and post-Saddam Iraq could serve as a key litmus test for the future of democratic reform in the region. However, it is also highly unlikely that regime change will come quickly to the moderate authoritarian states of the region that are also US allies. Decisionmakers in Washington may be able to dictate the political future of Iraq, but even America's mighty arsenal of political, economic, and military power cannot alter the basic demographic and environmental trends in the region.

Notes

1. Alex Marshall, ed., *The State of World Population 1997* (New York: United Nations Population Fund, 1997), p. 70.

2. "The World Bank: Middle East and North Africa Data Profile," *The World Bank Group Country Data* (2000), http://www.worldbank.org/data/countrydata/countrydata.html.

3. The leading scholar in this area is Thomas Homer-Dixon. For example, see his recent book (coedited with Jessica Blitt), *Ecoviolence: Links Among Environment, Population, and Security* (New York: Rowman & Littlefield, 1998), which focuses on Chiapas, Gaza, South Africa, Pakistan, and Rwanda.

4. Mostafa Dolatyar and Tim S. Gray, *Water Politics in the Middle East: A Context for Conflict or Co-operation?* (New York: St. Martin's Press, 2000), p. 6.

5. *Human Development Report: Consumption for Human Development* (New York: United Nations Development Programme, Oxford Univ. Press, 1998), p. 55; "Water Woes Around the World," MSNBC, 9 September 2002, http://www.msnbc.com/news/802693.asp.

6. Thomas Naff, "Conflict and Water Use in the Middle East," in *Water in the Arab World: Perspectives and Prognoses,* ed. Peter Rogers and Peter Lydon (Cambridge, Mass.: Harvard Univ. Press, 1994), p. 273.

7. Alwyn R. Rouyer, *Turning Water into Politics: The Water Issue in the Palestinian-Israeli Conflict* (New York: St. Martin's Press, 2000), p. 7.

8. Ulrich Küffner, "Contested Waters: Dividing or Sharing?" in *Water in the Middle East: Potential for Conflicts and Prospects for Cooperation,* ed. Waltina Scheumann and Manuel Schiffler (New York: Springer, 1998), p. 71.

9. Miriam R. Lowi, *Water and Power: The Politics of a Scarce Resource in the Jordan River Basin* (Cambridge, Eng.: Cambridge Univ. Press, 1993), p. 1.

10. Ibid., pp. 2ff.

11. Thomas Homer-Dixon and Jessica Blitt, "Introduction: A Theoretical Overview," in *Ecoviolence: Links Among Environment, Population,and Scarcity,* ed. Thomas Homer-Dixon and Jessica Blitt (New York: Rowman & Littlefield, 1998), p. 6.

12. Thomas Homer-Dixon and Valerie Percival, "The Case of Senegal-Mauritania," in *Environmental Scarcity and Violent*

Conflict: Briefing Book (Washington: American Association for the Advancement of Science and the University of Toronto, 1996), pp. 35–38.

13. Douglas Borer witnessed this agricultural problem while visiting rural areas on the Bataan peninsula in late 1985 and early 1986. The members of the New People's Army which he met were uninterested in Marxism, but they were very interested in ridding themselves of the Marcos regime. See Thomas Homer-Dixon and Valerie Percival, "The Case of the Philippines," ibid., p. 49.

14. Ted Gurr, *Why Men Rebel* (Princeton, N.J.: Princeton Univ. Press, 1970).

15. One need only look 90 miles southward from the Florida coast to find proof of this reality in Castro's Cuba.

16. Thus, Saddam Hussein was able to remain in power in Iraq until 2003 due to two essential factors. First, as noted in a recent article by James Quinlivan, Saddam had created "groups with special loyalties to the regime and the creation of parallel military organizations and multiple internal security agencies," that made Iraq essentially a "coup-proof" regime. (See James T. Quinlivan, "Coup-Proofing: Its Practice and Consequences in the Middle East," *International Security,* 24 [Fall 1999], 131–65.) Second, Saddam had convinced a significant portion of his people that the United States (and Britain) were responsible for their suffering. Thus, as long as these perceptions held and Saddam was able to command loyalty of the inner regime, his ouster from power by domestic sources remained unlikely.

17. Ewan W. Anderson, "Water: The Next Strategic Resource," in *The Politics of Scarcity: Water in the Middle East,"* ed. Joyce R. Starr and Daniel C. Stoll (Boulder, Colo.: Westview Press, 1988), p. 1.

18. Hussein A. Amery and Aaron T. Wolf, "Water, Geography, and Peace in the Middle East," in *Water in the Middle East: A Geography of Peace,* ed. Hussein A. Amery and Aaron T. Wolf (Austin: Univ. of Texas Press, 2000).

19. Nubit Kliot, *Water Resources and Conflict in the Middle East* (London and New York: Routledge, 1994), p. v, as quoted in Dolatyar and Gray, p. 9.

20. Rouyer, p. 9.

21. Martin Sherman, *The Politics of Water in the Middle East: An Israeli Perspective on the Hydro-Political Aspects of the Conflict* (New York: St. Martin's Press, 1999), p. xi.

22. Jochen Renger, "The Middle East Peace Process: Obstacles to Cooperation Over Shared Waters," in *Water in the Middle East: Potential for Conflict and Prospects for Cooperation,* ed. Waltina Scheumann and Manuel Schiffler (New York: Springer, 1998), p. 50.

23. Thus, even though the Saudi government is much more Islamized in religious terms than that of the Iraqis or Syrians, as long as the Saudi government is pro-US and serves US interests in supplying cheap oil, it receives the benevolent "moderate" label, while the more secularized Iraqis and Syrians have been labeled with the prerogative labels "radical" or "rogue-states."

24. Tony Allan, "Watersheds and Problemsheds: Explaining the Absence of Armed Conflict over Water in the Middle East," in *MERNIA: Middle East Review of International Affairs Journal,* 2 (March 1998), http://biu.ac.il/SOC/besa/meria/journal/1998/issue1/jv2n1a7.html.

25. Ibid.

26. Ibid.

27. Ibid.

28. Prices from 26 June 2001 quoted at http://www.usafutures.com/commodityprices.htm.

29. "Agricultural Policy Reform in the WTO—The Road Ahead," in *ERS Agricultural Economics Report,* No. 802, ed. Mary E. Burfisher (Washington: US Department of Agriculture, May 2001), p. iii.

30. Ibid., p. 6.

31. For instance, as of 2004, Freedom House (http://www.freedomhouse.org/) classifies Algeria, Egypt, and Tunisia as "not free," and Turkey as only "partly free."

JASON J. MORRISSETTE is a doctoral candidate and instructor of record in the School of Public and International Affairs at the University of Georgia. He is currently writing his dissertation on the political economy of water scarcity and conflict.

DOUGLAS A. BORER (Ph.D., Boston University, 1993) is an Associate Professor in the Department of Defense Analysis at the Naval Postgraduate School. He recently served as Visiting Professor of Political Science at the US Army War College. Previously he was Director of International Studies at Virginia Tech, and he has taught overseas in Fiji and Australia. Dr. Borer is a former Fulbright Scholar at the University of Kebangsaan Malaysia, and has published widely in the areas of security, strategy, and foreign policy.

From *Parameters,* Winter 2004–2005, pp. 86–101. Published in 2005 by the U.S. Army War College.

The Irony of Climate

Archaeologists suspect that a shift in the planet's climate thousands of years ago gave birth to agriculture. Now climate change could spell the end of farming as we know it.

BRIAN HALWEIL

High in the Peruvian Andes, a new disease has invaded the potato fields in the town of Chacllabamba. Warmer and wetter weather associated with global climate change has allowed late blight—the same fungus that caused the Irish potato famine—to creep 4,000 meters up the mountainside for the first time since humans started growing potatoes here thousands of years ago. In 2003, Chacllabamba farmers saw their crop of native potatoes almost totally destroyed. Breeders are rushing to develop tubers resistant to the "new" disease that retain the taste, texture, and quality preferred by Andean populations.

Meanwhile, old-timers in Holmes County, Kansas, have been struggling to tell which way the wind is blowing, so to speak. On the one hand, the summers and winters are both warmer, which means less snow and less snowmelt in the spring and less water stored in the fields. On the other hand, there's more rain, but it's falling in the early spring, rather than during the summer growing season. So the crops might be parched when they need water most. According to state climatologists, it's too early to say exactly how these changes will play out—if farmers will be able to push their corn and wheat fields onto formerly barren land or if the higher temperatures will help once again to turn the grain fields of Kansas into a dust bowl. Whatever happens, it's going to surprise the current generation of farmers.

Asian farmers, too, are facing their own climate-related problems. In the unirrigated rice paddies and wheat fields of Asia, the annual monsoon can make or break millions of lives. Yet the reliability of the monsoon is increasingly in doubt. For instance, El Niño events (the cyclical warming of surface waters in the eastern Pacific Ocean) often correspond with weaker monsoons, and El Niños will likely increase with global warming. During the El Niño-induced drought in 1997, Indonesian rice farmers pumped water from swamps close to their fields, but food losses were still high: 55 percent for dryland maize and 41 percent for wetland maize, 34 percent for wetland rice, and 19 percent for cassava. The 1997 drought was followed by a particularly wet winter that delayed planting for two months in many areas and triggered heavy locust and rat infestations. According to

Bambang Irawan of the Indonesian Center for Agricultural Socio-Economic Research and Development, in Bogor, this succession of poor harvests forced many families to eat less rice and turn to the less nutritious alternative of dried cassava. Some farmers sold off their jewelry and livestock, worked off the farm, or borrowed money to purchase rice, Irawan says. The prospects are for more of the same: "If we get a substantial global warming, there is no doubt in my mind that there will be serious changes to the monsoon," says David Rhind, a senior climate researcher with NASA's Goddard Institute for Space Studies.

Archaeologists believe that the shift to a warmer, wetter, and more stable climate at the end of the last ice age was key for humanity's successful foray into food production. Yet, from the American breadbasket to the North China Plain to the fields of southern Africa, farmers and climate scientists arc finding that generations-old patterns of rainfall and temperature are shifting. Farming may be the human endeavor most dependent on a stable climate—and the industry that will struggle most to cope with more erratic weather, severe storms, and shifts in growing season lengths. While some optimists are predicting longer growing seasons and more abundant harvests as the climate warms, farmers are mostly reaping surprises.

Toward the Unknown (Climate) Region

For two decades, Hartwell Allen, a researcher with the University of Florida in Gainesville and the U.S. Department of Agriculture, has been growing rice, soybeans, and peanuts in plastic, greenhouse-like growth chambers that allow him to play God. He can control—"rather precisely"—the temperature, humidity, and levels of atmospheric carbon. "We grow the plants under a daily maximum/minimum cyclic temperature that would mimic the real world cycle," Allen says. His lab has tried regimes of 28 degrees C day/18 degrees C night, 32/22, 36/26, 40/30, and 44/34. "We ran one experiment to 48/38, and got very few surviving plants," he says. Allen found that while a doubling of carbon dioxide and a slightly increased temperature stimulate seeds to germinate and the plants to grow larger and lusher, the higher temperatures are deadly when the plant starts producing pollen. Every stage of the process—pollen transfer, the growth of the tube that links the pollen to the seed, the viability of the pollen itself—is highly sensitive. "It's all or nothing, if pollination isn't successful," Allen notes. At temperatures above 36 degrees C during pollination, peanut yields dropped about six percent per degree of temperature increase. Allen is particularly concerned about the implications for places like India and West Africa, where peanuts are a dietary staple and temperatures during the growing season are already well above 32 degrees C: "In these regions the crops are mostly rain-fed. If global warming also leads to drought in these areas, yields could be even lower."

As plant scientists refine their understanding of climate change and the subtle ways in which plants respond, they are beginning to think that the most serious threats to agriculture will not be the most dramatic: the lethal heatwave or severe drought or endless deluge. Instead, for plants that humans have bred to thrive in specific climatic conditions, it is those subtle shifts in temperatures and rainfall during key periods in the crops' lifecycles that will be most disruptive. Even today, crop losses associated with background climate variability are significantly higher than those caused by disasters such as hurricanes or flooding.

John Sheehy at the International Rice Research Institute in Manila has found that damage to the world's major grain crops begins when temperatures climb above 30 degrees C during flowering. At about 40 degrees C, yields are reduced to zero. "In rice, wheat, and maize, grain yields are likely to decline by 10 percent for every 1 degree C increase over 30 degrees. We are already at or close to this threshold," Sheehy says, noting regular heat damage in Cambodia, India, and his own center in the Philippines, where the average temperature is now 2.5 degrees C higher than 50 years ago. In particular, higher night-time temperatures forced the plants to work harder at respiration and thus sapped their energy, leaving less for producing grain. Sheehy estimates that grain yields in the tropics might fall as much as 30 percent over the next 50 years, during a period when the region's already malnourished population is projected to increase by 44 percent. (Sheehy and his colleagues think a potential solution is breeding rice and other crops to flower early in the morning or at night so that the sensitive temperature process misses the hottest part of the day. But, he says, "we haven't been successful in getting any real funds for the work.") The world's major plants can cope with temperature shifts to some extent, but since the dawn of agriculture farmers have selected plants that thrive in stable conditions.

Climatologists consulting their computer climate models see anything but stability, however. As greenhouse gases trap more of the sun's heat in the Earth's atmosphere, there is also more energy in the climate system, which means more extreme swings—dry to wet, hot to cold. (This is the reason that there can still be severe winters on a warming planet, or that March 2004 was the third-warmest month on record after one of the coldest winters ever.) Among those projected impacts that climatologists have already observed in most regions: higher maximum temperatures and more hot days, higher minimum temperatures and fewer cold days, more variable and extreme rainfall events, and increased summer drying and associated risk of drought in continental interiors. All of these conditions will likely accelerate into the next century.

Cynthia Rosenzweig, a senior research scholar with the Goddard Institute for Space Studies at Columbia University, argues that although the climate models will always be improving, there are certain changes we can already predict with a level of confidence. First, most studies indicate "intensification of the hydrological cycle," which essentially means more droughts and floods, and more variable and extreme rainfall. Second, Rosenzweig says, "basically every study has shown that there will be increased incidence of crop pests." Longer growing seasons mean more generations of pests during the summer, while shorter and warmer winters mean that fewer adults, larvae, and eggs will die off.

Third, most climatologists agree that climate change will hit farmers in the developing world hardest. This is partly a result of geography. Farmers in the tropics already find themselves near the temperature limits for most major crops, so any warming is likely to push their crops over the top. "All increases in temperature, however small, will lead to decreases in production," says Robert Watson, chief scientist at the World Bank and former chairman of the Intergovernmental Panel on Climate Change. "Studies have consistently shown that agricultural regions in the developing world are more vulnerable, even before we consider the ability to cope," because of poverty, more limited irrigation technology, and lack of weather tracking systems. "Look at the coping strategies, and then it's a real double whammy," Rosenzweig says. In sub-Saharan Africa—ground zero of global hunger, where the number of starving people has doubled in the last 20 years—the current situation will undoubtedly be exacerbated by the climate crisis. (And by the 2080s, Watson says, projections indicate that even temperate latitudes will begin to approach the upper limit of the productive temperature range.)

Coping with Change

"Scientists may indeed need decades to be sure that climate change is taking place," says Patrick Luganda, chairman of the Network of Climate Journalists in the Greater Horn of Africa. "But, on the ground, farmers have no choice but to deal with the daily reality as best they can." Luganda says that several years ago local farming communities in Uganda could determine the onset of rains and their cessation with a fair amount of accuracy. "These days there is no guarantee that the long rains will start, or stop, at the usual time," Luganda says. The Ateso people in north-central Uganda report the disappearance of asisinit, a swamp grass favored for thatch houses because of its beauty and durability. The grass is increasingly rare because farmers have started to plant rice and millet in swampy areas in response to more frequent droughts. (Rice farmers in Indonesia coping with droughts have done the same.) Farmers have also begun to sow a wider diversity of crops and to stagger their plantings to hedge against abrupt climate shifts. Luganda adds that repeated crop failures have pushed many farmers into the urban centers: the final coping mechanism.

The many variables associated with climate change make coping difficult, but hardly futile. In some cases, farmers may need to install sprinklers to help them survive more droughts. In other cases, plant breeders will need to look for crop varieties that can withstand a greater range of temperatures. The good news is that many of the same changes that will help farmers cope with climate change will also make communities more self-sufficient and reduce dependence on the long-distance food chain.

Planting a wider range of crops, for instance, is perhaps farmers' best hedge against more erratic weather. In parts of Africa, planting trees alongside crops—a system called agroforestry that might include shade coffee and cacao, or leguminous trees with corn—might be part of the answer. "There is good reason to believe that these systems will be more resilient than a maize monoculture," says Lou Verchot, the lead scientist on climate change at the International Centre for Research in Agroforestry in Nairobi. The trees send their roots considerably deeper than the crops, allowing them to survive a drought that might damage the grain crop. The tree roots will also pump water into the upper soil layers where crops can tap it. Trees improve the soil as well: their roots create spaces for water flow and their leaves decompose into compost. In other words, a farmer who has trees won't lose everything. Farmers in central Kenya are using a mix of coffee, macadamia nuts, and cereals that results in as many as three marketable crops in a good year. "Of course, in any one year, the monoculture will yield more money," Verchot admits, "but farmers need to work on many years." These diverse crop mixes are all the more relevant since rising temperatures will eliminate much of the traditional coffee- and tea-growing areas in the Caribbean, Latin America, and Africa. In Uganda, where coffee and tea account for nearly 100 percent of agricultural exports, an average temperature rise of 2 degrees C would dramatically reduce the harvest, as all but the highest altitude areas become too hot to grow coffee.

In essence, farms will best resist a wide range of shocks by making themselves more diverse and less dependent on outside inputs. A farmer growing a single variety of wheat is more likely to lose the whole crop when the temperature shifts dramatically than a farmer growing several wheat varieties, or better yet, several varieties of plants besides wheat. The additional crops help form a sort of ecological bulwark against blows from climate change. "It will be important to devise more resilient agricultural production systems that can absorb and survive more variability," argues Fred Kirschenmann, director of the Leopold Center for Sustainable Agriculture at Iowa State University. At his own family farm in North Dakota, Kirschenmann has struggled with two years of abnormal weather that nearly eliminated one crop and devastated another. Diversified farms will cope better with drought, increased pests, and a range of other climate-related jolts. And they will tend to be less reliant on fertilizers and pesticides, and the fossil fuel inputs they require. Climate change might also be the best argument for preserving local crop varieties around the world, so that plant breeders can draw from as wide a palette as possible when trying to develop plants that can cope with more frequent drought or new pests.

Farms with trees planted strategically between crops will not only better withstand torrential downpours and parching droughts, they will also "lock up" more carbon. Lou Verchot says that the improved fallows used in Africa can lock up 10–20 times the carbon of nearby cereal monocultures, and 30 percent of the carbon in an intact forest. And building up a soil's stock of organic matter—the dark, spongy stuff in soils that stores carbon and gives them their rich smell—not only increases the amount of water the soil can hold (good for weathering droughts), but also helps bind more nutrients (good for crop growth).

Best of all, for farmers at least, systems that store more carbon are often considerably more profitable, and they might become even more so if farmers get paid to store carbon under the Kyoto Protocol. There is a plan, for instance, to pay farmers in Chiapas, Mexico, to shift from farming that involves regular forest clearing to agroforestry. The International Automobile Federation is funding the project as part of its commitment to reducing carbon emissions from sponsored sports car races. Not only that, "increased costs for fossil fuels will accelerate demand for renewable energies," says Mark Muller of the Institute for Agriculture and Trade Policy in Minneapolis, Minnesota, who believes that farmers will find new markets for biomass fuels like switchgrass that can be grown on the farm, as well as additional royalties from installing wind turbines on their farms.

However, "carbon farming is a temporary solution," according to Marty Bender of the Land Institute's Sunshine Farm in Salina, Kansas. He points to a recent paper in *Science* showing that even if America's soils were returned to their pre-plow carbon content—a theoretical maximum for how much carbon they could lock up—this would be equal to only two decades of American carbon emissions. "That is how little time we will be buying," Bender says, "despite the fact that it may take a hundred years of aggressive, national carbon farming and forestry to restore this lost carbon." (Cynthia Rosenzweig also notes that the potential to lock up carbon is limited, and that a warmer planet will reduce the amount of carbon that soils can hold: as land heats up, invigorated soil microbes respire more carbon dioxide.)

"We really should be focusing on energy efficiency and energy conservation to reduce the carbon emissions by our national economy," Bender concludes. That's why Sunshine Farm, which Bender directs, has been farming without fossil fuels, fertilizers, or pesticides in order to reduce its contribution to climate change and to find an inherently local solution to a global problem. As the name implies, Sunshine Farm runs essentially on sunlight. Homegrown sunflower seeds and soybeans become biodiesel that fuels tractors and trucks. The farm raises nearly three-fourths of the feed—oats, grain sorghum, and alfalfa—for its draft horses, beef cattle, and poultry. Manure and legumes in the crop rotation substitute for energy-gobbling nitrogen fertilizers. A 4.5-kilowatt photovoltaic array powers the workshop tools, electric fencing, water pumps, and chick brooding pens. The farm has eliminated an amount of energy equivalent to that used to make and transport 90 percent of its supplies. (Including the energy required to make the farm's machinery lowers the figure to 50 percent, still a huge gain over the standard American farm.)

The atmosphere, which sets the stage for climate dynamics, is a global commons. It has no gatekeeper—everyone can access it—and it has a limited capacity to absorb emissions before climate stability is undermined. Societies have developed a variety of gatekeeping solutions to help manage commons resources. One, privatization, would be difficult to apply to the atmosphere, although carbon-trading schemes might reduce emissions if caps are set low enough. Governments might also act as gatekeepers by means of treaties, such as the Kyoto Protocol, that limit emissions.

But these energy savings are only part of this distinctly local solution to an undeniably global problem, Bender says. "If local food systems could eliminate the need for half of the energy used for food processing and distribution, then that would save 30 percent of the fossil energy used in the U.S. food system," Bender reasons. "Considering that local foods will require some energy use, let's round the net savings down to 25 percent. In comparison, on-farm direct and indirect energy consumption constitutes 20 percent of energy use in the U.S. food system. Hence, local food systems could potentially save more energy than is used on American farms."

In other words, as climate tremors disrupt the vast intercontinental web of food production and rearrange the world's major breadbaskets, depending on food from distant suppliers will be more expensive and more precarious. It will be cheaper and easier to cope with local weather shifts, and with more limited supplies of fossil fuels, than to ship in a commodity from afar.

Agriculture is in third place, far behind energy use and chlorofluorocarbon production, as a contributor to climate warming. For farms to play a significant role, changes in cropping practices must happen on a large scale, across large swaths of India and Brazil and China and the American Midwest. As Bender suggests, farmers will be able to shore up their defenses against climate change, and can make obvious reductions in their own energy use which could save them money.

But the lasting solution to greenhouse gas emissions and climate change will depend mostly on the choices that everyone else makes. According to the London-based NGO Safe Alliance, a basic meal—some meat, grain, fruits, and vegetables—using imported ingredients can easily generate four times the greenhouse gas emissions as the same meal with ingredients from local sources. In terms of our personal contribution to climate change, eating local can be as important as driving a fuel-efficient car, or giving up the car for a bike. As politicians struggle to muster the will power to confront the climate crisis, ensuring that farmers have a less erratic climate in which to raise the world's food shouldn't be too hard a sell.

BRIAN HALWEIL is a senior researcher at Worldwatch Institute, and the author of *Eat Here: Reclaiming Homegrown Pleasures in a Global Supermarket.*

Avoiding Green Marketing Myopia

Ways to Improve Consumer Appeal for Environmentally Preferable Products

Jacquelyn A. Ottman, Edwin R. Stafford, and Cathy L. Hartman

In 1994, Philips launched the "EarthLight," a super energy-efficient compact fluorescent light (CFL) bulb designed to be an environmentally preferable substitute for the traditional energy-intensive incandescent bulb. The CFL's clumsy shape, however, was incompatible with most conventional lamps, and sales languished. After studying consumer response, Philips reintroduced the product in 2000 under the name "Marathon," to emphasize the bulb's five-year life. New designs offered the look and versatility of conventional incandescent light bulbs and the promise of more than $20 in energy savings over the product's life span compared to incandescent bulbs. The new bulbs were also certified by the U.S. Environmental Protection Agency's (EPA) Energy Star label. Repositioning CFL bulbs' features into advantages that resonated with consumer values—convenience, ease-of-use, and credible cost savings—ultimately sparked an annual sales growth of 12 percent in a mature product market.[1]

Philips' experience provides a valuable lesson on how to avoid the common pitfall of "green marketing myopia." Philips called its original entry "EarthLight" to communicate the CFL bulbs' environmental advantage. While noble, the benefit appealed to only the deepest green niche of consumers. The vast majority of consumers, however, will ask, "If I use 'green' products, what's in it for me?" In practice, green appeals are not likely to attract mainstream consumers unless they also offer a desirable benefit, such as cost-savings or improved product performance.[2] To avoid green marketing myopia, marketers must fulfill consumer needs and interests beyond what is good for the environment.

Although no consumer product has a zero impact on the environment, in business, the terms "green product" and "environmental product" are used commonly to describe those that strive to protect or enhance the natural environment by conserving energy and/or resources and reducing or eliminating use of toxic agents, pollution, and waste.[3] Paul Hawken, Amory Lovins, and L. Hunter Lovins write in their book *Natural Capitalism: Creating the Next Industrial Revolution* that greener, more sustainable products need to dramatically increase the productivity of natural resources, follow biological/cyclical production models, encourage dematerialization, and reinvest in and contribute to the planet's "natural" capital.[4] Escalating energy prices, concerns over foreign oil dependency, and calls for energy conservation are creating business opportunities for energy-efficient products, clean energy, and otherenvironmentally-sensitive innovations and products—collectively known as "cleantech"[5] (see the box on page 95). For example, Pulitzer Prize–winning author and *New York Times* columnist Thomas L. Friedman argues that government policy and industry should engage in a "geo-green" strategy to promote energy efficiency, renewable energy, and other cleantech innovations to help alleviate the nation's

dependency on oil from politically conflicted regions of the world.[6] Friedman asserts that such innovations can spark economic opportunity and address the converging global challenges of rising energy prices, terrorism, climate change, and the environmental consequences of the rapid economic development of China and India.

To exploit these economic opportunities to steer global commerce onto a more sustainable path, however, green products must appeal to consumers outside the traditional green niche.[7] Looking at sustainability from a green engineering perspective, Arnulf Grubler recently wrote in *Environment*, "To minimize environmental impacts by significant orders of magnitude requires the blending of good engineering with good economics as well as changing consumer preferences."[8] The marketing discipline has long argued that innovation must consider an intimate understanding of the customer,[9] and a close look at green marketing practices over time reveals that green products must be positioned on a consumer value sought by targeted consumers.

Drawing from past research and an analysis of the marketing appeals and strategies of green products that have either succeeded or failed in the marketplace over the past decade, some important lessons emerge for crafting effective green marketing and product strategies.[10] Based on the evidence, successful green products are able to appeal to mainstream consumers or lucrative market niches and frequently command price premiums by offering "non-green" consumer value (such as convenience and performance).

Green Marketing Myopia Defined

Green marketing must satisfy two objectives: improved environmental quality and customer satisfaction. Misjudging either or overemphasizing the former at the expense of the latter can be termed "green marketing myopia." In 1960, Harvard business professor Theodore Levitt introduced the concept of "marketing myopia" in a now-famous and influential article in the *Harvard Business Review*.[11] In it, he characterized the common pitfall of companies' tunnel vision, which focused on "managing products" (that is, product features, functions, and efficient production) instead of "meeting customers' needs" (that is, adapting to consumer expectations and anticipation of future desires). Levitt warned that a corporate preoccupation on products rather than consumer needs was doomed to failure because consumers select products and new innovations that offer benefits they desire. Research indicates that many green products have failed because of green marketing myopia—marketers' myopic focus on their products' "greenness" over the broader expectations of consumers or other market players (such as regulators or activists).

Green marketing must satisfy two objectives: improved environmental quality and customer satisfaction.

For example, partially in response to the 1987 Montreal Protocol, in which signatory countries (including the United States) agreed to phase out ozone-depleting chlorofluorocarbons (CFCs) by 2000, Whirlpool (in 1994) launched the "Energy Wise" refrigerator, the first CFC-free cooler and one that was 30 percent more efficient than the U.S. Department of Energy's highest standard.[12] For its innovation, Whirlpool won the "Golden Carrot," a $30 million award package of consumer rebates from the Super-Efficient Refrigerator Program, sponsored by the Natural Resources Defense Council and funded by 24 electric utilities. Unfortunately, Energy Wise's sales languished because the CFC-free benefit and energy-savings did not offset its $100 to $150 price premium, particularly in markets outside the rebate program, and the refrigerators did not offer additional features or new styles that consumers desired.[13] General Motors (GM) and Ford encountered similar problems when they launched their highly publicized EV-1 and Think Mobility electric vehicles, respectively, in the late 1990s to early 2000s in response to the 1990 zero-emission vehicle (ZEV) regulations adopted in California.[14] Both automakers believed their novel two-seater cars would be market successes (GM offered the EV-1 in a lease program, and Ford offered Think Mobility vehicles as rentals via the Hertz car-rental chain). Consumers, however, found electric vehicles' need for constant recharging with few recharging locations too inconvenient. Critics charged that the automakers made only token efforts to make electric cars a success, but a GM spokesperson recently explained, "We spent more than $1 billion to produce and market the vehicle, [but] fewer than 800 were leased."[15] Most drivers were not willing to drastically change their driving habits and expectations to accommodate electric cars, and the products ultimately were taken off the market.[16]

Aside from offering environmental benefits that do not meet consumer preferences, green marketing myopia can also occur when green products fail to provide credible, substantive environmental benefits. Mobil's Hefty photodegradable plastic trash bag is a case in point. Introduced in 1989, Hefty packages prominently displayed the term "degradable" with the explanation that a special ingredient promoted its decomposition into harmless particles in landfills "activated by exposure to the elements" such as sun, wind, and rain. Because most garbage is buried in landfills that allow limited exposure to the elements, making degradation virtually impossible, the claim enraged environmentalists. Ultimately, seven state attorneys general sued Mobil on charges of deceptive advertising and consumer fraud. Mobil removed the claim from its packaging and vowed to use extreme caution in making environmental claims in the future.[17]

Other fiascos have convinced many companies and consumers to reject green products. Roper ASW's 2002 "Green Gauge Report" finds that the top reasons consumers do not buy green products included beliefs that they require sacrifices—inconvenience, higher costs, lower performance—without significant environmental benefits.[18] Ironically, despite what consumers think, a plethora of green products available in the marketplace are in fact desirable because they deliver convenience, lower operating costs, and/or better performance. Often these are not marketed along with their green benefits, so consumers do not immediately recognize them as green and form misperceptions about their benefits. For instance, the appeal of premium-priced Marathon and other brands of CFL bulbs can be attributed to their energy savings and long life, qualities that make them convenient and economical over time. When consumers are convinced of the desirable "non-green" benefits of environmental products, they are more inclined to adopt them.

Other environmental products have also scored market successes by either serving profitable niche markets or offering mainstream appeal. Consider the Toyota Prius, the gas-electric hybrid vehicle that achieves about 44 miles per gallon of gasoline.[19] In recent years, Toyota's production has hardly kept pace with the growing demand, with buyers enduring long waits and paying thousands above the car's sticker price.[20] Consequently, other carmakers have scrambled to launch their own hybrids.[21] However, despite higher gas prices, analysts assert that it can take 5 to 20 years for lower gas expenses to offset many hybrid cars' higher prices. Thus, economics alone cannot explain their growing popularity.

Analysts offer several reasons for the Prius' market demand. Initially, the buzz over the Prius got a boost at the 2003 Academy Awards when celebrities such as Cameron Diaz, Harrison Ford, Susan Sarandon, and Robin Williams abandoned stretch limousines and oversized sport utility vehicles, arriving in Priuses to symbolize support for reducing America's dependence on foreign oil.[22] Since then, the quirky-looking Prius' badge of "conspicuous conservation" has satisfied many drivers' desires to turn heads and make a statement about their social responsibility, among them Google founders Larry Page and Sergey Brin, columnist Arianna Huffington, comic Bill Maher, and Charles, Prince of Wales.[23] The Prius ultimately was named *Motor Trend's* Car of the Year in 2004. The trendy appeal of the Prius illustrates that some green products can leverage consumer desires for being distinctive. Others say the Prius is just fun to drive—the dazzling digital dashboard that offers continuous feedback on fuel efficiency and other car operations provides an entertaining driving experience. More recently, however, the Prius has garnered fans for more practical reasons. A 2006 Maritz Poll finds that owners purchased hybrids because of the convenience of fewer fill-ups, better performance, and the enjoyment of driving the latest technology.[24] In some states, the Prius and other high-mileage hybrid vehicles, such as Honda's Insight, are granted free parking and solo-occupancy access to high occupancy vehicle (HOV) lanes.[25] In sum, hybrid vehicles offer consumers several desirable benefits that are not necessarily "green" benefits.

Many environmental products have become so common and widely distributed that many consumers may no longer recognize them as green because they buy them for non-green reasons. Green household products, for instance, are widely available at supermarkets and discount retailers, ranging from energy-saving Tide Coldwater laundry detergent to non-toxic Method and Simple Green cleaning products. Use of recycled or biodegradable paper products (such as plates, towels, napkins, coffee filters, computer paper, and other goods) is also widespread. Organic and rainforest-protective "shade grown" coffees are available at Starbucks and other specialty stores and supermarkets. Organic baby food is expected to command 12 percent market share in 2006 as parents strive to protect their children's mental and physical development.[26] Indeed, the organic food market segment has increased 20 percent annually since 1990, five times faster than the conventional food market, spurring the growth of specialty retailers such as Whole Foods Market and Wild Oats. Wal-Mart, too, has joined this extensive distribution of organic products.[27] Indeed, Wal-Mart has recently declared that in North American stores, its non-farm-raised fresh fish will be certified by the Marine Stewardship Council as sustainably harvested.[28]

Super energy-efficient appliances and fixtures are also becoming popular. Chic, front-loading washing machines, for example, accounted for 25 percent of the market in 2004, up from 9 percent in 2001.[29] EPA's Energy Star label, which certifies that products consume up to 30 percent less energy than comparable alternatives, is found on products ranging from major appliances to light fixtures to entire buildings (minimum efficiency standards vary from product to product). The construction industry is becoming increasingly green as government and industry demand office buildings that are "high

Emerging Age of Cleantech

In a 1960 Harvard Business Review article, Harvard professor Theodore Levitt introduced the classic concept of "marketing myopia" to characterize businesses' narrow vision on product features rather than consumer benefits.[1] The consequence is that businesses focus on making better mousetraps rather than seeking better alternatives for controlling pests. To avoid marketing myopia, businesses must engage in "creative destruction," described by economist Joseph Schumpeter as destroying existing products, production methods, market structures and consumption patterns, and replacing them with ways that better meet ever-changing consumer desires.[2] The dynamic pattern in which innovative upstart companies unseat established corporations and industries by capitalizing on new and improved innovations is illustrated by history. That is, the destruction of Coal Age technologies by Oil Age innovations, which are being destroyed by Information Age advances and the emerging Age of Cleantech—clean, energy- and resource-efficient energy technologies, such as those involving low/zero-emissions, wind, solar, biomass, hydrogen, recycling, and closed-loop processes.[3]

Business management researchers Stuart Hart and Mark Milstein argue that the emerging challenge of global sustainability is catalyzing a new round of creative destruction that offers "unprecedented opportunities" for new environmentally sensitive innovations, markets, and products.[4] Throughout the twentieth century, many technologies and business practices have contributed to the destruction of the very ecological systems on which the economy and life itself depends, including toxic contamination, depletion of fisheries and forests, soil erosion, and biodiversity loss. Recent news reports indicate, however, that many companies and consumers are beginning to respond to programs to help conserve the Earth's natural resources, and green marketing is making a comeback.[5] The need for sustainability has become more acute economically as soaring demand, dwindling supplies, and rising prices for oil, gas, coal, water, and other natural resources are being driven by the industrialization of populous countries, such as China and India. Politically, America's significant reliance on foreign oil has become increasingly recognized as a security threat. Global concerns over climate change have led 141 countries to ratify the Kyoto Protocol, the international treaty requiring the reduction of global warming gases created through the burning of fossil fuels. Although the United States has not signed the treaty, most multinational corporations conducting business in signatory nations are compelled to reduce their greenhouse gas emissions, and many states (such as California) and cities (such as Chicago and Seattle) have or are initiating their own global warming gas emission reduction programs.[6] State and city-level policy incentives and mandates, such as "renewable portfolio standards," requiring utilities to provide increasing amounts of electricity from clean, renewable sources such as wind and solar power, are also driving cleaner technology markets.

While some firms have responded grudgingly to such pressures for more efficient and cleaner business practices, others are seizing the cleantech innovation opportunities for new twenty-first-century green products and technologies for competitive advantage. Toyota, for instance, plans to offer an all-hybrid fleet in the near future to challenge competitors on both performance and fuel economy.[7] Further, Toyota is licensing its technology to its competitors to gain profit from their hybrid sales as well. General Electric's highly publicized "Ecomagination" initiative promises a greener world with a plan to double its investments (to $1.5 billion annually) and revenues (to $20 billion) from fuel-efficient diesel locomotives, wind power, "clean" coal, and other cleaner innovations by 2010.[8] Cleantech is attracting investors looking for the "Next Big Thing," including Goldman Sachs and Kleiner Perkins Caufield & Byers.[9] Wal-Mart, too, is testing a sustainable 206,000-square foot store design in Texas that deploys 26 energy-saving and renewable-materials experiments that could set new standards in future retail store construction.[10] In sum, economic, political, and environmental pressures are coalescing to drive cleaner and greener technological innovation in the twenty-first century, and companies that fail to adapt their products and processes accordingly are destined to suffer from the consequences of marketing myopia and creative destruction.

1. T. Levitt, "Marketing Myopia," *Harvard Business Review* 28, July–August (1960): 24–47.
2. See J. Schumpeter, *The Theory of Economic Development* (Cambridge: Harvard University Press, 1934); and J. Schumpeter, *Capitalism, Socialism and Democracy* (New York: Harper Torchbooks, 1942).
3. "Alternate Power: A Change Is in the Wind," *Business Week,* 4 July 2005, 36–37.
4. S. L. Hart and M. B. Milstein, "Global Sustainability and the Creative Destruction of Industries," *MIT Sloan Management Review* 41, Fall (1999): 23–33.
5. See for example T. Howard, "Being Eco-Friendly Can Pay Economically; 'Green Marketing' Sees Growth in Sales, Ads," *USA Today,* 15 August 2005; and E. R. Stafford, "Energy Efficiency and the New Green Marketing," *Environment,* March 2003, 8–10.
6. J. Ball, "California Sets Emission Goals That Are Stiffer than U.S. Plan," *Wall Street Journal,* 2 June 2005; and J. Marglis, "Paving the Way for U.S. Emissions Trading," *Grist Magazine,* 14 June 2005, www.climatebiz.com/sections/news_print.dfm?NewsID=28255.
7. Bloomberg News, "Toyota Says It Plans Eventually to Offer an All-Hybrid Fleet," 14 September 2005, http://www.nytimes.com/2005/09/14/automobiles/14toyota.html.
8. J. Erickson, "U.S. Business and Climate Change: Siding with the Marketing?" *Sustainability Radar,* June, www.climatebiz.com/sections/new_ print.cfm?NewsID=28204.
9. *Business Week,* note 3 above.
10. Howard, note 5 above.

performance" (for example, super energy- and resource-efficient and cost-effective) and "healthy" for occupants (for example, well-ventilated; constructed with materials with low or no volatile organic compounds [VOC]). The U.S. Green Building Council's "Leadership in Energy and Environmental Design" (LEED) provides a rigorous rating system and green building checklist that are rapidly becoming the standard for environmentally sensitive construction.[30]

Home buyers are recognizing the practical long-term cost savings and comfort of natural lighting, passive solar heating, and heat-reflective windows, and a 2006 study sponsored by home improvement retailer Lowe's found nine out of ten builders surveyed are incorporating energy-saving features into new homes.[31] Additionally, a proliferation of "green" building materials to serve the growing demand has emerged.[32] Lowe's competitor The Home Depot is testing an 'EcoOptions' product line featuring natural fertilizers and mold-resistant drywall in its Canadian stores that may filter into the U.S. market.[33] In short, energy efficiency and green construction have become mainstream.

The diversity and availability of green products indicate that consumers are not indifferent to the value offered by environmental benefits. Consumers are buying green—but not necessarily for environmental reasons. The market growth of organic foods and energy-efficient appliances is because consumers desire their perceived safety and money savings, respectively.[34] Thus, the apparent paradox between what consumers say and their purchases may be explained, in part, by green marketing myopia—a narrow focus on the greenness of products that blinds companies from considering the broader consumer and societal desires. A fixation on products' environmental merits has resulted frequently in inferior green products (for example, the original EarthLight and GM's EV-1 electric car) and unsatisfying consumer experiences. By contrast, the analysis of past research and marketing strategies finds that successful green products have avoided green marketing myopia by following three important principles: "The Three Cs" of consumer value positioning, calibration of consumer knowledge, and credibility of product claims.

Consumer Value Positioning

The marketing of successfully established green products showcases non-green consumer value, and there are at least five desirable benefits commonly associated with green products: efficiency and cost effectiveness; health and safety; performance; symbolism and status; and convenience. Additionally, when these five consumer value propositions are not inherent in the green product, successful green marketing programs bundle (that is, add to the product design or market offering) desirable consumer value to broaden the green product's appeal. In practice, the implication is that product designers and marketers need to align environmental products' consumer value (such as money savings) to relevant consumer market segments (for example, cost-conscious consumers).

Efficiency and Cost Effectiveness

As exemplified by the Marathon CFL bulbs, the common inherent benefit of many green products is their potential energy and resource efficiency. Given sky-rocketing energy prices and tax incentives for fuel-efficient cars and energy-saving home improvements and appliances, long-term savings have convinced cost-conscious consumers to buy green.

Recently, the home appliance industry made great strides in developing energy-efficient products to achieve EPA's Energy Star rating. For example, Energy Star refrigerators use at least 15 percent less energy and dishwashers use at least 25 percent less energy than do traditional models.[35] Consequently, an Energy Star product often commands a price premium. Whirlpool's popular Duet front-loading washer and dryer, for example, cost more than $2,000, about double the price of conventional units; however, the washers can save up to 12,000 gallons of water and $110 on electricity annually compared to standard models (Energy Star does not rate dryers).[36]

Laundry detergents are also touting energy savings. Procter & Gamble's (P&G) newest market entry, Tide Cold-water, is designed to clean clothes effectively in cold water. About 80 to 85 percent of the energy used to wash clothes comes from heating water. Working with utility companies, P&G found that consumers could save an average of $63 per year by using cold rather than warm water.[37] Adopting Tide Coldwater gives added confidence to consumers already washing in cold water. As energy and resource prices continue to soar, opportunities for products offering efficiency and savings are destined for market growth.

Health and Safety

Concerns over exposure to toxic chemicals, hormones, or drugs in everyday products have made health and safety important choice considerations, especially among vulnerable consumers, such as pregnant women, children, and the elderly.[38] Because most environmental products are grown or designed to minimize or eliminate the use of toxic agents and adulterating processes, market positioning on consumer safety and health can achieve broad appeal among health-conscious consumers. Sales of organic foods, for example, have grown considerably in the wake of public fear over "mad cow" disease, antibiotic-laced meats, mercury in fish, and genetically modified foods.[39] Mainstream appeal of organics is not derived from marketers promoting the advantages of free-range animal ranching and pesticide-free soil. Rather, market positioning of organics as flavorful, healthy alternatives to factory-farm foods has convinced consumers to pay a premium for them.

A study conducted by the Alliance for Environmental Innovation and household products-maker S.C. Johnson found that consumers are most likely to act on green messages that strongly connect to their personal environments.[40] Specifically, findings suggest that the majority of consumers prefer such environmental household product benefits as "safe to use around children," "no toxic ingredients," "no chemical residues," and "no strong fumes" over such benefits as "packaging can be recycled" or "not tested on animals." Seventh Generation, a brand of non-toxic and environmentally-safe household products, derived its name from the Iroquois belief that, "In our every deliberation, we must consider the impact of our decisions on the next seven generations." Accordingly, its products promote the family-oriented value of making the world a safer place for the next seven generations.

Indoor air quality is also a growing concern. Fumes from paints, carpets, furniture, and other décor in poorly ventilated "sick buildings" have been linked to headaches, eye, nose, and throat irritation, dizziness, and fatigue among occupants. Consequently, many manufacturers have launched green products to reduce indoor air pollution. Sherwin Williams, for example, offers "Harmony," a line of interior paints that is low-odor, zero-VOC, and silica-free. And Mohawk sells EverSet Fibers, a carpet that virtually eliminates the need for harsh chemical cleaners because its design allows most stains to be removed with water. Aside from energy efficiency, health and safety have been key motivators driving the green building movement.

Performance

The conventional wisdom is that green products don't work as well as "non-green" ones. This is a legacy from the first generation of environmentally sensitive products that clearly were inferior. Consumer perception of green cleaning agents introduced in health food stores in the 1960s and 1970s, for example, was that "they cost twice as much to remove half the grime."[41] Today, however, many green products are designed to perform better than conventional ones and can command a price premium. For example, in addition to energy efficiency, front-loading washers clean better and are gentler on clothes compared to conventional top-loading machines because they spin clothes in a motion similar to clothes driers and use centrifugal force to pull dirt and water away from clothes. By contrast, most top-loading washers use agitators to pull clothes through tanks of water, reducing cleaning and increasing wear on clothes. Consequently, the efficiency and high performance benefits of top-loading washers justify their premium prices.

Market positioning on consumer safety and health can achieve broad appeal among health-conscious consumers.

Homeowners commonly build decks with cedar, redwood, or pressure-treated pine (which historically was treated with toxic agents such as arsenic). Wood requires stain or paint and periodic applications of chemical preservatives for maintenance. Increasingly, however, composite deck material made from recycled milk jugs and wood fiber, such as Weyerhaeuser's ChoiceDek, is marketed as the smarter alternative. Composites are attractive, durable, and low maintenance. They do not contain toxic chemicals and never need staining or chemical preservatives. Accordingly, they command a price premium— as much as two to three times the cost of pressure-treated pine and 15 percent more than cedar or redwood.[42]

Likewise, Milgard Windows' low emissivity SunCoat Low-E windows filter the sun in the summer and reduce heat loss in the winter. While the windows can reduce a building's overall energy use, their more significant benefit comes from helping to create a comfortable indoor radiant temperature climate and protecting carpets and furniture from harmful ultraviolet rays. Consequently, Milgard promotes the improved comfort and performance of its SunCoat Low-E windows over conventional windows. In sum, "high performance" positioning can broaden green product appeal.

Symbolism and Status

As mentioned earlier, the Prius, Toyota's gas-electric hybrid, has come to epitomize "green chic." According to many automobile analysts, the cool-kid cachet that comes with being an early adopter of the quirky-looking hybrid vehicle trend continues to partly motivate sales.[43] Establishing a green chic appeal, however, isn't easy. According to popular culture experts, green marketing must appear grass-roots driven and humorous without sounding preachy. To appeal to young people, conservation and green consumption need the unsolicited endorsement of high-profile celebrities and connection to cool technology.[44] Prius has capitalized on its evangelical following and high-tech image with some satirical ads, including a television commercial comparing the hybrid with Neil Armstrong's moon landing ("That's one small step on the accelerator, one giant leap for mankind") and product placements in popular Hollywood films and sitcoms (such as *Curb Your Enthusiasm*). More recently, Toyota has striven to position its "hybrid synergy drive" system as a cut above other car makers' hybrid technologies with witty slogans such as, "Commute with Nature," "mpg:)," and "There's Nothing Like That New Planet Smell."[45] During the 2006

Super Bowl XL game, Ford launched a similarly humorous commercial featuring Kermit the Frog encountering a hybrid Escape sports utility vehicle in the forest, and in a twist, changing his tune with "I guess it *is* easy being green!"[46]

In business, where office furniture symbolizes the cachet of corporate image and status, the ergonomically designed "Think" chair is marketed as the chair "with a brain and a conscience." Produced by Steelcase, the world's largest office furniture manufacturer, the Think chair embodies the latest in "cradle to cradle" (C2C) design and manufacturing. C2C, which describes products that can be ultimately returned to technical or biological nutrients, encourages industrial designers to create products free of harmful agents and processes that can be recycled easily into new products (such as metals and plastics) or safely returned to the earth (such as plant-based materials).[47] Made without any known carcinogens, the Think chair is 99 percent recyclable; it disassembles with basic hand tools in about five minutes, and parts are stamped with icons showing recycling options.[48] Leveraging its award-winning design and sleek comfort, the Think chair is positioned as symbolizing the smart, socially responsible office. In sum, green products can be positioned as status symbols.

Convenience

Many energy-efficient products offer inherent convenience benefits that can be showcased for competitive advantage. CFL bulbs, for example, need infrequent replacement and gas-electric hybrid cars require fewer refueling stops—benefits that are highlighted in their marketing communications. Another efficient alternative to incandescent bulbs are light-emitting diodes (LEDs): They are even more efficient and longer-lasting than CFL bulbs; emit a clearer, brighter light; and are virtually unbreakable even in cold and hot weather. LEDs are used in traffic lights due to their high-performance convenience. Recently, a city in Idaho became a pioneer by adopting LEDs for its annual holiday Festival of Lights. "We spent so much time replacing strings of lights and bulbs," noted one city official, "[using LEDs] is going to reduce two-thirds of the work for us."[49]

To encourage hybrid vehicle adoption, some states and cities are granting their drivers the convenience of free parking and solo-occupant access to HOV lanes. A Toyota spokesperson recently told the *Los Angeles Times,* "Many customers are telling us the carpool lane is the main reason for buying now."[50] Toyota highlights the carpool benefit on its Prius Web site, and convenience has become an incentive to drive efficient hybrid cars in traffic-congested states like California and Virginia. Critics have charged, however, that such incentives clog carpool lanes and reinforce a "one car, one person" lifestyle over alternative transportation. In response, the Virginia legislature has more recently enacted curbs on hybrid drivers use of HOV lanes during peak hours, requiring three or more people per vehicle, except for those that have been grandfathered in.[51]

Solar power was once used only for supplying electricity in remote areas (for example, while camping in the wilderness or boating or in homes situated off the power grid). That convenience, however, is being exploited for other applications. In landscaping, for example, self-contained solar-powered outdoor evening lights that recharge automatically during the day eliminate the need for electrical hookups and offer flexibility for reconfiguration. With society's increasing mobility and reliance on electronics, solar power's convenience is also manifest in solar-powered calculators, wrist watches, and other gadgets, eliminating worries over dying batteries. Reware's solar-powered "Juice Bag" backpack is a popular portable re-charger for students, professionals, and outdoor enthusiasts on the go. The Juice Bag's flexible, waterproof solar panel has a 16.6-volt capacity to generate 6.3 watts to recharge PDAs, cell phones, iPods, and other gadgets in about 2 to 4 hours.[52]

Bundling

Some green products do not offer any of the inherent five consumer-desired benefits noted above. This was the case when energy-efficient and CFC-free refrigerators were introduced in China in the 1990s. While Chinese consumers preferred and were willing to pay about 15 percent more for refrigerators that were "energy-efficient," they did not connect the environmental advantage of "CFC-free" with either energy efficiency or savings. Consequently, the "CFC-free" feature had little impact on purchase decisions.[53] To encourage demand, the CFC-free feature was bundled with attributes desired by Chinese consumers, which included energy efficiency, savings, brand/quality, and outstanding after-sales service.

According to popular culture experts, green marketing must appear grass-roots driven and humorous without sounding preachy.

Given consumer demand for convenience, incorporating time-saving or ease-of-use features into green products can further expand their mainstream acceptance. Ford's hybrid Escape SUV comes with an optional 110-volt AC power outlet suitable for work, tailgating, or camping. Convenience has also enhanced the appeal of Interface's recyclable FLOR carpeting, which is marketed as "practical, goof-proof, and versatile." FLOR comes in modular square tiles with four peel-and-stick dots on the back for easy installation (and pull up for altering, recycling, or washing with water in the sink). Modularity offers versatility to assemble tiles for a custom look. Interface promotes the idea that its carpet tiles can be changed and reconfigured in minutes to dress up a room for any occasion. The tiles come in pizza-style boxes for storage, and ease of use is FLOR's primary consumer appeal.

Finally, Austin (Texas) Energy's "Green Choice" program has led the nation in renewable energy sales for the past three years.[54] In 2006, demand for wind energy outpaced supply so that the utility resorted to selecting new "Green Choice" subscribers by lottery.[55] While most utilities find it challenging to sell green electricity at a premium price on its environmental merit, Austin Energy's success comes from bundling three benefits that appeal to commercial power users: First, Green Choice customers are recognized in broadcast media for their corporate responsibility; second, the green power is marketed as "home grown," appealing to Texan loyalties; and third, the program offers a fixed price that is locked in for 10 years. Because wind power's cost is derived primarily from the construction of wind farms and is not subject to volatile fossil fuel costs, Austin Energy passes its inherent price stability onto its Green Choice customers. Thus, companies participating in Green Choice enjoy the predictability of their future energy costs in an otherwise volatile energy market.

In summary, the analysis suggests that successful green marketing programs have broadened the consumer appeal of green products by convincing consumers of their "non-green" consumer value. The lesson for crafting effective green marketing strategies is that planners need to identify the inherent consumer value of green product attributes (for example, energy efficiency's inherent long-term money savings) or bundle desired consumer value into green products (such as fixed pricing of wind power) and to draw marketing attention to this consumer value.

Calibration of Consumer Knowledge

Many of the successful green products in the analysis described here employ compelling, educational marketing messages and slogans that connect green product attributes with desired consumer value. That is, the marketing programs successfully calibrated consumer knowledge to recognize the green product's consumer benefits. In many instances, the environmental benefit was positioned as secondary, if mentioned at all. Changes made in EPA's Energy Star logo provide an example, illustrating the program's improved message calibration over the years. One of Energy Star's early marketing messages, "EPA Pollution Preventer," was not only ambiguous but myopically focused on pollution rather than a more mainstream consumer benefit. A later promotional message, "Saving The Earth. Saving Your Money." better associated energy efficiency with consumer value, and one of its more recent slogans, "Money Isn't All You're Saving," touts economic savings as the chief benefit. This newest slogan also encourages consumers to think implicitly about what else they are "saving"—the logo's illustration of the Earth suggests the answer, educating consumers that "saving the Earth" can also meet consumer self-interest.

The connection between environmental benefit and consumer value is evident in Earthbound Farm Organic's slogan, "Delicious produce is our business, but health is our bottom line," which communicates that pesticide-free produce is flavorful and healthy. Likewise, Tide Coldwater's "Deep Clean. Save Green." slogan not only assures consumers of the detergent's cleaning performance, but the term "green" offers a double meaning, connecting Tide's cost saving with its environmental benefit. Citizen's solar-powered Eco-Drive watch's slogan, "Unstoppable Caliber," communicates the product's convenience and performance (that is, the battery will not die) as well as prestige. Table 1 on page 89 shows other successful marketing messages that educate consumers of the inherent consumer value of green.

Some compelling marketing communications educate consumers to recognize green products as "solutions" for their personal needs *and* the environment.[56] When introducing its Renewal brand, Rayovac positioned the reusable alkaline batteries as a solution for heavy battery users and the environment with concurrent ads touting "How to save $150 on a CD player that costs $100" and "How to save 147 batteries from going to landfills." Complementing the money savings and landfill angles, another ad in the campaign featured sports star Michael Jordan proclaiming, "More Power. More Music. And More Game Time." to connect Renewal batteries' performance to convenience.[57] In practice, the analysis conducted here suggests that advertising that draws attention to how the environmental product benefit can deliver desired personal value can broaden consumer acceptance of green products.

Credibility of Product Claims

Credibility is the foundation of effective green marketing. Green products must meet or exceed consumer expectations by delivering their promised consumer value and providing substantive environmental benefits. Often, consumers don't have the expertise or ability to verify green products' environmental and consumer values, creating misperceptions and skepticism. As exemplified in the case of Mobil's Hefty photodegradable plastic trash bag described earlier, green marketing that touts a product's or a company's environmental credentials can spark the scrutiny of advocacy groups or regulators. For example, although it was approved by the U.S. Food and Drug Administration, sugar substitute

Splenda's "Made from sugar, so it tastes like sugar" slogan and claim of being "natural" have been challenged by the Sugar Association and Generation Green, a health advocacy group, as misleading given that its processing results in a product that is "unrecognizable as sugar."[58]

To be persuasive, past research suggests that green claims should be specific and meaningful.[59] Toyota recognizes the ambiguity of the term "green" and discourages its use in its marketing of its gas-electric hybrid cars. One proposed slogan, "Drive green, breathe blue" was dismissed in favor of specific claims about fuel efficiency, such as "Less gas in. Less gasses out."[60] Further, environmental claims must be humble and not over-promise. When Ford Motor Company publicized in *National* *Geographic* and other magazines its new eco-designed Rouge River Plant that incorporated the world's largest living roof of plants, critics questioned the authenticity of Ford's environmental commitment given the poor fuel economy of the automaker's best-selling SUVs.[61] Even the Prius has garnered some criticism for achieving considerably less mileage (approximately 26 percent less according to *Consumer Reports*) than its government sticker rating claims, although the actual reduced mileage does not appear to be hampering sales.[62] Nonetheless, green product attributes need to be communicated honestly and qualified for believability (in other words, consumer benefits and environmental effectiveness claims need to be compared with comparable alternatives or likely usage scenarios). For example, Toyota includes an "actual mileage may vary" disclaimer in Prius advertising. When Ford's hybrid Escape SUV owners complained that they were not achieving expected mileage ratings, Ford launched the "Fuel-Economy School" campaign to educate drivers about ways to maximize fuel efficiency.[63] Further, EPA is reconsidering how it estimates hybrid mileage ratings to better reflect realistic driving conditions (such as heavy acceleration and air conditioner usage).[64]

Table 1 Marketing Messages Connecting Green Products with Desired Consumer Value

Value	Message and business/product
Efficiency and cost effectiveness	"The only thing our washer will shrink is your water bill." —ASKO
	"Did you know that between 80 and 85 percent of the energy used to wash clothes comes from heating the water? Tide Coldwater—The Coolest Way to Clean." —Tide Coldwater Laundry Detergent "mpg:)" —Toyota Prius
Health and safety	"20 years of refusing to farm with toxic pesticides. Stubborn, perhaps. Healthy, most definitely." —Earthbound Farm Organic
	"Safer for You and the Environment." —Seventh Generation Household Cleaners
Performance	"Environmentally friendly stain removal. It's as simple as H_2O." —Mohawk EverSet Fibers Carpet
	"Fueled by light so it runs forever. It's unstoppable. Just like the people who wear it." —Citizen Eco-Drive Sport Watch
Symbolism	"Think is the chair with a brain and a conscience." —Steelcase's Think Chair
	"Make up your mind, not just your face." —The Body Shop
Convenience	"Long life for hard-to-reach places." —General Electric's CFL Flood Lights
Bundling	"Performance and luxury fueled by innovative technology." —Lexus RX400h Hybrid Sports Utility Vehicle

Source: Compiled by J.A. Ottman, E.R. Stafford, and C.L. Hartman, 2006.

Third Party Endorsements and Eco-Certifications

Expert third parties with respected standards for environmental testing (such as independent laboratories, government agencies, private consultants, or nonprofit advocacy organizations) can provide green product endorsements and/or "seals of approval" to help clarify and bolster the believability of product claims.[65] The "Energy Star" label, discussed earlier, is a common certification that distinguishes certain electronic products as consuming up to 30 percent less energy than comparable alternatives. The U.S. Department of Agriculture's "USDA Organic" certifies the production and handling of organic produce and dairy products.

Green Seal and Scientific Certification Systems emblems certify a broad spectrum of green products. Green Seal sets specific criteria for various categories of products, ranging from paints to cleaning agents to hotel properties, and for a fee, companies can have their products evaluated and monitored annually for certification. Green Seal-certified products include Zero-VOC Olympic Premium interior paint and Johnson Wax professional cleaners. Green Seal has also certified the Hyatt Regency in Washington, DC, for the hotel's comprehensive energy and water conservation, recycling programs, and environmental practices. By contrast, Scientific Certification Systems (SCS) certifies specific product claims or provides a detailed "eco-profile" for a product's environmental impact for display on product labels for a broad array of products, from agricultural products to fisheries to construction. For example, Armstrong hard surface flooring holds SCS certification, and SCS works with retailers like The Home Depot to monitor its vendors' environmental claims.[66]

Although eco-certifications differentiate products and aid in consumer decisionmaking, they are not without controversy. The science behind eco-seals can appear subjective and/or complex, and critics may take issue with certification criteria.[67] For example, GreenOrder, a New York-based environmental consulting firm, has devised a scorecard to evaluate cleantech products marketed in General Electric's "Ecomagination" initiative, which range from fuel-efficient aircraft engines to wind turbines to water treatment technologies. Only those passing GreenOrder's criteria are marketed as Ecomagination products, but critics have questioned GE's inclusion of "cleaner coal" (that is, coal gasification for cleaner burning and sequestration of carbon dioxide emissions) as an "Ecomagination" product.[68]

Although eco-certifications differentiate products and aid in consumer decisionmaking, they are not without controversy.

Consequently, when seeking endorsements and eco-certifications, marketers should consider the environmental tradeoffs and complexity of their products and the third parties behind endorsements and/or certifications: Is the third party respected? Are its certification methodologies accepted by leading environmentalists, industry experts, government regulators, and other key stakeholders? Marketers should educate their customers about the meaning behind an endorsement or an eco-seal's criteria. GE recognizes that its cleaner coal technology is controversial but hopes that robust marketing and educational outreach will convince society about cleaner coal's environmental benefits.[69] On its Web site, GE references U.S. Energy Information Administration's statistics that coal accounts for about 24 percent of the world's total energy consumption, arguing that coal will continue to be a dominant source of energy due to its abundance and the increasing electrification of populous nations such as China and India.[70] In response to GE's commitment to clean coal, Jonathan Lash, president of the World Resources Institute, said, "Five years ago, I had to struggle to suppress my gag response to terms like 'clean coal,' but I've since faced the sobering reality that every two weeks China opens a new coal-fired plant. India is moving at almost the same pace. There is huge environmental value in developing ways to mitigate these plants' emissions."[71]

Word-of-Mouth Evangelism and the Internet

Increasingly, consumers have grown skeptical of commercial messages, and they're turning to the collective wisdom and experience of their friends and peers about products.[72] Word-of-mouth or "buzz" is perceived to be very credible, especially as consumers consider and try to comprehend complex product innovations. The Internet, through e-mail and its vast, accessible repository of information, Web sites, search engines, blogs, product ratings sites, podcasts, and other digital platforms, has opened significant opportunities for tapping consumers' social and communication networks to diffuse credible "word-of-mouse" (buzz facilitated by the Internet) about green products. This is exemplified by one of the most spectacular product introductions on the Web: Tide Coldwater.

In 2005, Proctor & Gamble partnered with the non-profit organization, the Alliance to Save Energy (ASE), in a "viral marketing" campaign to spread news about the money-saving benefits of laundering clothes in cold water with specially formulated Tide Coldwater.[73] ASE provided credibility for the detergent by auditing and backing P&G's claims that consumers could save an average of $63 a year if they switched from warm to cold water washes. ASE sent e-mail promotions encouraging consumers to visit Tide.com's interactive Web site and take the "Coldwater Challenge" by registering to receive a free sample. Visitors could calculate how much money they would save by using the detergent, learn other energy-saving laundry tips, and refer e-mail addresses of their friends to take the challenge as well. Tide.com offered an engaging map of the United States where, over time, visitors could track and watch their personal networks grow across the country when their friends logged onto the site to request a free sample.

Given the immediacy of e-mail and the Internet, word-of-mouse is fast becoming an important vehicle for spreading credible news about new products. According to the Pew Internet & American Life Project, 44 percent of online U.S. adults (about 50 million Americans) are "content creators," meaning that they contribute to the Internet via blogs, product recommendations, and reviews.[74] To facilitate buzz, however, marketers need to create credible messages, stories, and Web sites about their products that are so compelling, interesting, and/or entertaining that consumers will seek the information out and forward it to their friends and family.[75] The fact that P&G was able to achieve this for a low-involvement product is quite remarkable.

International online marketing consultant Hitwise reported that ASE's e-mail campaign increased traffic at the Tide Coldwater Web site by 900 percent in the first week, and then tripled that level in week two.[76] Within a few months, more than one million Americans accepted the "Coldwater Challenge," and word-of-mouse cascaded through ten degrees of separation across all 50 states and more than 33,000 zip codes.[77] In October 2005, Hitwise reported that Tide.com ranked as the twelfth most popular site by market share of visits in the "Lifestyle—House and Garden" category.[78] No other laundry detergent brand's Web site has gained a significant Web presence in terms of the number of visits.

P&G's savvy implementation of "The Three Cs"—consumer value positioning on money savings, calibration of consumer knowledge about cold wash effectiveness via an engaging Web site, and credible product messages dispatched by a respected non-profit group and consumers' Internet networks—set the stage for Tide Coldwater's successful launch.

The Future of Green Marketing

Clearly, there are many lessons to be learned to avoid green marketing myopia (see the box)—the short version of all this is that effective green marketing requires applying good marketing principles to make green products desirable for consumers. The question that remains, however, is, what is green marketing's future? Historically, green marketing has been a misunderstood concept. Business scholars have viewed it as a "fringe" topic, given that environmentalism's acceptance of limits and conservation does not mesh well with marketing's traditional axioms of "give customers what they want" and "sell as much as you can." In practice, green marketing myopia has led to ineffective products and consumer reluctance. Sustainability, however, is destined to dominate twenty-first century commerce. Rising energy prices, growing pollution and resource consumption in Asia, and political pressures to address climate change are driving innovation toward healthier, more-efficient, high-performance products. In short, all marketing will incorporate elements of green marketing.

As the authors of *Natural Capitalism* argue, a more sustainable business model requires "product dematerialization"—that is, commerce will shift from the "sale of goods" to the "sale of services" (for example, providing illumination rather than selling light bulbs).[79] This model is illustrated, if unintentionally, by arguably the twenty-first century's hottest product—Apple's iPod. The iPod gives consumers the convenience to download, store, and play tens of thousands of songs without the environmental impact of manufacturing and distributing CDs, plastic jewel cases, and packaging.

Innovations that transform material goods into efficient streams of services could proliferate if consumers see them as desirable. To encourage energy and water efficiency, Electrolux piloted a "pay-per-wash" service in Sweden in 1999 where consumers were given new efficient washing machines for a small home installation fee and then were charged 10 Swedish kronor (about $1) per use. The machines were connected via the Internet to a central database to monitor use, and Electrolux maintained ownership and servicing of the washers. When the machines had served their duty, Electrolux took them

Summary of Guideposts for the "Three C'S"

Evidence indicates that successful green products have avoided green marketing myopia by following three important principles: consumer value positioning, calibration of consumer knowledge, and the credibility of product claims.

Consumer Value Positioning

- Design environmental products to perform as well as (or better than) alternatives.
- Promote and deliver the consumer-desired value of environmental products and target relevant consumer market segments (such as market health benefits among health-conscious consumers).
- Broaden mainstream appeal by bundling (or adding) consumer-desired value into environmental products (such as fixed pricing for subscribers of renewable energy).

Calibration of Consumer Knowledge

- Educate consumers with marketing messages that connect environmental product attributes with desired consumer value (for example, "pesticide-free produce is healthier"; "energy-efficiency saves money"; or "solar power is convenient").

- Frame environmental product attributes as "solutions" for consumer needs (for example, "rechargeable batteries offer longer performance").
- Create engaging and educational Internet sites about environmental products' desired consumer value (for example, Tide Coldwater's interactive Web site allows visitors to calculate their likely annual money savings based on their laundry habits, utility source (gas or electricity), and zip code location).

Credibility of Product Claims

- Employ environmental product and consumer benefit claims that are specific, meaningful, unpretentious, and qualified (that is, compared with comparable alternatives or likely usage scenarios).
- Procure product endorsements or eco-certifications from trustworthy third parties, and educate consumers about the meaning behind those endorsements and eco-certifications.
- Encourage consumer evangelism via consumers' social and Internet communication networks with compelling, interesting, and/or entertaining information about environmental products (for example, Tide's "Coldwater Challenge" Web site included a map of the United States so visitors could track and watch their personal influence spread when their friends requested a free sample).

---

back for remanufacturing. Pay-per-wash failed, however, because consumers were not convinced of its benefits over traditional ownership of washing machines.[80] Had Electrolux better marketed pay-per-wash's convenience (for example, virtually no upfront costs for obtaining a top-of-the-line washer, free servicing, and easy trade-ins for upgrades) or bundled pay-per-wash with more desirable features, consumers might have accepted the green service. To avoid green marketing myopia, the future success of product dematerialization and more sustainable services will depend on credibly communicating and delivering consumer-desired value in the marketplace. Only then will product dematerialization steer business onto a more sustainable path.

Notes

1. G. Fowler, "'Green Sales Pitch Isn't Moving Many Products," *Wall Street Journal,* 6 March 2002.

2. See, for example, K. Alston and J. P. Roberts, "Partners in New Product Development: SC Johnson and the Alliance for Environmental Innovation," *Corporate Environmental Strategy* 6, no. 2: 111–28.

3. See, for example, J. Ottman, *Green Marketing: Opportunity for Innovation* (Lincolnwood [Chicago]: NTC Business Books, 1997).

4. P. Hawken, A. Lovins, and L. H. Lovins, *Natural Capitalism: Creating the Next Industrial Revolution* (Boston: Little, Brown, and Company, 1999).

5. See, for example, *Business Week,* "Alternate Power: A Change in the Wind," 4 July 2005, 36–37.

6. See T. L. Friedman, "Geo-Greening by Example," *New York Times,* 27 March 2005; and T. L. Friedman, "The New 'Sputnik' Challenges: They All Run on Oil," *New York Times,* 20 January 2006.

7. There is some debate as to how to define a "green consumer." Roper ASW's most recent research segments American consumers by their propensity to purchase environmentally sensitive products into five categories, ranging from "True Blue Greens," who are most inclined to seek out and buy green on a regular basis (representing 9 percent of the population), to "Basic Browns," who are the least involved group and believe environmental indifference is mainstream (representing 33 percent of the population); see Roper ASW "Green Gauge Report 2002:Americans Perspective on Environmental Issues—Yes . . . But," November 2002, http://www.windustry. com/conferences/november2002/nov2002_proceedings/ plenary/greenguage2002.pdf(accessed 7 February 2006). Alternatively, however, some marketers view green consumers as falling into three broad segments concerned with preserving the planet, health consequences of environmental problems, and animal welfare; see Ottman, note 3 above, pages 19–44. Because environmental concerns are varied, ranging from resource/energy conservation to wildlife protection to air quality, marketing research suggests that responses to green advertising appeals vary by consumer segments. For example, in one study, young college-educated students were found to be drawn to health-oriented green appeals, whereas working adults were more responsive toward health, waste, and energy

appeals; see M. R. Stafford, T. F. Stafford, and J. Chowdhury, "Predispositions Toward Green Issues: The Potential Efficacy of Advertising Appeals," *Journal of Current Issues and Research in Advertising* 18, no. 2 (1996): 67–79. One of the lessons from the study presented here is that green products must be positioned on the consumer value sought by targeted consumers.

8. A. Grubler, "Doing More with Less: Improving the Environment through Green Engineering," *Environment 48,* no. 2 (March 2006): 22–37.

9. See, for example, L.A. Crosby and S. L. Johnson, "Customer-Centric Innovation," *Marketing Management* 15, no. 2 (2006): 12–13.

10. The methodology for this article involved reviewing case descriptions of green products discussed in the academic and business literature to identify factors contributing to consumer acceptance or resistance. Product failure was defined as situations in which the green product experienced very limited sales and ultimately was either removed from the marketplace (such as General Motor's EV1 electric car and Electrolux's "pay-per-wash" service) or re-positioned in the marketplace (such as Philips' "EarthLight"). Product success was defined as situations in which the green product attained consumer acceptance and was widely available at the time of the analysis. Particular attention centered on the market strategies and external market forces of green products experiencing significant growth (such as gas-electric hybrid cars and organic foods), and the study examined their market context, pricing, targeted consumers, product design, and marketing appeals and messages.

11. See T. Levitt, "Marketing Myopia," *Harvard Business Review* 28, July–August (1960): 24–47.

12. A. D. Lee and R. Conger, "Market Transformation: Does it Work? The Super Energy Efficient Refrigerator Program," *ACEEE Proceedings,* 1996, 3.69–3.80.

13. Ibid.

14. The California Air Resources Board (CARB) adopted the Low-Emission Vehicle (LEV) regulations in 1990. The original LEV regulations required the introduction of zero-emission vehicles (ZEVs) in 1998 as 2 percent of all vehicles produced for sale in California, and increased the percentage of ZEVs from 2 percent to 10 percent in 2003. By 1998, significant flexibility was introduced through partial ZEV credits for very-low-emission vehicles. For a review, see S. Shaheen, "California's Zero-Emission Vehicle Mandate," *Institute of Transporation Studies,* Paper UCD-ITS-RP-04-14, 2 September 2004.

15. C. Palmeri, "Unplugged," *Business Week,* 20 March 2006, 12.

16. "Think Tanks," *Automotive News,* 6 March 2006, 42; J. Ottman, "Lessons from the Green Graveyard," *Green@Work,* April 2003, 62–63.

17. J. Lawrence, "The Green Revolution: Case Study," *Advertising Age,* 29 January 1991, 12.

18. See Roper ASW, note 7 above.

19. "Fuel Economy: Why You're Not Getting the MPG You Expect," *Consumer Reports,* October 2005, 20–23.

20. J. O'Dell, "Prices Soar for Hybrids with Rights to Fast Lane," *Los Angeles Times,* 27 August 2005.

21. M. Landler and K. Bradsher, "VW to Build Hybrid Minivan with Chinese," *New York Times,* 9 September 2005.

22. K. Carter, "'Hybrid' Cars Were Oscars' Politically Correct Ride," *USA Today,* 31 March 2003.

23. See, for example, H. W. Jenkins, "Dear Valued Hybrid Customer . . . ," *Wall Street Journal,* 30 November 2005; E. R. Stafford, "Conspicuous Conservation," *Green@Work,* Winter 2004, 30–32. A recent Civil Society Institute poll found that 66 percent of survey participants agreed that driving fuel efficient vehicles was "patriotic"; see Reuters, "Americans See Fuel Efficient Cars as 'Patriotic,'" 18 March 2005, http://www.planetark.com/avantgo/dailynewsstory.cfm?newsid=29988.

24. "Rising Consumer Interest in Hybrid Technology Confirmed by Maritz Research," PRNewswire, 5 January 2006.

25. O'Dell, note 20 above.

26. J. Fetto, "The Baby Business," *American Demographics,* May 2003, 40.

27. See D. McGinn, "The Green Machine," *Newsweek,* 21 March 2005, E8–E12; and J. Weber, "A Super-Natural Investing Opportunity," *Business 2.0,* March 2005, 34.

28. A. Murray, "Can Wal-Mart Sustain a Softer Edge?" *Wall Street Journal,* 8 February 2006.

29. C. Tan, "New Incentives for Being Green," *Wall Street Journal,* 4 August 2005.

30. For an overview of the Leadership in Energy and Environmental Design Green Building Rating System, see http://www.usgbc.org. The 69-point LEED rating system addresses energy and water use, indoor air quality, materials, siting, and innovation and design. Buildings can earn basic certification or a silver, gold, or platinum designation depending on the number of credits awarded by external reviewers. Critics charge, however, that the costly and confusing administration of the LEED system is inhibiting adoption of the program and impeding the program's environmental objectives; see A. Schendler and R. Udall, "LEED is Broken; Let's Fix It," *Grist Magazine,* 16 October 2005, http://www.grist.com/comments/soapbox/2005/10/26/leed/index1.html.

31. GreenBiz.com, "Survey: Home Builders Name Energy Efficiency as Biggest Industry Trend," 26 January 2006, http://www.greenerbuildings.com/news_details.cfm?NewsID=30221.

32. D. Smith, "Conservation: Building Grows Greener in Bay Area," *San Francisco Chronicle,* 1 June 2005.

33. E. Beck, "Earth-Friendly Materials Go Mainstream," *New York Times,* 5 January 2006, 8.

34. J. M. Ginsberg and P. N. Bloom, "Choosing the Right Green Marketing Strategy," *MIT Sloan Management Journal,* Fall 2004: 79–84.

35. Tan, note 29 above.

36. Tan, note 29, above.

37. C. C. Berk, "P&G Will Promote 'Green' Detergent," *Wall Street Journal,* 19 January 2005.

38. K. McLaughlin, "Has Your Chicken Been Drugged?" *Wall Street Journal,* 2 August 2005; and E. Weise, "Are Our Products Our Enemy?" *USA Today,* 13 August 2005.

39. McLaughlin, ibid.

40. Alston and Roberts, note 2 above.

41. R. Leiber, "The Dirt on Green Housecleaners," *Wall Street Journal,* 29 December 2005.

42. M. Alexander, "Home Improved," *Readers Digest,* April 2004, 77–80.

43. For example, see D. Leonhardt, "Buy a Hybrid, and Save a Guzzler," *New York Times,* 8 February 2006.

44. See, for example, D. Cave, "It's Not Sexy Being Green (Yet)," *New York Times,* 2 October 2005.

45. G. Chon, "Toyota Goes After Copycat Hybrids; Buyers are Asked to Believe Branded HSD Technology is Worth the Extra Cost," *Wall Street Journal,* 22 September 2005.

46. B. G. Hoffman, "Ford: Now It's Easy Being Green," *Detroit News,* 31 January 2006.

47. See W. McDonough and M. Braungart, *Cradle to Cradle: Remaking the Way We Make Things* (New York: North Point Press, 2002).

48. R. Smith, "Beyond Recycling: Manufacturers Embrace 'C2C' Design," *Wall Street Journal,* 3 March 2005.

49. K. Hafen, "Preston Festival Goes LED," *Logan Herald Journal,* 21 September 2005.

50. O'Dell, note 20 above.

51. A. Covarrubias, "In Carpool Lanes, Hybrids Find Cold Shoulders," *Los Angeles Times,* 10 April 2006.

52. M. Clayton, "Hot Stuff for a Cool Earth," *Christian Science Monitor,* 21 April 2005.

53. See Ogilvy & Mather Topline Report, *China Energy-Efficient CFC-Free Refrigerator Study* (Beijing: Ogilvy & Mather, August 1997); E. R. Stafford, C. L. Hartman, and Y. Liang, "Forces Driving Environmental Innovation Diffusion in China: The Case of Green-freeze," *Business Horizons* 9, no. 2 (2003): 122–35.

54. J. Baker, Jr., K. Denby, and J. E. Jerrett, "Market-based Government Activities in Texas," *Texas Business Review,* August 2005, 1–5.

55. T. Harris, "Austinites Apply to Save With Wind Power," *KVUE News,* 13 February 2006.

56. J. Ottman, note 3 above.

57. J. Ottman, note 3 above.

58. Generation Green, "Splenda Letter to Federal Trade Commission," 13 January 2005, http://www.generationgreen. org/2005_01-FTC-letter.htm (accessed 7 February 2006).

59. J. Davis, "Strategies for Environmental Advertising," *Journal of Consumer Marketing* 10, no. 2 (1993): 23–25.

60. S. Farah, "The Thin Green Line," CMO Magazine, 1 December 2005, http://www.cmomagazine.com/read/120105/green_line. html (accessed 9 February 2006).

61. Ibid.

62. See Jenkins, note 23 above.

63. See Farah, note 60 above.

64. M. Maynard, "E.P.A. Revision is Likely to Cut Mileage Ratings," *New York Times,* 11 January 2006.

65. For a more comprehensive overview of eco-certifications and labeling, see L. H. Gulbrandsen, "Mark of Sustainability? Challenges for Fishery and Forestry Eco-labeling," *Environment* 47, no. 5 (2005): 8–23.

66. For a comprehensive overview of other eco-certifications, see Consumers Union's Web site at http://www.eco-labels. org/home.cfm.

67. Gulbrandsen, note 65 above, pages 17–19.

68. Farah, note 60 above.

69. Farah, note 60 above.

70. See GE Global Research, *Clean Coal,* http://ge.com/research/ grc_2_1_3.html (accessed 16 April 2006).

71. A. Griscom Little, "It Was Just My Ecomagination," *Grist Magazine,* 10 May 2005, http://grist.org/news/ muck/2005/05/10/little-ge/index.html.

72. E. Rosen, *The Anatomy of Buzz: How to Create Word-of-Mouth Marketing* (New York: Doubleday, 2000).

73. Viral marketing is a form of "word-of-mouse" buzz marketing defined as "the process of encouraging honest communication among consumer networks, and it focuses on email as the channel." See J. E. Phelps, R. Lewis, L. Mobilio, D. Perry, and N. Raman, "Viral Marketing or Electronic Word-of-Mouth Advertising: Examining Consumer Responses and Motivation to Pass Along Email," *Journal of Advertising Research* 44, no. 4 (2004): 333–48.

74. G. Ramsey, "Ten Reasons Why Word-of-Mouth Marketing Works," Online Media Daily, 23 September 2005, http:// publications.mediapost.com/index.cfm?fuseaction=Articles. san&s=34339&Nid=15643&p=114739 (accessed 16 February 2006).

75. See Rosen, note 72 above.

76. Ramsey, note 74 above.

77. Tide press release, "ColdWater Challenge Reaches One Million," http://www.tide.com/tidecoldwater/challenge.html (accessed 13 September 2005).

78. L. Prescott, "Case Study: Tide Boosts Traffic 9-fold," iMedia Connection, 30 November 2005, http://www.imdiaconnection. com/content/7406.asp.

79. Hawken, Lovins, and Lovins, note 4 above; see also A. B. Lovins, L. H. Lovins, and P. Hawken, "A Road Map for Natural Capitalism," *Harvard Business Review,* May–June 1999, 145–58.

80. J. Makower, "Green Marketing: Lessons from the Leaders," Two Steps Forward, September 2005, http://makower.typepad. com/joel_makower/2005/09/green_marketing.html.

JACQUELYN A. OTTMAN is president of J. Ottman Consulting, Inc. in New York and author of *Green Marketing: Opportunity for Innovation,* 2nd edition (NTC Business Books, 1997). She can be reached at jaottman@greenmarketing.com. **EDWIN R. STAFFORD** is an associate professor of marketing at Utah State University, Logan. He researches the strategic marketing and policy implications of clean technology (also known as "cleantech") and is the co-principal investigator for a $1 million research grant from the U.S. Department of Energy on the diffusion of wind power in Utah. He may be reached at ed.stafford@ usu.edu. **CATHY L. HARTMAN** is a professor of marketing at Utah State University, Logan. Her research centers on how interpersonal influence and social systems affect the diffusion of ideas and clean products and technology. She is principal investigator on a $1 million U.S. Department of Energy grant for developing wind power in the state of Utah. She can be contacted at cathy.hartman@usu.edu.

From *Environment,* June 2006, pp. 23–36. Reprinted by permission of the Helen Dwight Reid Educational Foundation. Published by Heldref Publications, 1319 Eighteenth St., NW, Washington, DC 20036-1802. Copyright © 2006. www.heldref.org

Energy: Present and Future Problems

Unit Selections

Key Points to Consider

- Many politicians, from Governor Arnold Schwarzenegger on down have suggested that hydrogen-powered cars are the way of the future. Creating zero harmful emissions, they sound almost too good to be true. What is the future of this power option?

- Why are some energy experts rethinking the long moratorium on the development of new nuclear power plants to create electricity? Have the lessons of Three Mile Island and Chernobyl been forgotten or ignored, or are there new technologies that make nuclear energy more attractive?

- How can conservation measures actually increase the economic benefits of energy production? What impact would improving energy efficiency have on fossil fuel consumption and on potential climate change?

- What are some of the major benefits of such alternate energy sources as solar power and wind power? Do these energy alternatives really have a chance at competing with fossil fuels for a share of the global energy market?

- Nuclear energy gives off no harmful byproducts . . . other than spent nuclear fuel. While everyone acknowledges that the waste is incredibly toxic, there has not been a great effort to deal with this problem. Should nuclear power be an important part of our energy portfolio? What is the cost? And what kind of subsides presently exist? Would these need to continue?

Student Web Site

www.mhcls.com/online

Internet References

Further information regarding these Web sites may be found in this book's preface or online.

Alliance for Global Sustainability (AGS)
http://globalsustainability.org/

Alternative Energy Institute, Inc.
http://www.altenergy.org

Energy and the Environment: Resources for a Networked World
http://zebu.uoregon.edu/energy.html

Institute for Global Communication/EcoNet
http://www.igc.org/

Nuclear Power Introduction
http://library.thinkquest.org/17658/pdfs/nucintro.pdf

U.S. Department of Energy
http://www.energy.gov

The dramatic climate changes we are experiencing and can expect to continue can be tied directly to the development and exploitation of inexpensive fossil fuel-based energy sources. Energy is cheap and readily available. A man working at near peak capacity generates enough energy to power a 75-watt light bulb. The electrical cost to do the same is about 10¢ per day. Cheap energy allowed for our explosive population growth over the last several centuries and is responsible for the increased levels of greenhouse gases in our atmosphere. We are presently adding a staggering 6,000 million metric tons of CO_2 to the atmosphere annually, which is expected to increase to 8,000 million metric tons by 2030 (Energy Information Administration). The long-term effects of increasing the CO_2 levels in the atmosphere are documented in "Climate Change 2007" in Unit 1. What is clear is that immediate and forceful action is necessary to reduce our CO_2 emissions. In addition to the damage to the environment, our continued reliance on fossil fuels places additional burdens on our society. The United States has an enormous trade deficit with the Middle East nations, a region that controls well over 30% of the known petroleum reserves. Our dependence on these reserves leaves us economically vulnerable. Certainly the most dramatic illustration of our vulnerabilities is the war in Iraq and the nuclear tensions with Iran.

A number of alternatives to fossil fuels have been proposed. All have benefits, and all have limitations. If a magic bullet existed, we would have exploited it already. The fact is that there's no simple answer to our energy demands. What is needed is a creative, flexible, and intelligent energy policy for the 21st century. In this unit, a number of different topics are explored. Energy requirements for homes and buildings consume the largest fraction of our energy needs, but transportation—particularly automobiles—are not far behind. The first article, "Gassing Up with Hydrogen" by Sunita Satyapal, John Petrovic, and George Thomas explores the future of hydrogen fuel cells as an alterative to the traditional internal combustion engine. Hydrogen is a super-clean fuel, with no harmful emissions. The obstacles to a hydrogen-based economy are many, however. Producing hydrogen is costly. A nationwide network of hydrogen fueling stations would have to be implemented. Advances in other technologies, such as electric cars, would have to be minimal to keep hydrogen at the forefront. The cost of producing a hydrogen car would have to come down by orders of magnitude from the present price tag of close to $1 million. And finally, as discussed in the article, a safe and economical method of storage would have to be developed. Unfortunately, a nation driven by hydrogen cars is not in the immediate future.

The next two articles explore a far more promising development, namely electricity generation from wind. Once expensive compared to traditional energy sources, wind power is now cheaper than any other form except for hydroelectric. The negatives to wind power certainly exist. Wind is fickle and doesn't always blow. It is also noisy and kills birds. but compared to other energy sources, these drawbacks are relatively minor. Initial installation costs are high, but wind turbines can be erected quickly and require very little maintenance. But wind's negatives pale in comparison to problems with other energy sources. Ever larger wind turbines are being constructed, and these are more efficient and spin more slowly, reducing bird deaths. As prices drop, wind energy will continue to become an important part of our overall energy mix.

Article 16 is the "The Future of Nuclear Power," written by the MIT working group. It addresses the controversial proposal of expanding our nuclear power option. Nuclear power has no harmful emissions and does not require importation of the fuel source. It is essentially 'renewable' because our nuclear resources are extensive. But the negatives to nuclear power weigh heavy. First, it is costly. There are huge upfront costs that make it viable only with large federal subsides. Second, nuclear power plants will always be vulnerable to a terrorist attack. Third, the fuel is extremely dangerous, and great care is required to insure that it is handled correctly. And finally, there is the political thorny issue of disposal. In 100 years, we will probably have answers to the problems of nuclear disposal, but for now, serious questions remain.

The article "Personalized Energy" by Stephen M. Millett, looks at our energy needs from a different perspective. Instead of viewing energy generation in terms of large, centralized power plants, Millett asks whether one facet of our energy portfolio might not be more decentralized power sources. By turning to more localized sources, energy solutions can be developed that are more tailored to local communities' strengths.

Gassing Up with Hydrogen

Researchers are working on ways for fuel-cell vehicles to hold the hydrogen gas they need for long-distance travel.

SUNITA SATYAPAL, JOHN PETROVIC AND GEORGE THOMAS

On a late summer day in Paris in 1783, Jacques Charles did something astonishing. He soared 3,000 feet above the ground in a balloon of rubber-coated silk bags filled with lighter-than-air hydrogen gas. Terrified peasants destroyed the balloon soon after it returned to earth, but Charles had launched a quest that researchers two centuries later are still pursuing: to harness the power of hydrogen, the lightest element in the universe, for transportation.

Burned or used in fuel cells, hydrogen is an appealing option for powering future automotive vehicles for several reasons. Domestic industries can make it from a range of chemical feedstocks and energy sources (for instance, from renewable, nuclear and fossil-fuel sources), and the nontoxic gas could serve as a virtually pollution-free energy carrier for machines of many kinds. When it burns, it releases no carbon dioxide, a potent greenhouse gas. And if hydrogen is fed into a fuel-cell stack—a battery-like device that generates electricity from hydrogen and oxygen—it can propel an electric car or truck with only water and heat as by-products [see "On the Road to Fuel-Cell Cars," by Steven Ashley; *Scientific American,* March 2005]. Fuel-cell-powered vehicles could offer more than twice the efficiency of today's autos. Hydrogen could therefore help ease pressing environmental and societal problems, including air pollution and its health hazards, global climate change and dependence on foreign oil imports.

Yet barriers to gassing up cars with hydrogen are significant. Kilogram for kilogram, hydrogen contains three times the energy of gasoline, but today it is impossible to store hydrogen gas as compactly and simply as the conventional liquid fuel. One of the most challenging technical issues is how to efficiently and safely store enough hydrogen onboard to provide the driving range and performance that motorists demand. Researchers must find the "Goldilocks" storage solutions that are "just right." Storage devices should hold sufficient hydrogen to support today's minimum acceptable travel range—300 miles—on a tank of fuel in a volume of space that does not compromise passenger or luggage room. They should release it at the required flow rates for acceleration on the highway and operate at practical temperatures. They should be refilled or recharged in a few minutes and

come with a competitive price tag. Current hydrogen storage technologies fall far short of these goals.

Researchers worldwide in the auto industry, government and academia are expending considerable effort to overcome these limitations. The International Energy Agency's Hydrogen Implementing Agreement, signed in 1977, is now the largest international group focusing on hydrogen storage, with more than 35 researchers from 13 countries. The International Partnership for the Hydrogen Economy, formed in 2003, now includes 17 governments committed to advancing hydrogen and fuel-cell technologies. And in 2005 the U.S. Department of Energy set up a National Hydrogen Storage Project with three Centers of Excellence and many industry, university and federal laboratory efforts in both basic and applied research. Last year alone this project provided more than $30 million to fund about 80 research projects.

Infrastructural Hurdles

One obstacle to the wide adoption of hydrogen fuel-cell cars and trucks is the sheer size of the problem. U.S. vehicles alone consume 383 million gallons of gasoline a day (about 140 billion

---

Overview/Hydrogen Storage

- One of the biggest obstacles to future fuel-cell vehicles is how engineers will manage to stuff enough hydrogen onboard to provide the 300-mile minimum driving range that motorists demand.
- Typically hydrogen is stored in pressurized tanks as a highly compressed gas at ambient temperature, but the tanks do not hold enough gas. Liquid-hydrogen systems, which operate at cryogenic temperatures, also suffer from significant drawbacks.
- Several alternative high-density storage technologies are under development, but none is yet up to the challenge.

gallons annually), which accounts for about two thirds of the total national oil consumption. More than half of that petroleum comes from overseas. Clearly, the nation would need to invest considerable capital to convert today's domestic auto industry to fuel-cell vehicle production and the nation's extensive gasoline refining and distribution network to one that handles vast quantities of hydrogen. The fuel-cell vehicles themselves would have to become cheap and durable enough to compete with current technology while offering equivalent performance. They also must address safety concerns and a lingering negative public perception—people still remember the 1937 *Hindenburg* airship tragedy and associate it with hydrogen, despite some credible evidence that the airship's flammable skin was the crucial factor in the ignition of the blaze.

Why is it so difficult to store enough hydrogen onboard a vehicle? At room temperature and atmospheric pressure (one atmosphere is about 14.5 pounds per square inch, or psi), hydrogen exists as a gas with an energy density about $\frac{1}{3,000}$ that of liquid gasoline. A 20-gallon tank containing hydrogen gas at atmospheric pressure would propel a standard car only about 500 feet. So engineers must increase the density of stored hydrogen in any useful onboard hydrogen containment system.

A 300-mile minimum driving range is one of the principal operational aims of an industry-government effort—the FreedomCAR and Fuel Partnership—to develop advanced technology for future automobiles. Engineers employ a useful rule of thumb in making such calculations: a gallon of gasoline is equal, on an energy basis, to one kilogram (2.2 pounds) of hydrogen. Whereas today's average automobile needs about 20 gallons of gasoline to travel at least 300 miles, the typical fuel-cell vehicle would need only about eight kilograms of hydrogen because of its greater operational efficiency. Depending on the vehicle type and size, some models would require less hydrogen to go that far, some more. Tests of about 60 hydrogen-fueled prototypes from several automakers have so far demonstrated driving ranges of 100 to 190 miles.

Aiming for a practical goal that could be achievable by 2010 (when some companies expect the first production fuel-cell cars to hit the road), researchers compare the performance of various storage technologies against the "6 weight percent" benchmark. That is, a fuel storage system in which 6 percent of its total weight is hydrogen. For a system weighing a total of 100 kilograms (a reasonable size for a vehicle), six kilograms would be stored hydrogen. Although 6 percent may not seem like much, achieving that level will be extremely tough; less than 2 percent is the best possible today—using storage materials that operate at relatively low pressures. Further, keeping the system's total volume to about that of a standard automotive gasoline tank will be even more difficult, given that much of its allotted space will be taken up by the tanks, valves, tubing, regulators, sensors, insulation and anything else that is required to hold the six kilograms of hydrogen. Finally, a useful system must release hydrogen at rates fast enough for the fuel-cell and electric motor combination to provide the power and acceleration that drivers expect.

Containing Hydrogen

At present, most of the several hundred prototype fuel-cell vehicles store hydrogen gas in high-pressure cylinders, similar to scuba tanks. Advanced filament-wound, carbon-fiber composite technology has yielded strong, lightweight tanks that can safely contain hydrogen at pressures of 5,000 psi (350 times atmospheric pressure) to 10,000 psi (700 times atmospheric pressure). Simply raising the pressure does not proportionally increase the hydrogen density, however. Even at 10,000 psi, the best achievable energy density with current high-pressure tanks (39 grams per liter) is about 15 percent of the energy content of gasoline in the same given volume. Today's high-pressure tanks can contain only about 3.5 to 4.5 percent of hydrogen by weight. Ford recently introduced a prototype "crossover SUV" called Edge that is powered by a combination plug-in hybrid/fuel-cell system that stores 4.5 kilograms of hydrogen fuel in a 5,000-psi tank to achieve a total maximum range of 200 miles.

High-pressure tanks would be acceptable in certain transportation applications, such as transit buses and other large vehicles that have the physical size necessary to accommodate storage for sufficient hydrogen, but it would be difficult to manage in cars. Also, the current cost of such tanks is 10 or more times higher than what is competitive for autos.

Liquefying stored hydrogen can improve its energy density, packing the most hydrogen into a given volume of any existing option. Like any gas, hydrogen that is cooled sufficiently condenses into a liquid, which at atmospheric pressure occurs around –253 degrees Celsius. Liquid hydrogen exhibits a density of 71 grams per liter, or about 30 percent of the energy density of gasoline. The hydrogen weight densities achievable by these systems depend on the containment and insulation equipment they use.

Liquefied hydrogen has important drawbacks, though. First, its very low boiling point necessitates cryogenic equipment and special precautions for safe handling. In addition, because it operates at low temperature, the containers have to be insulated extremely well. Finally, liquefying hydrogen takes more energy than compressing the gas to high pressures. This requirement drives up the cost of the fuel and reduces the overall energy efficiency of the cryocooling process.

Nevertheless, one carmaker is pushing this technology onto the road. BMW plans to introduce a vehicle this year called Hydrogen 7, which will incorporate an internal-combustion engine capable of running on either gasoline (for 300 miles) or on liquid hydrogen for 125 miles. Hydrogen 7 will be sold on a limited basis to selected customers in the U.S. and other countries with local access to hydrogen refueling stations.

Chemical Compaction

Searching for promising ways to raise energy density, scientists may be able to take advantage of the chemistry of hydrogen itself. In their pure gas and liquid phases, hydrogen molecules contain two bound atoms each. But when hydrogen atoms are chemically bound to certain other elements, they can be packed even closer together than in liquid hydrogen. The principal aim of hydrogen storage research now is finding the materials that can pull off this trick.

Some researchers are focusing on a class of substances called reversible metal hydrides, which were discovered by accident in 1969 at the Philips Eindhoven Labs in the Netherlands. Investigators found that a samarium-cobalt alloy exposed to pressurized hydrogen gas would absorb hydrogen, somewhat like a sponge soaks up water. When the pressure was then removed, the hydrogen within the alloy reemerged; in other words, the process was reversible.

Intensive research followed this discovery. In the U.S., scientists James Reilly of Brookhaven National Laboratory and Gary Sandrock of Inco Research and Development Center in Suffern, N.Y., pioneered the development of hydride alloys with finely tuned hydrogen absorption properties. This early work formed the basis for today's widely used nickel-metal hydride batteries. The density of hydrogen in these alloys can be very high: 150 percent more than liquid hydrogen, because the hydrogen atoms are constrained between the metal atoms in their crystal lattices.

Many properties of metal hydrides are well suited to automobiles. Densities surpassing that of liquid hydrogen can be achieved at relatively low pressures, in the range of 10 to 100 times atmospheric pressure. Metal hydrides are also inherently stable, so they require no extra energy to maintain storage, although heat is required to release the stored gas. But their Achilles' heel is mass. They weigh too much for practical onboard storage. Metal hydride researchers have so far attained a maximum hydrogen capacity of 2 percent of the total material weight (2 weight percent). This level translates into a 1,000-pound hydrogen storage system (for a 300-mile driving range), which is clearly too heavy for today's 3,000-pound car.

Metal hydride studies currently concentrate on materials with inherently high hydrogen content, which researchers then modify to meet the hydrogen storage system requirements of operating temperatures in the neighborhood of 100 degrees C, pressures from 10 to 100 atmospheres and delivery rates sufficient to support rapid vehicle acceleration. In many cases, materials that contain useful proportions of hydrogen are a bit too stable in that they require substantially higher temperatures to release the hydrogen. Magnesium, for example, forms magnesium hydride with 7.6 weight percent hydrogen but must be heated to above 300 degrees C for release to occur. If a practical system is to rely on waste heat from a fuel-cell stack (about 80 degrees C) to serve as the "switch" to liberate hydrogen from a metal hydride, then the trigger temperature must be lower.

Destabilized Hydrides

Chemists John J. Vajo and Gregory L. Olson of HRL Laboratories in Malibu, Calif., as well as researchers elsewhere are exploring a clever approach to overcoming the temperature problem. Their "destabilized hydrides" combine several substances to alter the reaction pathway so that the resulting compounds release the gas at lower temperatures.

Destabilized hydrides are part of a class of hydrogen-containing materials called complex hydrides. Chemists long thought that many of these compounds were not optimal for refueling a vehicle, because they were irreversible—once the hydrogen was freed by decomposition of the compounds, the materials would require reprocessing to return them to a hydrogenated state. Chemists Borislav Bogdanovic and Manfred Schwickardi of the Max Planck Institute of Coal Research in Mulheim, Germany, however, stunned the hydride research community in 1996 when they demonstrated that the complex hydride sodium alanate becomes reversible when a small amount of titanium is added. This work triggered a flurry of activity during the past decade. HRL's lithium borohydride destabilized with magnesium hydride, for example, holds around 9 percent of hydrogen by weight reversibly and features a 200 degree C operating temperature. This improvement is notable, but its operating temperature is still too high and its hydrogen release rate too slow for automotive applications. Nevertheless, the work is promising.

Although current metal hydrides have limitations, many automakers see them as the most viable low-pressure approach in the near- to mid-term. Toyota and Honda engineers, for example, are planning a so-called hybrid approach in a system that combines a solid metal hydride with moderate pressure (significantly lower than 10,000 psi), which they predict could achieve a driving range of more than 300 miles. General Motors has teams of storage experts, including Scott Jorgensen, who are supporting research on a wide range of metal hydride systems worldwide (including in Russia, Canada and Singapore). GM is also collaborating with Sandia National Laboratories on a four-year, $10-million effort to produce a prototype complex metal hydride system.

Hydrogen Carriers

Other hydrogen storage options have the potential to work well in cars, but they suffer a penalty in the refueling step. In general, these chemical hydride substances need industrial processing to reconstitute the spent material. The step requires off-board regeneration; that is, once hydrogen stored onboard a vehicle is released, a leftover by-product must be reclaimed at a service station and regenerated in a chemical plant.

More than 20 years ago Japanese researchers studied this approach using, for example, the decalin-naphthalene system. When decalin ($C_{10}H_{18}$) is heated, it converts chemically to naphthalene (a pungent-smelling compound with the formula $C_{10}H_8$) by changing the nature of its chemical bonds, which liberates five hydrogen molecules. Hydrogen gas thus bubbles out of the liquid decalin as it transforms into naphthalene. Exposing naphthalene to moderate hydrogen gas pressures reverses the process; it absorbs hydrogen and changes back to decalin (6.2 weight percent for the material alone). Research chemists Alan Cooper and Guido Pez of Air Products and Chemicals in Allentown, Pa., are investigating a similar technique using organic (hydrocarbon-based) liquids. Other scientists, including S. Thomas Autrey and his co-workers at the Pacific Northwest National Laboratory and chemistry professor Larry G. Sneddon of the University of Pennsylvania, are working on new liquid carriers, such as aminoboranes, that can store large amounts of hydrogen and release it at moderate temperatures.

Designer Materials

Yet another approach to the hydrogen storage problem centers on lightweight materials with very high surface areas to which hydrogen molecules stick (or adsorb). As one might expect, the amount of hydrogen retained on any surface correlates with the material's surface area. Recent developments in nanoscale engineering have yielded a host of new high-surface-area materials, some with more than 5,000 square meters of surface area per gram of material. (This amount equates to about three acres of surface area within just a teaspoon of powder.) Carbon-based materials are particularly interesting because they are lightweight, can be low cost and can form a variety of nanosize structures: carbon nanotubes, nanohorns (hornlike tubes), fullerenes (ball-shaped molecules) and aerogels (ultraporous solids). One relatively cheap material, activated carbon, can store up to about 5 weight percent hydrogen.

These carbon structures all share a common limitation, however. Hydrogen molecules bond very weakly with the carbon atoms, which means that the high-surface-area materials must be kept at or near the temperature of liquid nitrogen, −196 degrees C. In contrast to hydride research, in which scientists are struggling to lower the hydrogen binding energy, carbon researchers are exploring ways to raise the binding energy by modifying the surfaces of materials or by adding metal dopants that may alter their properties. These investigators employ theoretical modeling of carbon structures to discover promising systems for further study.

Beyond the carbon-based approaches, another fascinating nanoscale engineering concept is a category of substances called metal-organic materials. A few years ago Omar Yaghi, then a chemistry professor at the University of Michigan at Ann Arbor and now at the University of California, Los Angeles, invented these so-called metal-organic frameworks, or MOFs. Yaghi and his co-workers showed that this new class of highly porous, crystalline materials could be produced by linking inorganic compounds together with organic "struts". The resulting MOFs are synthetic compounds with elegant-looking structures and physical characteristics that can be controlled to provide various desired functions. These heterogeneous structures can have very large surface areas (as high as 5,500 square meters per gram), and researchers can tailor chemical sites on them for optimal binding to hydrogen. To date, investigators have demonstrated MOFs that exhibit hydrogen capacities of 7 weight percent at −196 degrees C. They continue to work on boosting this performance.

Although current progress on hydrogen storage methods is encouraging, finding the "just right" approach may take time, requiring sustained, innovative research and development efforts. Over the centuries, the basic promise—and challenge—of using hydrogen for transportation has remained fundamentally unchanged: Holding onto hydrogen in a practical, lightweight container allowed Jacques Charles to travel across the sky in his balloon during the last decades of the 18th century. Finding a similarly suitable container to store hydrogen in automobiles will permit people to travel across the globe in the coming decades of the 21st century without fouling the sky above.

SUNITA SATYAPAL, JOHN PETROVIC and GEORGE THOMAS work in the U.S. Department of Energy's applied research and development program in hydrogen storage technology. Satyapal, who has held various positions in academia and industry, serves as the team leader for the DOE's applied hydrogen storage R&D activities. Petrovic, a laboratory fellow (retired) of Los Alamos National Laboratory, is a consultant for the DOE and a fellow of both the American Ceramic Society and the American Society for Materials international. Thomas, currently a consultant with the DOE, has more than 30 years of experience studying the effects of hydrogen on metals at Sandia National Laboratories. These views are those of the authors only; they do not reflect the positions of the U.S. Department of Energy.

Wind Power
Obstacles and Opportunities

Martin J. Pasqualetti

To know the wind is to respect nature. You ride with the wind when it fills your sails, but pay its power no heed and risk inconvenience, expense, even death. Drive through calm air in Los Angeles one moment only to encounter 30 minutes later Santa Anas whipping wildfires across mountaintops and pushing tractor-trailers into ditches. Lounge on the beach on Kauai one day, but find yourself huddling for protection the next day as a hurricane levels entire forests.[1] Sit on the porch during a quiet and muggy Oklahoma night when suddenly a mass of debris, once a house, swirls past, before dropping nearby as a pile of kindling and shattered dreams. More than any other force of nature, we have little defense against the wind. The wind keeps us on our toes.

If we cannot control the wind, perhaps we can put it to our use. It is a challenge with which we have had some success. Historically, we have used the wind to help us with work that would otherwise fall heavily upon our own backs. The wind helped humans explore the world; they had no other energy source. It continues to help us prepare foods and pump water. In some places, the wind is such a part of daily life that in its absence, silence blankets the landscape and puts us out of sorts. When it picks up again, flags flutter, well water rises, and grains are again ground to flour.

The wind machines humans developed were among the earliest icons of civilization. We can see them in early sketches from the Orient, scrolls from Persia, paintings from the Low Countries, photographs from the Dust Bowl, and even movies from Hollywood. Putting the wind to work was our first conscious use of solar power.

Perhaps the most widespread use of wind machines, at least in the United States, has been to pump water. Dotting the Great Plains by the hundreds of thousands, farm windmills—along with grain silos—were once as characteristic of the landscape as coal spoils were of Appalachia. Spinning whenever air moved, they brought to the surface the water that allowed ranches and settlements to flourish in an area otherwise too dry for either to exist for long. Most of these ingenious whirling devices eventually gave way to powerful compact motors that ran on fossil fuels, and as a result, wind energy landscapes largely disappeared. Before they all were removed, some folks preserved a few of them, drawn to their quaint beauty and the nostalgia they

evoked as symbols of a Great Plains lifestyle. By then, however, most people considered the era of wind machines dead.

As it has happened, the epitaphs were premature. Today, wind machines are back. It has not been a quiet resurrection but rather one with substantial notoriety and publicity, plus a controversial mix of support and resistance. The new devices look and act little like their ancestors: Instead of the creaking, wooden machines of the past, those of the new species are made of metal and fiberglass—and are bigger, quieter, sleeker, and more powerful than ever. Instead of pumping water, the moving blades spin generators housed with an assemblage of gears in the nacelle, which is located behind the hub where all the blades meet. Instead of a stream of water, modern wind machines are pumping a stream of electrons, a product proving to be a valuable asset to farmers who are trying to address present day economic realities of living off the land.

The new appearance and mechanics of wind machines reflects their different role. Instead of producing mechanical power for the purposes of pumping and grinding, the new machines convert mechanical energy into electricity. Instead of being erected here and there in splendidly independent isolation, many are being clustered in symmetrically interdependent neighborhoods, designed to work together as parts of a larger organism. Nor are they just generating electricity: Unexpectedly, modern wind machines are prompting us to consider how best to weigh the energy we need against the environmental quality we want. All the while they are continuing their transformation from public indifference to public curiosity, from an overlooked energy supplier to alternative energy's "holy grail," one possible way to get most of what we want and little of which we do not.

An Old Resource with a New Mission

Compared to the variety of uses that stretch back millennia, converting wind energy to electricity is a recent application. Although a few people were trying to accomplish this at the same time Thomas Edison opened his coal-fired Pearl Street generating plant in the latter years of the nineteenth century, it would be another 80 years before such proof-of-concept machines would

Table 1 Historical Wind Turbines

Turbine, Country	Date in Service	Diameter (meters)	Swept Area (meters)	Power (kilowatts)	Specific Power (kilowatts per square meter)	Number of Blades	Tower Height (meters)
Poul la Cour, Denmark	1891	23	408	18	0.04	4	—
Smith-Putnam, United States	1941	53	2,231	1,250	0.56	2	34
F. L. Smidth, Denmark	1941	17	237	50	0.21	3	24
F. L. Smidth, Denmark	1942	24	456	70	0.15	3	24
Gedser, Denmark	1957	24	452	200	0.44	3	25
Hütter, Germany	1958	34	908	100	0.11	2	22

Source: P. Gipe, *Wind Energy Comes of Age* (New York: John Wiley & Sons, 1995), 78.

evolve into the commercial generators that started sprouting in the California landscape in 1981 (see Table 1). Indeed, the beginning of the modern era of wind power bore few similarities to earlier water pumping. The vision of modern wind power was much grander in scale, one that has evolved to row upon row of machines spreading over hundreds of acres, contributing enough electricity to power an entire city but—and this is the big difference—without undesirable side effects that accompanied the use of conventional resources.

Obviously, the machinery of the late twentieth century differs both in form and function from the equipment that nineteenth century ranchers and farmers developed to help them wrest a living from the dry lands that predominate west of the 100th meridian. For them, it was enough that machines were turning when the air was on the move. In the new era, such simple fulfillment is not enough; wind power today is viewed less from the living room and more from the boardroom. Wind power is big business, and the managers of that business must be sophisticated not just in the ways of making money, but in several disciplines that lead to success. Even something as seemingly innocent as turbine placement can no longer be considered just from the perspectives of convenience, necessity, or whim. Instead, the new wind barons must understand meteorology, metallurgy, physics, aerodynamics, capacity factors, land ownership, planning, zoning, and the influence of public perception.

Wind power's popularity is widespread and growing, a result of its increasing profitability and the perceived environmental benefits it engenders. It also results from the simple fact that, unlike fossil and nuclear fuels, wind is a widely available, familiar element of the environment. The first step is seemingly the easiest: finding it.

Wind and the Family Farm

The initial step in developing any resource, be it gold or wind, is locating it. While it is often an uncomplicated step, not every place is attractive. Just like gold, wind is not evenly distributed

in the richness of its product. Subtropical deserts, such as the Sonoran Desert surrounding Phoenix, are created by persistent high pressure and are often unsuitable for wind development. For obvious reasons, forested areas are unattractive for wind turbines, as are equatorial areas with their characteristically light and variable winds. The rest of the world is more promising, although detailed data collection must precede full-fledged capital investment.

Finding windy places is a relatively easy step: Unlike fossil or nuclear fuels, it does not require drilling rigs, seismic gear, or Geiger counters. Wind power, in most places, is simply "out there." This means that the most obvious early task of wind prospectors is to determine where winds are strong enough. Once identified, such areas reveal several common characteristics, including exposed terrain, colliding air masses, and, in particular circumstances, topographic funneling, as through mountain passes. In the United States, several areas meet such criteria, including sites in California, southeastern Washington, central Wyoming, east-central New Mexico, and most notably the Great Plains (see Figure 1). In Europe, strong winds are significant along the west coast of Ireland, Great Britain, the eastern North Sea, the southern Baltic Sea, the Pyrénées Mountains, and the Rhone Valley (see Figure 2). Many of these areas are "nuggets" that we are plucking first, and they have been stimulating further prospecting and the development of grand plans for the future. By the end of 2003, the total installed capacity in the European Union was 28,440 megawatts (MW).[2]

Like the gold rush of the 1850s, the modern wind rush started in California. California still leads the way, with 2,042 MW installed by January 2004, principally in four locations: Altamont Pass, on the edge of the Central Valley east of San Francisco; Tehachapi Pass at the southern end of the Sierra Nevada; San Gorgonio Pass, near Palm Springs; and in the rolling hills between San Francisco and Sacramento. Although first, the primacy of California is certainly temporary: With a potential for 6,770 MW, it ranks only seventeenth among the 50 states in potential (see Table 2).[3]

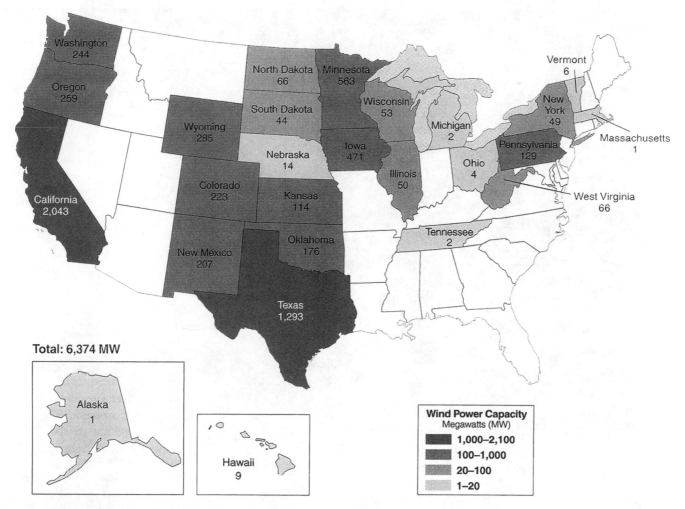

Figure 1 2003 year-end installed wind power capacity (in megawatts).

Source: American Wind Energy Association, *Wind Energy Projects Throughout the United States of America,* http://www.awea.org/projects (accessed 30 June 2004).

Among other states attracting interest, Texas has been the pacesetter with more than $1 billion in new wind investment and 1,293 MW installed, mostly in the western part of the state near such communities as Big Springs and McCamey. Coincidentally, many of these developments are positioned among the now derelict oil equipment that helped bring great wealth to this part of the state and has underpinned many of its towns and cities. At the end of January 2004, California, Texas, and 24 additional states held within their borders an installed capacity of 6,374 MW.[4]

Another 2,000 MW has been proposed for development in the near future, with some of the largest projects planned for the states of Washington and Massachusetts. However, the greatest potential remains where wind machines once so dominated the landscape—midway between these extremes in the Great Plains.

The wind of the Great Plains is as obvious as is its treeless expanse. When Francisco Vásquez de Coronado and his men crossed this region searching for the Seven Cities of Cíbola in 1540, they found no gold but two other resources instead. The most obvious and most useful to Coronado were the great

herds of bison—totaling perhaps 50 million head—that were scattered across a million square miles of grassland. Not only did they provide food, but their droppings helped guide the expedition across otherwise indistinct landscapes. The other resource was the wind, but three centuries would pass before it was appreciated.

By the late 1800s, the general pattern had reversed; bison were being decimated for sport, and wind power was lifting water for irrigation. Today, only this second resource remains, yet it is being used for a different mission: to generate electricity and make money.[5] It is a realistic ambition: The winds of the Great Plains are so abundant that the energy potential from just three states (North Dakota, Texas, and Kansas), were they fully developed, would match the electrical needs of the entire country. These and several other Great Plains states hold the largest expanse of class 4 (400–500 watts per square meter) lands in the country (see the box).[6]

Although weather on the Great Plains is often viewed as being inhospitable—farming families there endure swirling snow in winter and blowing dust in summer—attitudes toward the frequent tempests are lately bending in a new direction.

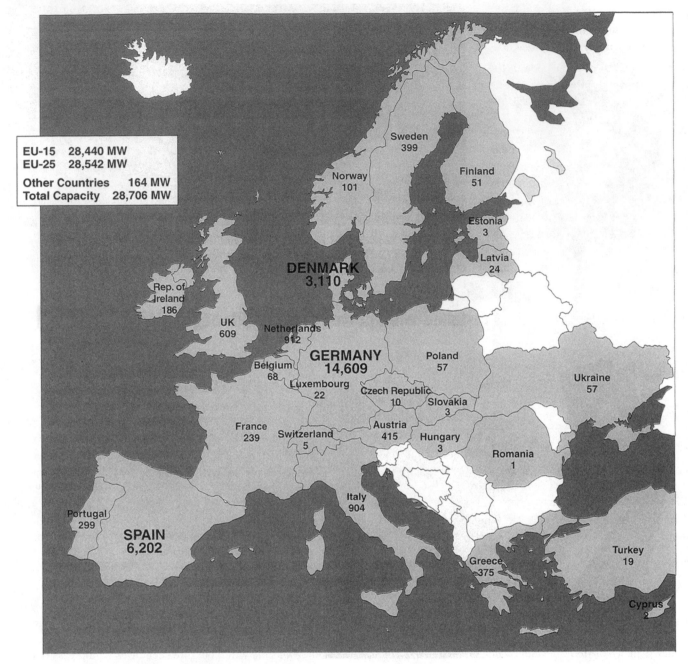

Figure 2 2003 year-end european installed wind power capacity (in megawatts).

Source: Adapted from a map compiled by the European Wind Energy Association, http://www.ewea.org, 3 February 2004.

Always alert for new sources of income to ease their financial volatility, locals are turning to wind developers with equanimity and even enthusiasm. They are finding that the same winds that strip soil from the fields and bury houses in snow can fuel rural economic development.[7]

Construction of a typical 100 MW wind farm produces more than 50,000 days (approximately 419,020 manhours) of employment. In Prowers County, Colorado, the recently completed wind development is each year providing $764,000 in new revenues, $917,000 in school general funds, $203,000 in school bond funds, $189,000 to the Prowers Medical Center, and $189,000 in additional revenue to the county tax base.[8] Meanwhile, a 250 MW project in Iowa is providing $2 million in property taxes and $5.5 million in operation and maintenance income. The leases on offer to farmers in this area commonly provide yearly royalties of more than $2,000 per turbine: For a single Iowa project, local farmers are receiving $640,000 annually. Other projects return $4,000–$5,000 per turbine. In some cases, a one-megawatt turbine could generate revenues for the owner of $150,000 per year once the debt for purchase is repaid.[9] Though appreciable at the individual level, such royalties are only a small part of the variable costs for wind developers (see Figure 3). To them, such largesse seems good business. To farmers, such revenues can mean the difference between bankruptcy and prosperity.

Table 2 Top 20 States for Wind Energy Potential (measured by annual energy potential in billions of kilowatt hours)

1	North Dakota	1,200
2	Texas	1,190
3	Kansas	1,070
4	South Dakota	1,030
5	Montana	1,020
6	Nebraska	868
7	Wyoming	747
8	Oklahoma	725
9	Minnesota	657
10	Iowa	551
11	Colorado	481
12	New Mexico	435
13	Idaho	73
14	Michigan	65
15	New York	62
16	Illinois	61
17	California	59
18	Wisconsin	58
19	Maine	56
20	Missouri	52

Note. As of July 2004, reevaluation of the wind potential has been done for 28 states by the U.S. Department of Energy's Windpowering America program; reevaluation of additional states is still in process. Once complete, the numbers for each state might change, as might the relative rankings. In many cases, the potential will increase. The Great Plains will continue to dominate the rest of the country in terms of potential.

Source: Pacific Northwest Laboratory, *An Assessment of the Available Wind Land Area and Wind Energy Potential in the Contiguous United States* (Richland, WA: Pacific Northwest National Laboratory, 1991).

In spite of a weighty list of environmental attributes, wind power carries some unexpectedly heavy baggage.

In spite of promising trends, it would be an overstatement to claim that wind power offers an energy panacea or a reversal of the national trend toward increasingly concentrated generation of electricity: It is, at least so far, a relatively small enterprise. Taken together, all the wind developments in the country contribute less than 1 percent of our current needs.[10] However, there is a real attraction to wind power's promise for the future: Its estimated generating potential in the United States alone is 10,777 billion kilowatts per hour (kWh) annually, or three times the electricity generated in the entire country today.[11] In recognition of such potential for pollution-free electricity, the U.S. government is sponsoring a program called Wind Powering America (WPA) to tap more deeply this vast natural resource. WPA's new goal is to increase to 30 the number of states with more than 100 megawatts of wind-generating capacity by 2010.[12] The program also aims to increase rural economic development and, to some degree, local energy independence.

The Environmental Irony

Wind power attracts many adherents from the environmental community. These organizations focus on its solar roots, emphasizing that it requires no mining, drilling, or pumping, no pipelines, port facilities, or supply trains. It produces no air pollution or radioactive waste, and it neither dirties water nor requires water for cooling. Wind power is relatively benign, simple, modular, affordable, and domestic. It is, in short, an environmental golden goose.

However, in spite of such a weighty list of attributes, wind power carries some unexpectedly heavy baggage. In England, anti-wind epithets have been particularly colorful: Developers have suffered their machines being called everything—including "lavatory brushes in the air" for their busy top ends. This leads us to an irony of wind power: While we usually consider wind power environmentally friendly, most of the objections to its expansion have had environmental origins.

We can follow the thread of such reactions most clearly to Palm Springs, California, in the mid-1980s. Soon after installing thousands of turbines in windy San Gorgonio Pass just north of the city limits, developers were battered with complaints that the machines interfered with television reception, produced annoying and inconsistent noise, posed risks to wildlife and aircraft, and represented incompatible land-use practices.[13] The most troubling, bitter, and outraged complaint, however, was that the wind machines destroyed the aesthetic appeal of the landscape, thereby threatening the very attribute that most attracts tourists to the area's fancy resorts.[14]

Developers and bureaucrats of the day, all of whom had expected a warmer welcome, were startled by such reactions. It was apparent to everyone with hopes for contributions from wind power that any future success would have to rest on greater environmental compatibility and a more complete respect for public attitudes and opinions. The initial experience provoked industry musing as to what actions might better attract support. Soon, manufacturers began making improvements to design and engineering. Locally, the concerns over wind impacts led to stricter planning rules and more uniform standards. These adjustments softened the problems, but they could not eliminate them. Turbines remained unavoidably visible and the center of a classic example of incompatible land use. The very characteristic that had long kept residential development in the San Gorgonio Pass minimal—the strong wind—was the same characteristic that prompted developers to fill it with machines. There was very little compromise potential.

So strong was the backlash against wind development that the City of Palm Springs sued the U.S. Bureau of Land Management and the County of Riverside, claiming that developers had not

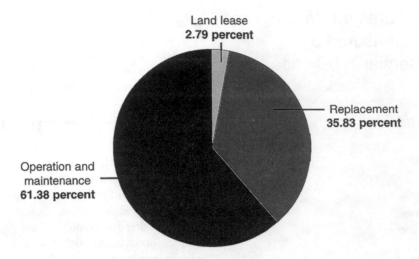

Figure 3 Variable costs of wind energy projects.

Source: E. DeMeo and B. Parsons, "Some Common Misconceptions about Wind Power," presented at the All States Wind Summit, Austin, TX, 22 May 2003. See U.S. Department of Energy, *State Wind Energy Handbook,* http://www.eere.energy.gov/windpoweringamerica/pdfs/wpa/34600_wind_handbook.pdf (accessed 30 June 2004), 90.

Wind Classes

Developers need to know the average wind speed at a particular site to design and build the most appropriate turbines. It would be no more prudent to size the turbines for the slowest speed than it would be to size them from the fastest, but infrequently occurring speed. One can get a good idea of this relationship by using the Weibull distribution, a plot of frequency against speed. This distribution helps identify various classes of wind (see the table). The higher the number, the stronger the average speed. A good wind speed is 7 meters per second (mps); 20 mps may be excessive and cause damage to equipment. Turbine manufacturers have a "rated wind speed" for all models and sizes of turbines they sell. Typical rated wind speed requirements are in the range of 8–13 mps, but many machines will produce some power with much slower speeds.[1] Currently, developers are concentrating on class 4 and above as the most promising areas.

1. For an excellent and detailed description of these and other principles, see P. Gipe, *Wind Energy* Comes of Age (New York: John Wiley & Sons, 1995).

Wind Power Classification

Wind power class	Resource potential	Wind power density at 50 meters (in watts per square meter)	Wind speed at 50 meters (in meters per second)	Wind speed at 50 meters (in miles per hour)
2	Marginal	200–300	5.6–6.4	12.5–14.3
3	Fair	300–400	6.4–7.0	14.3–15.7
4	Good	400–500	7.0–7.5	15.7–16.8
5	Excellent	500–600	7.5–8.0	16.8–17.9
6	Outstanding	600–800	8.0–8.8	17.9–19.7
7	Superb	800–1600	8.8–11.1	19.7–24.8

followed proper environmental procedures. Although the suit was eventually abandoned, it was not before the local jurisdictions, including Palm Springs and Riverside County, enacted a long list of required adjustments, stipulating (for example) height limitations, the use of nonglinting paint, reporting mechanisms for endangered species, and the establishment of decommissioning bonds.

A few years later, in an unpredictable turnaround, attitudes changed. This happened once Palm Springs, led by its mayor Sonny Bono, began eyeing wind machines as generators of tax revenue, as well as electricity. With a financial windfall in mind, the city annexed several square miles of land in the middle of the windiest part of the pass, thereby enlarging the city limits and sweeping additional tax revenues into the municipal treasury. Also, counter to the early intuition and opposition of city officials, the wind turbines have become something of a tourist attraction. Organized tours are available, images of wind farms adorn many local postcards, and brochures advertise the Palm Springs wind industry. Even Hollywood producers have incorporated the striking wind energy landscapes in movies and advertisements. These changes reflect the progression in public attitudes toward greater acceptance, although a closer look still finds disgruntled residents who have the original objections. As they point out, we can paint them, size them, sculpt them, and engineer them to a fine edge, but we cannot make them disappear.

The Aesthetic Core

Reactions to wind power tend to be both quick and subjective. While one group fights intrusion, another is organizing visits for enthusiastic tourists. Where one person loathes turbines, that person's neighbors find them fascinating. Whichever reaction prevails in any given location, wind turbines cannot be ignored, for they do not fit naturally upon the land. They are, to apply Massachusetts Institute of Technology historian Leo Marx's famous phrase, "machines in the garden."[15]

The spread of wind power encounters the most strident opposition where it interferes with local land use.

Wind power's development contained a surprise: Among its corps of supporters, no one anticipated the need to defend wind projects. Why did no one foresee objections? We can only speculate, but it seems that the advantages were considered by adherents to be so obvious, especially when compared to nuclear power, that developing a defensive strategy for this new technology seemed superfluous. Supporters failed to recognize how opposite is the signature between the two: Nuclear power is compact and quiet, whereas wind power is expansive and obvious. Reflecting on this difference, resistance to nuclear power accumulated slowly only after a long educational process that culminated with accidents at Three Mile Island and Chernobyl, while resistance to wind power was immediate and instinctive.

Although this difference suggests the heft of visual aesthetics in shaping public opinion, it masks two other ingredients of equal importance. One is the immobility of the resource: Wind moves, windy sites do not. In this way, wind differs from coal and most other fuels, because its nature does not allow it to be extracted and transported for use at a distant site. For wind power to be successful, turbines must be installed where sufficient wind resources exist or not at all. Thus, just like two other resources—geothermal energy and hydropower—the site-specific nature of

wind developments intrinsically invites conflicts with existing or planned land uses. This is true even in deserts, the common dumping ground of society.[16]

The second ingredient in helping form public attitudes toward wind power is the landscape itself. Simply put, some landscapes are more valued than others. Place turbines in sensitive areas—perhaps along the coast or in a national park—and prepare for an uproar. Place them out of view or in low-value areas—sanitary landfills, for example—and opposition diminishes.

These characteristics produce wind power's most intractable challenges. First, owing to resource immobility and the subjectivity of its aesthetic impact, total mitigation is impossible. Second, because environmental competition changes from place to place and from one time to another, generic solutions are few and elusive. Third, because nothing can make turbines invisible, little we do will make them more acceptable to those perceiving land-use interference. There is no escaping the essence of wind turbines: They will always be spinning, pulsing, exoskeletal contraptions that naturally attract the eye.

The foregoing notwithstanding, the future of wind power remains both promising and substantial, if we can identify and follow the appropriate path. Two general strategies suggest themselves: work to bend public opinion in favor of wind power, or install the turbines out of view. The first approach is under way but slow. The second approach can be quicker and would seem to hold promise, but it is being met with mixed results, especially when projects are proposed for offshore locations—the newest tactic to avoid public criticism and maximize profits.

Larger and Larger Turbines

Wind turbines are getting larger and larger. What is driving this trend? To answer this question, we need to know that movement obtains its impetus from the sun; as solar energy strikes the surface of the Earth, it creates differences in pressure. The wind, in turn, moves "downhill" along the pressure gradients that are formed, from higher to lower pressure. Speed increases as the horizontal distance between different pressures is shortened. Wind also typically accelerates when it is constricted, as when it moves through a mountain pass. The faster the wind moves, the more energy it carries, but it is not in a linear function. Rather, it increases with the cube of the wind speed, usually written x^3. This means that a wind speed of 8 meters per second (mps), yields 314 (8^3) watts for every square meter exposed to the wind, while at 16 mps, we get 2,509 (16^3) watts per square meter, again eight times as much. This relationship puts a premium on sites having the strongest winds. This relationship also explains why the area "swept" by the turbine blades is so important and why the wind industry has been striving fervently to increase the scale of the equipment it installs. A one-megawatt turbine at a typical European site would produce enough electricity annually to meet the needs of 700 typical European households.

117

Moving Offshore

The spread of wind power encounters the most strident opposition where it interferes with local land use. Tourism, recreation, entertainment, resorts, and a host of other outdoor activities create most of the challenge because their function is to help people escape reality. For relief from this dilemma, developers are looking for sites offshore, and they have been finding them, especially in the shallow, wind-swept waters of the Irish Sea and North Sea. Denmark, which is characteristically leading the way with this strategy, has already installed and activated several fields of this type. Other projects are in place or planned off Ireland, the United Kingdom, the Netherlands, and several other countries.

In addition to prohibiting wind projects in populated areas, positioning them offshore offers several operational advantages. For example, winds passing over water tend to be stronger than those passing over land; offshore placement removes no land from existing or planned uses; any noise produced at sea is muffled by that of the surf; road use is largely a moot issue; and negotiation with multiple landowners is unnecessary. Nonetheless, offshore placement requires some tradeoffs. For example, offshore equipment is more costly to construct and maintain, and it inherently increases the potential for conflict with any recreational use of the seashore. It also tends to encourage the installation of larger turbines (see the box).

Strictly from a public perspective, offshore placement has the presumed advantage of mitigating complaints about aesthetic intrusion. It has not, however, turned out to be the expected universal remedy. Indeed, moving offshore is increasing rather than diminishing the enmity of wind power in some quarters, especially in the northeastern United States.

Tempted by the strong offshore winds of coastal Massachusetts and responding to the hostility to wind developments witnessed in California, several entrepreneurs advocated placing wind turbines on the shallow offshore banks. The proposal itself may go down as the most foolhardy miscalculation in renewable energy history. The problem, as usual, is incompatible use of space. Called Cape Wind, the project is proposed for Nantucket Sound, a site between the popular vacation spots on Cape Cod and the exclusive holiday retreats of Martha's Vineyard and Nantucket Island. Like development near Palm Springs, Cape Wind is colliding with the wishes of a prosperous and politically astute residential corps bent on protecting existing scenic and recreational qualities that it has come to cherish.

Riding a Roller Coaster

Wind energy has experienced a wild ride over the past 20 years, one where initial enthusiasm soured quickly with the perception that generous incentives and lax oversight were allowing

---

The Cape Wind Project: A Wind Power Lightning Rod

Cape Wind is a proposed $500–$750 million wind development project for Horseshoe Shoal in Nantucket Sound. If approved, the turbines will come within 5 miles of land, spread over an area of 24 square miles, and consist of 130, 417-foot wind turbines connected to a central service platform that includes a helicopter pad and crew quarters. Each turbine blade will be 164 feet long with a total diameter of 328 feet. Each turbine will have a base diameter of 16 feet and an above-water profile taller than the Statue of Liberty.

The proposal has become a lightning rod for the wind industry. The Alliance to Protect Nantucket Sound (the Alliance),[1] which strongly opposes the project, has been accumulating arguments against it. They point out that

- each turbine will have about 150 gallons of hydraulic oil, and the service platform will have at least 30,000 gallons of dielectric oil, and diesel fuel;
- the project will be within the flight path of thousands of small planes; and
- the turbines will pose a navigation hazard to the commercial ferry lines in the area.

These and other objections, however, take a secondary position to the Alliance's primary objection, that of aesthetic intrusion. The Alliance claims that the turbines will be visible for farther than 20 miles, that they will be lighted at night, and that they will flicker with changing sun angle. The Alliance has developed many computer visualizations of how they would appear. (Some proponents might point out that

the Visualizations illustrate how inconsequential the turbines would appear from the beach.)

The pro-development side has not been idle. Cape Wind has its own Web site,[2] which identifies the many benefits of the project, including that it offsets the need for 113 million gallons of oil yearly and creates approximately 600–1,000 new jobs. They also refer to many studies attesting to the benefits of such projects as Cape Wind. One of the most recent references the positive impacts on sea creatures around the wind turbines off the southern Swedish coast.[3] Other Web sites provide many testimonials to the good sense of offshore wind power.[4] The controversy over Cape Wind's offshore proposal is just the beginning of many other anticipated projects along the East Coast, such as off Long Island.[5]

1. http://www.saveoursound.org.
2. http://www.capewind.org.
3. L. Nordstrom, "Windmills off Swedish Coast are Providing Unexpected Benefit for Marine Life, Scientists Say," *Environmental News Network,* 11 February 2004, http://www.enn.com/news/200402-11/s_13011 .asp.
4. windfarm@cleanpowernow.org; http://www.safewind. info.
5. See the Safe Wind Coalition's Web site, http://www. safewind.info/wind_farms_where.htm; and the Long Island Offshore Wind Initiative's site at http://www. lioffshorewindenergy.org/.

---

virtually any wind farm development, no matter how carelessly designed or operated, to be financially tenable. An undertow that quickly started to pull against the early currents of promise was a perception that wind developments were being installed without sufficient public notification, due consideration, or individual benefit. By the late 1980s, it was clear that improved turbines and business situations were going to be necessary if wind power was to develop a significant position in the alternative energy mix in the United States or abroad.

Some of the earliest advances first came into view in Europe, where even casual inspection spotted substantial differences from early installations in California. For example, instead of large clusters of turbines spread haphazardly upon the land, deployment of European turbines was more sensitively organized into smaller groupings that were carefully integrated into the landscape. This was partly a result of a higher sensitivity to existing conditions and partly a measured response to the experiences in California that had dulled the promise of wind power and threatened its future.[17]

Reflecting improvements and continued support, wind power's trajectory is once again upward. Today it is the fastest-growing renewable energy resource in the world.[18] Wind power is especially popular outside the United States: In countries like Germany, it is welcomed, encouraged, and promoted as one way to reduce greenhouse gas emissions. In Denmark, the value of wind power to the economy now exceeds that of its economic mainstay, ham. Spain's development of wind power is currently growing at a faster pace than it is in any other country.

The roller coaster ride is not over, however: Even amid news of improvements and quickened growth, wind power continues to have its critics. The more determined of these opponents work to keep wind machines from their property and out of their view. They hire public relations experts, make abundant use of the Internet to promote their view and attract adherents, and invite the support of prominent citizens to their cause. The group Save Our Sound is perhaps the most visible example of such techniques.[19] Such determined resistance was never envisioned when the champions of wind power came calling more than two decades ago. Today, despite progress in assuaging public apprehensions, a measure of uncertainty still hangs over wind's future.

From Incentives to Independence

What is to be made of the many incentives that wind power enjoys? Tax incentives, utility portfolio standards, feed-in laws,[20] and many other aids currently help make it an economically viable alternative energy provider. Some would say that the requirement for these incentives demonstrates that wind power is not a legitimate competitor for our energy dollars. Others might argue that the mere existence of these aids suggests how narrow the economic gap is between a present need for subsidies and independent viability. While its increasingly competitive status results partly from a rising cost of conventional energy, it also reflects the declining costs of all

alternatives, including wind. The message is this: Even without incentives, wind power has been moving toward economic independence, and it seems destined to reach parity with conventional sources soon.

It is often the smallest margin of help that wins the day for an emerging technology. One way to demonstrate this is to examine the impact of higher conventional energy cost. In one study of 12 Midwestern states, where electricity sold at 4.5 cents per kWh, the regional potential for cost-effective wind power was about 7 percent of current total generation in the United States.[21] If the market would support a price of 5.0 cents per kWh, however, the potential would grow to 177 percent of current generation. If one additional penny is added to the price, the potential blossoms to 14 times current levels.[22]

Until conventional energy makes this inevitable jump, wind operators need another way to bridge the gap. This brings us to the U.S. Production Tax Credit (PTC). This credit originally provided for an inflation-adjusted 1.5 cents per kilowatt-hour for electricity generated with wind turbines. With PTC now at about 1.9 cents, wind projects are economically favored. In its absence, however, development of new projects virtually ceases. This occurred, for instance, when the credit expired at the end of 2001, before it was reinstated some months later. PTC lapsed again on 31 December 2003, and discussions in Congress are once again under way as to whether to extend it for a five-year period.[23]

Without such a credit, the U.S. wind industry will suffer. According to Craig Cox, executive director of the Interwest Energy Alliance, "The lapse of the PTC has created uncertainty in the wind energy marketplace, and interest in new developments has slowed."[24] Renewal of the credit is part of the $31 billion energy bill that stalled in Congress at the end of 2003, again putting the wind energy industry back on its "roller coaster" in the United States. The world's major wind turbine manufacturer, Vestas Group, delayed its decision to build a wind turbine plant in Oregon because of the uncertainty of the credit. Ultimately, such uncertainty spreads to all phases of wind energy development, not just deployment of turbines. "We've been looking to establish a manufacturing facility in the U.S. but have not done that only because of the boom and bust cycle of the wind energy industry in the U.S.," Scott Kringen of Vestas told Reuters.[25] Other spokespersons have made similar observations: "Today, a wide range of U.S. companies are interested in the wind industry, but many are staying on the sidelines because of the on-again, off-again nature of the market produced by frequent expirations of the PTC," said Randall Swisher, executive director of the Washington, DC-based American Wind Energy Association.[26] Most countries offer more stable, longer-term policy support for wind than does the United States, and they use mechanisms that are inherently more pluralistic and egalitarian. This helps explain why wind power is on such a fast track in countries such as Germany, Denmark, and Spain.

Also playing an important role in helping wind power gain a competitive advantage are Renewable Energy Credits, or "green tags."[27] These tags result from laws currently in force in 13 states that require electricity providers to include a prescribed amount

Wind Power and Bird Mortality

There is a persistent public impression that birds and wind-mills don't mix very well. Particularly for the smaller turbines, the spinning blades are hard to see during the day and are invisible at night. Many of those who campaign against wind power expansion cite this concern as part of their argument.

Concerns about turbine-related bird mortality stem largely from the experience at Altamont Pass, California, where approximately 7,000 wind turbines are located on rolling grassland 50 miles east of San Francisco Bay.[1] Between 1989 and 1991, 182 dead birds were found in study plots associated with wind turbines, including approximately 39 golden eagles killed per year by the turbines.[2] Golden eagles, red-tailed hawks, and American kestrels had higher mortality than more common American ravens and turkey vultures.[3] Deaths of eagles and potential danger to endangered California condors are the biggest issues at Altamont Pass. Bird mortality at comparably sized wind facilities has been reported as being similar or lower than those at Altamont Pass.[4]

While such fatalities are regrettable, there is serious question as to whether they are sufficient to slow or halt the use of wind power. One environmental group, The Defenders of Wildlife, recommends that bird mortality should be "kept in perspective."[5] For comparison glass windows kill 100–900 million birds per year; house cats, 100 million; cars and trucks, 50–100 million; transmission line collisions, up to 175 million; agriculture, 67 million; and hunting, more than 100 million.[6] Clean Power Now, an advocacy group encouraging wind development in Nantucket Sound, answers the question "Do wind turbines kill birds?" by stating "Very few and not always."[7] Altamont Pass, where much of the concern for avian safety originated, appears to be more of the exception than the rule. Data show the actual numbers killed in the pass do not exceed one bird per turbine per year, and for raptors, reported kill rates are 0.05 per turbine per year.[8] Nevertheless, the wind power industry has made several adjustments. For example, perch guards are being installed and a program to replace the old machines with modern turbines on high monopoles is ongoing (One modern turbine replaces seven older machines).[9] More study on this matter would be welcome.

1. W. G. Hunt, R. E. Jackman, T. L. Hunt, D. E. Driscoll, and L. Culp, *A Population Study of Golden Eagles in the Altamont Pass Wind Resource Area: Population Trend Analysis 1997,* report prepared for the National Renewable Energy Laboratory (NREL), Subcontract XAT-6-16459-01 (Santa Cruz, CA: Predatory Bird Research Group, University of California, 1998).
2. S. Orloff and A. Flannery, *Wind Turbine Effects on Avian Activity, Habitat Use, and Mortality in Altamont Pass and Solano County WRAs,* report prepared by BioSystems Analysis, Inc. for the California Energy Commission, 1992.
3. C. G. Thelander and L. Rugge, *Avian Risk Behavior and Fatalities at the Altamont Pass Wind Resource Area,* report prepared for the National Renewable Energy Laboratory: SR-500-27545, (Santa Cruz, CA: Predatory Bird Research Group, University of California, 2000).
4. M. D. McCrary et al., *Summary of Southern California Edison's Bird Monitoring Studies in the San Gorgonio Pass,* unpublished data; and R. L. Anderson, J. Tom, N. Neumann, J. A. Cleckler, and J. A. Brownell, *Avian Monitoring and Risk Assessment at Tehachapi Pass Wind Resource Area, California* (Sacramento, CA: California Energy Commission, 1996).
5. Defenders of Wildlife, *Renewable Energy: Wind Energy Resources, Principles and Recommendations,* http://www.defenders.org/habitat/renew/wind.html.
6. Curry & Kerlinger, *What Kills Birds?* http://www.currykerlinger.com/birds.htm.
7. Clean Power Now, *Do Wind Turbines Kill Birds?* http://www.cleanpowernow.org/birdkills.php.
8. P. Kerlinger, *An Assessment of the Impacts of Green Mountain Power Corporation's Wind Power Facility on Breeding and Migrating Birds in Searsburg, Vermont, July 1996–July 1998,* report prepared for the Vermont Department of Public Service, NREL/SR-500-28591 (Golden, CO: NREL, March 2002), page 64.
9. R. C. Curry and P. Kerlinger, *Avian Mitigation Plan: Kenetech Model Wind Turbines, Altamont Pass WRA, California,* report presented at the National Avian Wind Power Planning Meeting III, San Diego, CA, May 1998, page 26.

of renewable electricity in the electric power-supply portfolio they offer to their customers. Electricity providers meet this requirement through several possible approaches. They can generate the necessary amount of renewable electricity themselves, purchase it from someone else, or buy credits from other providers who have excess. The green tags rely almost entirely on private market forces. Taken together with production tax credits and various industry improvements, they are helping wind power continue its trend toward independent profitability. Such status, coupled with reduced public resistance, will move wind power from the realm of alternative to the position of mainstream energy resource. This might be possible if we can move from NIMBY to PIMBY.

From NIMBY to PIMBY

When plans encounter resistance, developers usually make amended suggestions to attract greater support. This would seem to be a sensible tactic for wind developers, but it was not a part of planning for wind power in the mid-1980s. Instead, a naive impression prevailed that wind power would attract unquestioned support. However, the public resisted the blatant

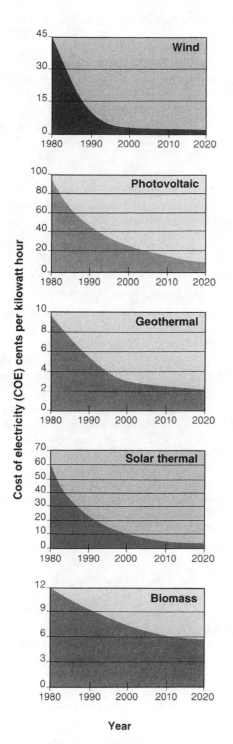

Figure 4 Renewable Energy Cost Trends.

Note: These graphs are reflections of historical cost trends, not precise annual historical data.
Source: National Renewable Energy Laboratory, 2002.

notices of lawsuits for their trouble. NIMBY, even then a battlescarred acronym (for "Not In My Back Yard"), emerged in the headlines.

The ingredients mixed in the cauldron of subsequent wind power development made for a rich and complex brew. On the positive and promising side, developers learned to appreciate the power of public opinion and to work to inform it more completely. This also applied to regulators and policymakers. All parties began ascribing primacy to cooperation over imposition. By the mid-1990s, wind companies successfully improved efficiency and design, and jurisdictional authorities made zoning codes more appropriate if not more restrictive. The use of focus groups and public hearings became common elements in wind development planning procedures. As a result, controversy and press attention subsided, and projects continued to come on line with little fanfare or public notice in Iowa, Kansas, Minnesota, and Texas.

Then came Cape Wind, and much of the old debate began anew. Developers who had forgotten or never fully appreciated the power of public opinion started retreating. Despite many improvements and increasing experience, Cape Wind planners had devoted insufficient attention to considering the combination of factors that make one place unique from another. They reasoned that if offshore installations were meeting with success in Europe, why should they not find acceptance in "green" Massachusetts? However, they failed to realize the poor comparability between the mindset of people in the United States, who live in a spacious and largely post-industrial country, and their European contemporaries who have been living with industrial landscapes, greater population density, and much less personal space for centuries. In making their calculations, they neglected to note that the coastal areas of Massachusetts, heavily utilized for recreation, is not comparable with the lightly settled coastal areas of Europe.

In many ways, the Cape Wind episode is an East Coast version of the California experience 20 years earlier. Admittedly, the setting is different—desert versus ocean—but the underlying problem is the same: wind turbines—immovable and numerous—interfering with the aesthetics of a valued recreational resource.

The experience of Cape Wind suggests the need for a fresh approach to wind power development. The key element of this new approach is simplicity itself: Avoid sites having a high potential for conflict. Making this assessment would involve two steps. The first step would be to assign sites "compatibility rankings," starting with the most compatible sites.

- Rank #1 properties would be those where it is not only suitable but overtly requested for wind development, such as farms in Iowa or Kansas.
- Rank #2 properties would likely be acceptable, such as in southeastern Washington.
- Rank #3 properties might be acceptable in certain circumstances, such as near Palm Springs.
- Rank #4 properties would be completely off-limits, for example, on the top of Mt. Rushmore.

placement of wind turbines on the landscape. It was a particularly unexpected experience because California was known as a state where many of the most ardent environmentalists held forth. Instead of receiving congratulatory handshakes, wind developers (and various government officials) received

Ranks would be determined according to points assigned to site-specific characteristics, including lines of site, type and color tone of terrain, ownership, bird flyways, endangered species, competitive economic value, transmission lines/corridors, protected status (such as national parks), economic development, energy security, and so forth. This should be part of the process of environmental impact assessment, and it should be initiated at any location with a strong, class 4 or above wind resource. Without such rankings, the current ad hoc and contentious approach will continue. It would be akin to a general plan for a city: Variances could be granted, but there would be a broad guidance document in place.

Part two of this plan would be to concentrate our attention on Rank #1 sites. In the United States, this means the Great Plains. There are two simple reasons to emphasize this region. First, the United States' greatest wind resource is there. Second, the small-scale farmers in the area generally welcome the turbines. The message is this: When contentious sites breed contempt, avoid them, at least for now, even if the resource base is attractive and the load centers are nearby. Admit that the wind power alternative is uniquely visible and interferes with scenic vistas and cease trying to force-feed developments down the throats of a resistant public. This is not good for the future of wind power.

On the other hand, there are places where wind power development is welcome. Small farms of the Great Plains have been losing ground for decades to consolidation and the vagaries of weather; they need an economic boost to stay viable. The owners of these farms have put out the welcome mat for wind developers in places such as along Buffalo Ridge, on the border between northwest Iowa and Minnesota, and even farther west in places like Lamar, Colorado. As Chris Rundell, a local rancher, phrased it: "The wind farm has installed a new spirit of community in Lamar . . . it's intangible but very real." They are embracing a new acronym, PIMBY—Please In My Back Yard.

Seeing the wind development in the Great Plains in recent years is a continuation of history, if in a slightly different form: Where a century ago hundreds of thousands of farm windmills made the local agricultural life possible, wind power is again proving its worth to those who would live there. It is bringing needed cash into the local economy and slowing a multiyear trend of farm abandonment and consolidation. The same lands that early wind machines helped develop, new wind machines are helping preserve.[28]

Notes

1. See http://www.state.hi.us/dbedt/ert/wwg/windy.html#molokai.

2. European Wind Energy Association, "Wind Power Expands 23% in Europe but Still Is Only a 3-Member State Story," press release, 3 February 2004, www.ewea.org/documents/0203_EU2003 figures&x005F; final6.pdf.

3. Pacific Northwest Laboratory, *An Assessment of the Available Windy Land Area and Wind Energy Potential in the Contiguous United States* (Washington, DC: U.S. Department of Energy (DOE), 1991).

4. American Wind Energy Association, "Wind Energy: An Untapped Resource," fact sheet, 13 January 2004, http://www.awea.org/pubs/factsheets/WindEnergyAnUntappedResource.pdf.

5. P. Gipe, "More Than First Thought? Wind Report Stirs Minor Tempest," *Renewable Energy World* 6, no. 5 (2003), available at http://www.jxj.com/magsandj/rew/2003_05/wind_report.html. See also C. L. Archer and M. Z. Jacobson, "The Spatial and Temporal Distributions of U.S. Winds and Wind Power at 80 M Derived from Measurements," *Journal of Geophysical Research* 108, no. D9 (2003): 4289.

6. North Dakota has the capacity to produce 138,000 MW; Texas, 136,000; Kansas, 122,000; South Dakota, 117,000 MW; and Montana, 116,000 MW.

7. American Wind Energy Association; see also the National Wind Technology Center Web site (http://www.nrel.gov/wind/).

8. Craig Cox, senior associate, Interwest Energy Alliance, personal communication with author, 27 April 2004.

9. Paul Gipe, executive director, Ontario Sustainable Energy Association, personal communication with author, 20 April 2004.

10. American Wind Energy Association, note 4 above.

11. American Wind Energy Association, note 4 above.

12. Lawrence Flowers, technical director, National Renewable Energy Laboratory, personal communication with author, 22 June 2004.

13. M. J. Pasqualetti and E. Butler, "Public Reaction to Wind Development in California," *International Journal of Ambient Energy* 8, no. 3 (1987): 83–90.

14. M. J. Pasqualetti, "Accommodating Wind Power in a Hostile Landscape," in M. J. Pasqualetti, P. Gipe, and R. Righter, eds., *Wind Power in View: Energy Landscapes in a Crowded World,* (San Diego, CA: Academic Press, 2002), 153–71; M. J. Pasqualetti, "Morality, Space, and the Power of Wind-Energy Landscapes," *The Geographical Review* 90, no. 3 (2001): 381–94; and M. J. Pasqualetti, "Wind Energy Landscapes: Society and Technology in the California Desert," *Society and Natural Resources* 14, no. 8 (2001): 689–99.

15. L. Marx, *The Machine in the Garden* (Oxford, UK: Oxford University Press, 1964).

16. C. C. Reith and B. M. Thomson, eds., *Deserts As Dumps? The Disposal of Hazardous Materials in Arid Ecosystems* (Albuquerque: University of New Mexico Press, 1992).

17. Pasqualetti, "Wind Energy Landscapes: Society and Technology in the California Desert," note 14 above.

18. Total worldwide wind energy installations were 1,000 MW in 1985, 18,000 MW in 2000, and nearly 40,000 MW in 2003, growing at about 35 percent per annum. See Solarbuzz, *Fast Solar Energy Facts: Solar Energy Global,* http://www.solarbuzz.com/FastFactsIndustry.htm.

19. See http://www.saveoursound.org/windspin.html.

20. Today's support started with Public Utility Regulatory Policies Act (PURPA), the 1978 law that promoted alternative energy sources and energy efficiency by requiring utilities to buy power from independent companies that could produce power for less than what it would have cost for the utility to generate the power, the so-called "avoided cost." In the past 20 years, electricity feed-in laws have been popular in Denmark, Germany, Italy, France, Portugal, and Spain. Private generators, or producers, charge a feed-in tariff for the price-per-unit of electricity the suppliers or utility buy. The rate of the tariff is determined by the federal government. In other words, the government sets the price for electricity in the country.

Because the producer is guaranteed a price for the electricity, if he or she meets certain criteria, feed-in laws help attract new generation capacity. During the past decade, Germany became the world leader in wind development. Much of this success is due to *Stromeinspeisungsgesetz,* literally meaning the "Law on Feeding Electricity from Renewable Sources into the Public Network." The original Electricity Feed Law set the price for renewable electricity sources at 90 percent the retail residential price. In 2001, the German feed law was modified to a simple, fixed price for each renewable technology. See http://www.geni.org/globalenergy/policy/renewableenergy/electricityfeed-inlaws/germany/index.shtml.

21. Union of Concerned Scientists, "How Wind Power Works," briefing, http://www.ucsusa.org/CoalvsWind/brief.wind.html.

22. Ibid.

23. The U.S. Production Tax Cut, by its emphasis on the actual generation and transmission of electricity, not just the construction of equipment, provides additional incentives for greater technical efficiency.

24. See http://www.interwestenergy.org.

25. Reuters went on to report that the "Wind industry backers say the gaps have created a roller coaster in U.S. wind production growth because companies become fearful of investing in the alternative energy source. They say the tax-break gaps hamper wind power growth in the United States which grew last year at a rate of only 10 percent compared to global growth of 28 percent." See "Wind Power Tax Credit Expires in December," *Reuters World Environment News,* 27 November 2003, http://www.planetark.com/dailynewsstory.cfm/newsid/22956/newsDate/27-Nov2003/story.htm.

26. *First Quarter Report: Wind Industry Trade Group Sees Little To No Growth In 2004, Following Near-Record Expansion In 2003,* http://www.awea.org/news/news0405121qt.html

27. For more information about green tags for wind power, see http://www.sustainablemarketing.com/wind.php?google.

28. For several economic summaries, see S. Clemmer, *The Economic Development Benefits of Wind Power,* presentation at Harvesting Clean Energy Conference, Boise, Idaho, 10 February 2003, available at http://www.eere.energy.gov/windpoweringamerica/pdfs/wpa/34600_wind_handbook.pdf.

MARTIN J. PASQUALETTI is a professor of geography at Arizona State University in Tempe. His research interests include renewable energy and the landscape impacts of energy development and use. He is a coeditor of and contributor to *Wind Power in View: Energy Landscapes in a Crowded World* (Academic Press, 2002). He also contributed articles on wind power to the *Encyclopedia of Energy* (Academic Press, 2004). His research has appeared in *The Geographical Review and Society and Natural Resources.* He thanks Paul Gipe and Robert Righter for reading the manuscript for this article and offering many helpful suggestions. Pasqualetti may be reached at (480) 965-7533 or via e-mail at pasqualetti@asu.edu.

From *Environment,* September 2004, pp. 23–38. Reprinted by permission of the Helen Dwight Reid Educational Foundation. Published by Heldref Publications, 1319 Eighteenth St., NW, Washington, DC 20036-1802. Copyright © 2004. www.heldref.org

Whither Wind?

A *Journey through the Heated Debate over Wind Power*

CHARLES KOMANOFF

It was a place I had often visited in memory but feared might no longer exist. Orange slabs of calcified sandstone teetered overhead, while before me, purple buttes and burnt mesas stretched over the desert floor. In the distance I could make out southeast Utah's three snowcapped ranges—the Henrys, the Abajos, and, eighty miles to the east, the La Sals, shimmering into the blue horizon.

No cars, no roads, no buildings. Two crows floating on the late-winter thermals. Otherwise, stillness.

Abbey's country. But my country, too. Almost forty years after *Desert Solitaire,* thirty-five since I first came to love this Colorado River plateau, I was back with my two sons, eleven and eight. We had spent four sun-filled days clambering across slickrock in Arches National Park and crawling through the slot canyons of the San Rafael Reef. Now, perched on a precipice above Goblin Valley, stoked on endorphins and elated by the beauty before me, I had what might seem a strange, irrelevant thought: I didn't want windmills here.

Not that any windmills are planned for this Connecticut-sized expanse—the winds are too fickle. But wind energy is never far from my mind these days. As Earth's climate begins to warp under the accumulating effluent from fossil fuels, the increasing viability of commercial-scale wind power is one of the few encouraging developments.

Encouraging to me, at least. As it turns out, there is much disagreement over where big windmills belong, and whether they belong at all.

Fighting fossil fuels, and machines powered by them, has been my life's work. In 1971, shortly after getting my first taste of canyon country, I took a job crunching numbers for what was then a landmark expose of U.S. power plant pollution, *The Price of Power.* The subject matter was drier than dust—emissions data, reams of it, printed out on endless strips of paper by a mainframe computer. Dull stuff, but nightmarish visions of coal-fired smokestacks smudging the crystal skies of the Four Corners kept me working 'round the clock, month after month.

A decade later, as a New York City bicycle commuter fed up with the oil-fueled mayhem on the streets, I began working with the local bicycle advocacy group, Transportation Alternatives, and we soon made our city a hotbed of urban American anti-car activism. The '90s and now the '00s have brought other battles—"greening" Manhattan tenement buildings through energy efficiency and documenting the infernal "noise costs" of Jet Skis, to name two—but I'm still fighting the same fight.

Why? Partly it's knowing the damage caused by the mining and burning of fossil fuels. And there's also the sheer awfulness of machines gone wild, their groaning, stinking combustion engines invading every corner of life. But now the stakes are immeasurably higher. As an energy analyst, I can tell you that the science on global warming is terrifyingly clear: to have even a shot at fending off climate catastrophe, the world must reduce carbon dioxide emissions from fuel burning by at least 50 percent within the next few decades. If poor countries are to have any room to develop, the United States, the biggest emitter by far, needs to cut back by 75 percent.

Although automobiles, with their appetite for petroleum, may seem like the main culprit, the number one climate change agent in the U.S. is actually electricity. The most recent inventory of U.S. greenhouse gases found that power generation was responsible for a whopping 38 percent of carbon dioxide emissions. Yet the electricity sector may also be the least complicated to make carbon free. Approximately three-fourths of U.S. electricity is generated by burning coal, oil, or natural gas. Accordingly, switching that same portion of U.S. electricity generation to nonpolluting sources such as wind turbines, while simultaneously ensuring that our ever-expanding arrays of lights, computers, and appliances are increasingly energy efficient, would eliminate 38 percent of the country's CO_2 emissions and bring us halfway to the goal of cutting emissions by 75 percent.

To achieve that power switch entirely through wind power, I calculate, would require 400,000 windmills rated at 2.5 megawatts each. To be sure, this is a hypothetical figure, since it ignores such real-world issues as limits on power transmission and the intermittency of wind, but it's a useful benchmark just the same.

What would that entail?

To begin, I want to be clear that the turbines I'm talking about are huge, with blades up to 165 feet long mounted on towers rising several hundred feet. Household wind machines like the 100-foot-high Bergey 10-kilowatt BWC Excel with 11-foot blades, the mainstay of the residential and small business wind turbine market, may embody democratic self-reliance and other "small is beautiful" virtues, but we can't look to them to make a real dent in the big energy picture. What dictates the supersizing of windmills are two basic laws of wind physics: a wind turbine's energy potential is proportional to the square of the length of the blades, and to the cube of the speed at which the blades spin. I'll spare you the math, but the difference in blade lengths, the greater wind speeds higher off the ground, and the sophisticated controls available on industrial-scale turbines all add up to a market-clinching five-hundred-fold advantage in electricity output for a giant General Electric or Vestas wind machine.

How much land do the industrial turbines require? The answer turns on what "require" means. An industry rule of thumb is that to maintain adequate exposure to the wind, each big turbine needs space around it of about 60 acres. Since 640 acres make a square mile, those 400,000 turbines would need 37,500 square miles, or roughly all the land in Indiana or Maine.

On the other hand, the land actually occupied by the turbines—their "footprint"—would be far, far smaller. For example, each 3.6-megawatt Cape Wind turbine proposed for Nantucket Sound will rest on a platform roughly 22 feet in diameter, implying a surface area of 380 square feet—the size of a typical one-bedroom apartment in New York City. Scaling that up by 400,000 suggests that just six square miles of land—less than the area of a single big Wyoming strip mine—could house the bases for all of the windmills needed to banish coal, oil, and gas from the U.S. electricity sector.

Of course, erecting and maintaining wind turbines can also necessitate clearing land: ridgeline installations often require a fair amount of deforestation, and then there's the associated clearing for access roads, maintenance facilities, and the like. But there are also now a great many turbines situated on farmland, where the fields around their bases are still actively farmed.

Depending, then, on both the particular terrain and how the question is understood, the land area said to be needed for wind power can vary across almost four orders of magnitude. Similar divergences of opinion are heard about every other aspect of wind power, too. Big wind farms kill thousands of birds and bats . . . or hardly any, in comparison to avian mortality from other tall structures such as skyscrapers. Industrial wind machines are soft as a whisper from a thousand feet away, and even up close their sound level would rate as "quiet" on standard noise charts . . . or they can sound like "a grinding noise" or "the shrieking sound of a wild animal," according to one unhappy neighbor of an upstate New York wind farm. Wind power developers are skimming millions via subsidies, state-mandated quotas, and "green power" scams . . . or are boldly risking their own capital to strike a blow for clean energy against the fossil fuel Goliath.

Some of the bad press is warranted. The first giant wind farm, comprising six thousand small, fast-spinning turbines placed directly in northern California's principal raptor flyway, Altamont Pass, in the early 1980s rightly inspired the epithet "Cuisinarts for birds." The longer blades on newer turbines rotate more slowly and thus kill far fewer birds, but bat kills are being reported at wind farms in the Appalachian Mountains; as many as two thousand bats were hacked to death at one forty-four-turbine installation in West Virginia. And as with any machine, some of the nearly ten thousand industrial-grade windmills now operating in the U.S. may groan or shriek when something goes wrong. Moreover, wind power does benefit from a handsome federal subsidy; indeed, uncertainty over renewal of the "production tax credit" worth 1.9 cents per kilowatt-hour nearly brought wind power development to a standstill a few years ago.

At the same time, however, there is an apocalyptic quality to much anti-wind advocacy that seems wildly disproportionate to the actual harm, particularly in the overall context of not just other sources of energy and modern industry in general. New York State opponents of wind farms call their website "Save Upstate New York," as if ecological or other damage from wind turbines might administer the coup de grÃ¢ce to the state's rural provinces that decades of industrialization and pollution, followed by outsourcing, have not. In neighboring Massachusetts, a group called Green Berkshires argues that wind turbines" are enormously destructive to the environment," but does not perform the obvious comparison to the destructiveness of fossil fuelâ "based power. Although the intensely controversial Cape Wind project "poses an imminent threat to navigation and raises many serious maritime safety issues," according to the anti-wind Alliance to Protect Nantucket Sound, the alliance was strangely silent when an oil barge bound for the region's electric power plant spilled ninety-eight thousand gallons of its deadly, gluey cargo into Buzzards Bay three years ago.

Of course rhetoric is standard fare in advocacy, particularly the environmental variety with its salvationist mentality—environmentalists always like to feel they are "saving" this valley or that species. It all comes down to a question of what we're saving, and for whom. You can spend hours sifting through the anti-wind websites and find no mention at all of the climate crisis, let alone wind power's potential to help avert it.

In fact, many wind power opponents deny that wind power displaces much, if any, fossil fuel burning. Green Berkshires insists, for example, that "global warming [and] dependence on fossil fuels . . . will not be ameliorated one whit by the construction of these turbines on our mountains."

This notion is mistaken. It is true that since wind is variable, individual wind turbines can't be counted on to produce on demand, so the power grid can't necessarily retire fossil fuel generators at the same rate as it takes on windmills. The coal- and oil-fired generators will still need to be there, waiting for a windless day. But when the wind blows, those generators can spin down. That's how the grid works: it allocates electrons. Supply more electrons from one source, and other sources can supply fewer. And since system operators program the grid to draw from the lowest-cost generators first, and wind power's "fuel," moving air, is free, wind-generated electrons are given priority. It follows that more electrons from wind power mean proportionately fewer from fossil fuel burning.

What about the need to keep a few power stations burning fuel so they can instantaneously ramp up and counterbalance fluctuations in wind energy output? The grid requires this ballast, known as spinning reserve, in any event both because demand is always changing and because power plants of any type are subject to unforeseen breakdowns. The additional variability due to wind generation is slight—wind speeds don't suddenly drop from strong to calm, at least not for every turbine in a wind farm and certainly not for every wind farm on the grid. The clear verdict of the engineers responsible for grid reliability—a most conservative lot—is that the current level of wind power development will not require additional spinning reserve, while even much larger supplies of wind-generated electricity could be accommodated through a combination of energy storage technologies and improved models for predicting wind speeds.

With very few exceptions, then, wind output can be counted on to displace fossil fuel burning one for one. No less than other nonpolluting technologies like bicycles or photovoltaic solar cells, wind power is truly an anti-fossil fuel.

I made my first wind farm visit in the fall of 2005. I had seen big windmills up close in Denmark, and I had driven through the big San Gorgonio wind farm that straddles Highway I-10 near Palm Springs, California. But this trip last November had a mission. After years of hearing industrial wind turbines in the northeastern United States characterized as either monstrosities or crowns of creation, I wanted to see for myself how they sat on the land. I also wanted to measure the noise from the turning blades, so I brought the professional noise meter I had used in my campaign against Jet Skis.

Madison County occupies the broad middle of New York State, with the Catskill Mountains to the south, Lake Ontario to the northwest, and the Adirondacks to the northeast. Its rolling farms sustain seventy thousand residents and, since 2001, two wind farms, the 20-windmill Fenner Windpower Project in the western part of the county and the 7-windmill Madison Windpower Project twenty miles east.

At the time of my visit Fenner was the state's largest wind farm, although that distinction has since passed to the 120-windmill Maple Ridge installation in the Tug Hill region farther north. It was windy that day, though not unusually so, according to the locals. All twenty-seven turbines were spinning, presumably at their full 1.5-megawatt ratings. For me the sight of the turning blades was deeply pleasing. The windmills, sleek, white structures more than three hundred feet tall sprinkled across farmland, struck me as graceful and marvelously useful. I thought of a story in the *New York Times* about a proposed wind farm near Cooperstown, New York, in which a retiree said that seeing giant windmills near your house "would be like driving through oil derricks to get to your front door." To my eye, the Fenner turbines were anti-derricks, oil rigs running in reverse.

For every hour it was in full use, each windmill was keeping a couple of barrels of oil, or an entire half-ton of coal, in the ground. Of course wind turbines don't generate full power all the time because the wind doesn't blow at a constant speed. The Madison County turbines have an average "capacity factor," or annual output rate, of 34 percent, meaning that over the course of a year they generate about a third of the electricity they would produce if they always ran at full capacity. But that still means an average three thousand hours a year of full output for each turbine. Multiply those hours by the twenty-seven turbines at Fenner and Madison, and a good 200,000 barrels of oil or 50,000 tons of coal were being kept underground by the two wind farms each year—enough to cover an entire sixty-acre farm with a six-inch-thick oil slick or pile of coal.

The windmills, spinning easily at fifteen revolutions per minute—that's one leisurely revolution every four seconds—were clean and elegant in a way that no oil derrick or coal dragline could ever be. The nonlinear arrangement of the Fenner turbines situated them comfortably among the traditional farmhouses, paths, and roads, while at Madison, a grassy hillside site, the windmills were more prominent but still unaggressive. Unlike a ski run, say, or a power line cutting through the countryside, the windmills didn't seem like a violation of the landscape. The turning vanes called to mind a natural force—the wind—in a way that a cell phone or microwave tower, for example, most certainly does not.

They were also relatively quiet. My sound readings, taken at distances ranging from one hundred to two thousand feet from the tower base, topped out at 64 decibels and went as low as 45—the approximate noise range given for a small-town residential cul-de-sac on standard noise charts. It's fair to say that the wind turbines in Madison County aren't terribly noisy even from up close and are barely audible from a thousand feet or more away. The predominant sound was a low, not unpleasant hum, or hvoohmm, like a distant seashore, but perhaps a bit thicker.

Thinking back on that November day, I've come to realize that a windmill, like any large structure, is a signifier. Cell-phone towers signify the intrusion of quotidian life—the reminder to stop off at the 7-Eleven, the unfinished business at the office. The windmills I saw in upstate New York signified, for me, not just displacement of destructive fossil fuels, but acceptance of the conditions of inhabiting the Earth. They signified, in the words of environmental lawyer and MIT research affiliate William Shutkin, "the capacity of environmentalists—of citizens—to match their public positions with the private choices necessary to move toward a more environmentally and economically sustainable way of life."

The notion of choices points to another criticism of wind turbines: the argument that the energy they might make could be saved instead through energy-efficiency measures. The Adirondack Council, for example, in a statement opposing the 10-windmill Barton Mines project on a former mountaintop mine site writes, "If the Barton project is approved, we will gain 27 to 30 megawatts of new, clean power generation. Ironically, we could save more than 30 megawatts of power in the Adirondack Park through simple, proven conservation methods in homes and businesses."

The council's statement is correct, of course. Kilowatts galore could be conserved in any American city or town by swapping out incandescent light bulbs in favor of compact fluorescents, replacing inefficient kitchen appliances, and extinguishing "vampire" loads by plugging watt-sucking electronic devices into on-off power strips. If this notion sounds familiar, it's because it has been raised in virtually every power plant dispute since the 1970s. But the ground has shifted, now that we have such overwhelming proof that we're standing on the threshold of catastrophic climate change.

Those power plant debates of yore weren't about fuels and certainly not about global warming, but about whether to top off the grid with new megawatts of supply or with "negawatts"—watts that could be saved through conservation. It took decades of struggle by legions of citizen advocates and hundreds of experts (I was one) to embed the negawatt paradigm in U.S. utility planning. But while we were accomplishing that, inexorably rising fossil fuel use here and around the world was overwhelming Earth's "carbon sinks," causing carbon dioxide to accumulate in the atmosphere at an accelerating rate, contributing to disasters such as Hurricane Katrina and Europe's 2003 heat wave, and promising biblical-scale horrors such as a waning Gulf Stream and disappearing polar icepacks.

The energy arena of old was local and incremental. The new one is global and all-out. With Earth's climate, and the world as we know and love it, now imperiled, topping off the regional grid pales in comparison to the task at hand. In the new, ineluctable struggle to rescue the climate from fossil fuels, efficiency and "renewables" (solar and biomass as well as wind) must all be pushed to the max. Those thirty negawatts that lie untapped in the kitchens and TV rooms of Adirondack houses are no longer an alternative to the Barton wind farm—they're another necessity.

In this new, desperate, last-chance world—and it is that, make no mistake—pleas like the Adirondack Council's, which once would have seemed reasonable, now sound a lot like fiddling while the Earth burns. The same goes for the urgings by opponents of Cape Wind and other pending wind farms to "find a more suitable site"; those other suitable wind farm sites (wherever they exist) need to be developed in addition to, not instead of, Nantucket Sound, or Barton Mines, or the Berkshires.

There was a time when the idea of placing immense turbines in any of these places would have filled me with horror. But now, what horrifies me more is the thought of keeping them windmill free.

Part of the problem with wind power, I suspect, is that it's hard to weigh the effects of any one wind farm against the greater problem of climate change. It's much easier to comprehend the immediate impact of wind farm development than the less tangible losses from a warming Earth. And so the sacrifice is difficult, and it becomes progressively harder as rising affluence brings ever more profligate uses of energy.

Picture this: Swallowing hard, with deep regret for the change in a beloved landscape formerly unmarked in any obvious way by humankind, you've just cast the deciding affirmative vote to permit a wind farm on the hills outside your town. On the way home you see a new Hummer in your neighbor's driveway. How do you not feel like a self-sacrificing sucker?

Intruding the unmistakable human hand on any landscape for wind power is, of course, a loss in local terms, and no small one, particularly if the site is a verdant ridgeline. Uplands are not just visible markers of place but fragile environments, and the inevitable access roads for erecting and serving the turbines can be damaging ecologically as well as symbolically. In contrast, few if any benefits of the wind farm will be felt by you in a tangible way. If the thousands of tons of coal a year that your wind farm will replace were being mined now, a mile from your house, it might be a little easier to take. Unfortunately, our society rarely works that way. The bread you cast upon the waters with your vote will not come back to you in any obvious way—it will be eaten in Wyoming, or Appalachia. And you may just have to mutter an oath about the Hummer and use your moral imagination to console yourself about the ridge.

But what if the big push for wind power simply "provides more energy for people to waste?" as Carl Safina, an oceanographer who objects to the Cape Wind project, asked me recently. Safina is unusual among Cape Wind opponents, not just because he is a MacArthur Fellow and prize-winning author (*Song for the Blue Ocean, Voyage of the Turtle*), but because he is completely honest about the fact that his objections are essentially aesthetic.

"I believe the aesthetics of having a national seashore with a natural view of the blue curve of the planet are very important," he wrote in an e-mail from coastal Long Island, where he lives. "I think turbines and other structures should be sited in places not famed for natural beauty"—a statement that echoed my feelings about Utah's Goblin Valley.

"Six miles is a very short distance over open water," Safina continued, referring to the span from the public beach at Craigville on Cape Cod to the closest proposed turbine, "and a group of anything several hundred feet high would completely dominate the view." While the prominence of the turbines when seen from the shore is open to debate (the height of a Cape Wind tower from six miles would be just two-thirds of one degree, not quite half the width of your finger held at arm's length), there is no question that the wind turbines would, in his words, "put an end to the opportunity for people to experience an original view of a piece of the natural world in one of America's most famously lovely coastal regions."

Yet for all his fierce attachment to that view, Safina says he might give it up if doing so made a difference. "If there was a national energy strategy that would make the U.S. carbon neutral in fifty years," he wrote, "and if Cape Wind was integral and significant, that might be a worthwhile sacrifice." But the reality, as Safina described in words that could well have been mine, is that "Americans insist on wasting energy and needing more. We will affect the natural view of a famously beautiful piece of America's ocean and still not develop a plan to conserve energy."

Safina represents my position and, I imagine, that of others on both sides of the wind controversy when he pleads for federal action that could justify local sacrifice for the greater good. If Congress enacted an energy policy that harnessed the spectrum of cost-effective energy efficiency together with renewable energy, thereby ensuring that fossil fuel use shrank starting today, a windmill's contribution to climate protection might actually register, providing psychic reparation for an altered viewshed. And if carbon fuels were taxed for their damage to the climate, wind power's profit margins would widen, and surrounding communities could extract bigger tax revenues from wind farms. Then some of that bread upon the waters would indeed come back—in the form of a new high school, or land acquired for a nature preserve.

I t's very human to ask, "Why me? Why my ridgeline, my seascape, my viewshed?" These questions have been difficult to answer; there has been no framework—local or national—to guide wind farm siting by ranking potential wind power locales for their ecological and community suitability. That's a gap that the Appalachian Mountain Club is trying to bridge, using its home state of Massachusetts as a model.

According to AMC research director Kenneth Kimball, who heads the project, Massachusetts has ninety-six linear miles of "Class 4" ridgelines, where wind speeds average fourteen miles per hour or more, the threshold for profitability with current technology. Assuming each mile can support seven to nine large turbines of roughly two megawatts each, the state's uplands could theoretically host 1,500 megawatts of wind power. (Coastal areas such as Nantucket Sound weren't included in the survey.)

Kimball's team sorted all ninety-six miles into four classes of governance—Appalachian Trail corridor or similar lands where development is prohibited; other federal or state conservation lands; Massachusetts open space lands; and private holdings. They then overlaid these with ratings denoting conflicts with recreational, scenic, and ecological values. The resulting matrix suggests the following rankings of wind power suitability:

1. *Unsuitable*—lands where development is prohibited (Appalachian Trail corridors, for example) or "high conflict" areas: 24 miles (25 percent).
2. *Less than ideal*—federal or state conservation lands rated "medium conflict": 21 miles (22 percent).
3. *Conditionally favorable*—Conservation or open space lands rated "low conflict," or open space or private lands rated "medium conflict": 27 miles (28 percent).
4. *Most favorable*—Unrestricted private land and "low conflict" areas: 24 miles (25 percent).

Category 4 lands are obvious places to look for wind farm development. Category 3 lands could also be considered, says the AMC, if wind farms were found to improve regional air quality, were developed under a state plan rather than piecemeal, and were bonded to assure eventual decommissioning. If these conditions were met, then categories 3 and 4, comprising approximately fifty miles of Massachusetts ridgelines, could host four hundred wind turbines capable of supplying nearly 4 percent of the state's annual electricity—without grossly endangering wildlife or threatening scenic, recreational, or ecological values (e.g., critical habitat, roadless areas, rare species, old growth, steep slopes).

Whether that 4 percent is a little or a lot depends on where you stand and, equally, on where we stand as a society. You could call the four hundred turbines mere tokenism against our fuel-besotted way of life, and considering them in isolation, you'd be right. But you could also say this: Go ahead and halve the state's power usage, as could be done even with present-day technology, and "nearly 4 percent" doubles to 7–8 percent. Add the Cape Wind project and other offshore wind farms that might follow, and wind power's statewide share might reach 20 percent, the level in Denmark.

Moreover, the windier and emptier Great Plains states could reach 100 percent wind power or higher, even with a suitability framework like the AMC's, thereby becoming net exporters of clean energy. But even at 20 percent, Massachusetts would be doing its part to displace that 75 percent of U.S. electricity generated by fossil fuels. If you spread the turbines needed to achieve that goal across all fifty states, you'd be looking to produce roughly eight hundred megawatt-hours of wind output per square mile—just about what Massachusetts would be generating in the above scenario. And the rest of New England and New York could do the same, affording these "blue" states a voice in nudging the rest of the country greenward.

S o goes my notion, anyway. You could call it wind farms as signifiers, with their value transcending energy-share percentages to reach the realm of symbols and images. That is where we who love nature and obsess about the environment have lost the high ground, and where *Homo americanus* has been acting out his (and her) disastrous desires—opting for the "manly" SUV over the prim Prius, the macho powerboat over the meandering canoe, the stylish halogen lamp over the dorky compact fluorescent.

Throughout his illustrious career, wilderness champion David Brower called upon Americans "to determine that an untrammeled wildness shall remain here to testify that this generation had love for the next." Now that all wild things and all places are threatened by global warming, that task is more complex.

Could a windmill's ability to "derive maximum benefit out of the site-specific gift nature is providing—wind and open space," in the words of aesthetician Yuriko Saito, help Americans bridge the divide between pristine landscapes and sustainable ones? Could windmills help Americans subscribe to the "higher order of beauty" that environmental educator David Orr defines as something that "causes no ugliness somewhere else or at some later time"? Could acceptance of wind farms be our generation's way of avowing our love for the next?

I believe so. Or want to.

CHARLES KOMANOFF, an economic policy analyst and environmental activist, is the author of *Power Plant Cost Escalation.* He lives in New York City and advocates for energy efficiency, bicycle transportation, and urban revitalization.

The Rise of Renewable Energy

Solar cells, wind turbines and biofuels are poised to become major energy sources. New policies could dramatically accelerate that evolution.

DANIEL M. KAMMEN

No plan to substantially reduce greenhouse gas emissions can succeed through increases in energy efficiency alone. Because economic growth continues to boost the demand for energy—more coal for powering new factories, more oil for fueling new cars, more natural gas for heating new homes—carbon emissions will keep climbing despite the introduction of more energy-efficient vehicles, buildings and appliances. To counter the alarming trend of global warming, the U.S. and other countries must make a major commitment to developing renewable energy sources that generate little or no carbon.

Renewable energy technologies were suddenly and briefly fashionable three decades ago in response to the oil embargoes of the 1970s, but the interest and support were not sustained. In recent years, however, dramatic improvements in the performance and affordability of solar cells, wind turbines and biofuels—ethanol and other fuels derived from plants—have paved the way for mass commercialization. In addition to their environmental benefits, renewable sources promise to enhance America's energy security by reducing the country's reliance on fossil fuels from other nations. What is more, high and wildly fluctuating prices for oil and natural gas have made renewable alternatives more appealing.

We are now in an era where the opportunities for renewable energy are unprecedented, making this the ideal time to advance clean power for decades to come. But the endeavor will require a long-term investment of scientific, economic and political resources. Policymakers and ordinary citizens must demand action and challenge one another to hasten the transition.

Let the Sun Shine

Solar Cells, also known as photovoltaics, use semiconductor materials to convert sunlight into electric current. They now provide just a tiny slice of the world's electricity: their global generating capacity of 5,000 megawatts (MW) is only 0.15 percent of the total generating capacity from all sources. Yet sunlight could potentially supply 5,000 times as much energy as the world currently consumes. And thanks to technology improvements, cost declines and favorable policies in many states and nations, the annual production of photovoltaics has increased by more than 25 percent a year for the past decade and by a remarkable 45 percent in 2005. The cells manufactured last year added 1,727 MW to worldwide generating capacity, with 833 MW made in Japan, 353 MW in Germany and 153 MW in the U.S.

Solar cells can now be made from a range of materials, from the traditional multicrystalline silicon wafers that still dominate the market to thin-film silicon cells and devices composed of plastic or organic semiconductors. Thin-film photovoltaics are cheaper to produce than crystalline silicon cells but are also less efficient at turning light into power. In laboratory tests, crystalline cells have achieved efficiencies of 30 percent or more; current commercial cells of this type range from 15 to 20 percent. Both laboratory and commercial efficiencies for all kinds of solar cells have risen steadily in recent years, indicating that an expansion of research efforts would further enhance the performance of solar cells on the market.

Overview

- Thanks to advances in technology, renewable sources could soon become large contributors to global energy.
- To hasten the transition, the U.S. must significantly boost its R&D spending on energy.
- The U.S. should also levy a fee on carbon to reward clean energy sources over those that harm the environment.

Solar photovoltaics are particularly easy to use because they can be installed in so many places—on the roofs or walls of homes and office buildings, in vast arrays in the desert, even sewn into clothing to power portable electronic devices. The state of California has joined Japan and Germany in leading a global push for solar installations; the "Million Solar Roof" commitment is intended to create 3,000 MW of new generating capacity in the state by 2018. Studies done by my research group, the Renewable and Appropriate Energy Laboratory at the University of California, Berkeley, show that annual production of solar photovoltaics in the U.S. alone could grow to 10,000 MW in just 20 years if current trends continue.

The biggest challenge will be lowering the price of the photovoltaics, which are now relatively expensive to manufacture. Electricity produced by crystalline cells has a total cost of 20 to 25 cents per kilowatt-hour, compared with four to six cents for coal-fired electricity, five to seven cents for power produced by burning natural gas, and six to nine cents for biomass power plants. (The cost of nuclear power is harder to pin down because experts disagree on which expenses to include in the analysis; the estimated range is two to 12 cents per kilowatt-hour.) Fortunately, the prices of solar cells have fallen consistently over the past decade, largely because of improvements in manufacturing processes. In Japan, where 290 MW of solar generating capacity were added in 2005 and an even larger amount was exported, the cost of photovoltaics has declined 8 percent a year; in California, where 50 MW of solar power were installed in 2005, costs have dropped 5 percent annually.

Surprisingly, Kenya is the global leader in the number of solar power systems installed per capita (but not the number of watts added). More than 30,000 very small solar panels, each producing only 12 to 30 watts, are sold in that country annually. For an investment of as little as $100 for the panel and wiring, the system can be used to charge a car battery, which can then provide enough power to run a fluorescent lamp or a small black-and-white television for a few hours a day. More Kenyans adopt solar power every year than make connections to the country's electric grid. The panels typically use solar cells made of amorphous silicon; although these photovoltaics are only half as efficient as crystalline cells, their cost is so much lower (by a factor of at least four) that they are more affordable and useful for the two billion

---

5,000 megawatts
Global generating capacity of solar power

37 percent
Top efficiency of experimental solar cells

20 to 25 cents
Cost per kilowatt-hour of solar power

Growing Fast, but Still a Sliver

Solar cells, wind power and biofuels are rapidly gaining traction in the energy markets, but they remain marginal providers compared with fossil-fuel sources such as coal, natural gas and oil.

The Renewable Boom

Since 2000 the commercialization of renewable energy sources has accelerated dramatically. The annual global production of solar cells, also known as photovoltaics, jumped 45 percent in 2005. The construction of new wind farms, particularly in Europe, has boosted the worldwide generating capacity of wind power 10-fold over the past decade. And the production of ethanol, the most common biofuel, soared to 36.5 billion liters last year, with the lion's share distilled from American-grown corn.

Jen Christiansen; Sources; PV News, BTM Consult, AWEA, EWEA, F.O. Light and *BP Statistical Review of World Energy* 2006.

Hot Power from Mirrors

Solar-thermal systems, long used to provide hot water for homes or factories, can also generate electricity. Because these systems produce power from solar heat rather than light, they avoid the need for expensive photovoltaics.

Solar Concentrator

A solar-thermal array consists of thousands of dish-shaped solar concentrators, each attached to a Stirling engine that converts heat to electricity. The mirrors in the concentrator are positioned to focus reflected sunlight on the Stirling engine's receiver.

Stirling Engine

A high-performance Stirling engine shuttles a working fluid, such as hydrogen gas, between two chambers. The cold chamber is separated from the hot chamber by a regenerator that maintains the temperature difference between them. Solar energy from the receiver heats the gas in the hot chamber, causing it to expand and move the hot piston. This piston then reverses direction, pushing the heated gas into the cold chamber. As the gas cools, the cold piston can easily compress it, allowing the cycle to start a new. The movement of the pistons drives a turbine that generates electricity in an alternator.

Don Foley, Source: U.S. Department of Energy

people worldwide who currently have no access to electricity. Sales of small solar power systems are booming in other African nations as well, and advances in low-cost photovoltaic manufacturing could accelerate this trend.

Furthermore, photovoltaics are not the only fast-growing form of solar power. Solar-thermal systems, which collect sunlight to generate heat, are also undergoing a resurgence. These systems have long been used to provide hot water for homes or factories, but they can also produce electricity without the need for expensive solar cells. In one design, for example, mirrors focus light on a Stirling engine, a high-efficiency device containing a working fluid that circulates between hot and cold chambers. The fluid expands as the sunlight heats it, pushing a piston that, in turn, drives a turbine.

In the fall of 2005 a Phoenix company called Stirling Energy Systems announced that it was planning to build two large solar-thermal power plants in southern California. The company signed a 20-year power purchase agreement with Southern California Edison, which will buy the electricity from a 500-MW solar plant to be constructed in the Mojave Desert. Stretching across 4,500 acres, the facility will include 20,000 curved dish mirrors, each concentrating light on a Stirling engine about the size of an oil barrel. The plant is expected to begin operating in 2009 and could later be expanded to 850 MW. Stirling Energy Systems also signed a 20-year contract with San Diego Gas & Electric to build a 300-MW, 12,000-dish plant in the Imperial Valley. This facility could eventually be upgraded to 900 MW.

The financial details of the two California projects have not been made public, but electricity produced by present solar-thermal technologies costs between five and 13 cents per kilowatt-hour, with dish-mirror systems at the upper end of that range. Because the projects involve highly reliable technologies and mass production, however, the generation expenses are expected to ultimately drop closer to four to six cents per kilowatt-hour—that is, competitive with the current price of coal-fired power.

Blowing in the Wind

Wind power has been growing at a pace rivaling that of the solar industry. The worldwide generating capacity of wind turbines has increased more than 25 percent a year, on average, for the past decade, reaching nearly 60,000 MW in 2005. The growth has been nothing short of explosive in Europe—between 1994 and 2005, the installed wind power capacity in European Union nations jumped from 1,700 to 40,000 MW. Germany alone has more than 18,000 MW of capacity thanks to an aggressive construction program. The northern German state of Schleswig-Holstein currently meets one quarter of its annual electricity demand with more than 2,400 wind turbines, and in certain months wind power provides more than half the state's electricity. In addition,

---

60,000 megawatts
Global generating capacity of wind power

0.5 percent
Fraction of U.S. electricity produced by wind turbines

1.9 cents
Tax credit for wind power, per kilowatt-hour of electricity

Spain has 10,000 MW of wind capacity, Denmark has 3,000 MW, and Great Britain, the Netherlands, Italy and Portugal each have more than 1,000 MW.

In the U.S. the wind power industry has accelerated dramatically in the past five years, with total generating capacity leaping 36 percent to 9,100 MW in 2005. Although wind turbines now produce only 0.5 percent of the nation's electricity, the potential for expansion is enormous, especially in the windy Great Plains states. (North Dakota, for example, has greater wind energy resources than Germany, but only 98 MW of generating capacity is installed there.) If the U.S. constructed enough wind farms to fully tap these resources, the turbines could generate as much as 11 trillion kilowatt-hours of electricity, or nearly three times the total amount produced from all energy sources in the nation last year. The wind industry has developed increasingly large and efficient turbines, each capable of yielding 4 to 6 MW. And in many locations, wind power is the cheapest form of new electricity, with costs ranging from four to seven cents per kilowatt-hour.

The growth of new wind farms in the U.S. has been spurred by a production tax credit that provides a modest subsidy equivalent to 1.9 cents per kilowatt-hour, enabling wind turbines to compete with coal-fired plants. Unfortunately, Congress has repeatedly threatened to eliminate the tax credit. Instead of instituting a long-term subsidy for wind power, the lawmakers have extended the tax credit on a year-to-year basis, and the continual uncertainty has slowed investment in wind farms. Congress is also threatening to derail a proposed 130-turbine farm off the coast of Massachusetts that would provide 468 MW of generating capacity, enough to power most of Cape Cod, Martha's Vineyard and Nantucket.

The reservations about wind power come partly from utility companies that are reluctant to embrace the new technology and partly from so-called NIMBY-ism. ("NIMBY" is an acronym for Not in My Backyard.) Although local concerns over how wind turbines will affect landscape views may have some merit, they must be balanced against the social costs of the alternatives. Because society's energy needs are growing relentlessly, rejecting wind farms often means requiring the construction or expansion of fossil fuel-burning power plants that will have far more devastating environmental effects.

16.2 billion

Liters of ethanol produced in the U.S. in 2005

2.8 percent

Ethanol's share of all automobile fuel by volume

$2 billion

Annual subsidy for corn-based ethanol

Green Fuels

Researchers are also pressing ahead with the development of biofuels that could replace at least a portion of the oil currently consumed by motor vehicles. The most common biofuel by far in the U.S. is ethanol, which is typically made from corn and blended with gasoline. The manufacturers of ethanol benefit from a substantial tax credit: with the help of the $2-billion annual subsidy, they sold more than 16 billion liters of ethanol in 2005 (almost 3 percent of all automobile fuel by volume), and production is expected to rise 50 percent by 2007. Some policymakers have questioned the wisdom of the subsidy, pointing to studies showing that it takes more energy to harvest the corn and refine the ethanol than the fuel can deliver to combustion engines. In a recent analysis, though, my colleagues and I discovered that some of these studies did not properly account for the energy content of the by-products manufactured along with the ethanol. When all the inputs and outputs were correctly factored in, we found that ethanol has a positive net energy of almost five megajoules per liter.

We also found, however, that ethanol's impact on greenhouse gas emissions is more ambiguous. Our best estimates indicate that substituting corn-based ethanol for gasoline reduces greenhouse gas emissions by 18 percent, but the analysis is hampered by large uncertainties regarding certain agricultural practices, particularly the environmental costs of fertilizers. If we use different assumptions about

Plugging Hybrids

The environmental benefits of renewable biofuels would be even greater if they were used to fuel plug-in hybrid electric vehicles (PHEVs). Like more conventional gasoline-electric hybrids, these cars and trucks combine internal-combustion engines with electric motors to maximize fuel efficiency, but PHEVs have larger batteries that can be recharged by plugging them into an electrical outlet. These vehicles can run on electricity alone for relatively short trips; on longer trips, the combustion engine kicks in when the batteries no longer have sufficient juice. The combination can drastically reduce gasoline consumption: whereas conventional sedans today have a fuel economy of about 30 miles per gallon (mpg) and nonplug-in hybrids such as the Toyota Prius average about 50 mpg, PHEVs could get an equivalent or 80 to 160 mpg. Oil use drops still further if the combustion engines in PHEVs run on biofuel blends such as E85, which is a mixture of 15 percent gasoline and 85 percent ethanol.

If the entire U.S. vehicle fleet were replaced overnight with PHEVs, the nation's oil consumption would decrease by 70 percent or more, completely eliminating the need for petroleum imports. The switch would have equally profound implications for protecting the earth's fragile climate, not to mention the elimination of smog. Because most of the energy for cars would come from the electric grid instead of from fuel tanks, the environmental impacts would be concentrated in a few thousand power plants instead of in hundreds of millions of vehicles. This shift would focus the challenge of climate protection squarely on the task of reducing the greenhouse gas emissions from electricity generation.

PHEVs could also be the salvation of the ailing American auto industry. Instead of continuing to lose market share to foreign companies, U.S. automakers could become competitive again by retooling their factories to produce PHEVs that are significantly more fuel-efficient than the nonplug-in hybrids now sold by Japanese companies. Utilities would also benefit from the transition because most owners of PHEVs would recharge their cars at night, when power is cheapest, thus helping to smooth the sharp peaks and valleys in demand for electricity. In California, for example, the replacement of 20 million conventional cars with PHEVs would increase nighttime electricity demand to nearly the same level as daytime demand, making far better use of the grid and the many power plants that remain idle at night. In addition, electric vehicles not in use during the day could supply electricity to local distribution networks at times when the grid was under strain. The potential benefits to the electricity industry are so compelling that utilities may wish to encourage PHEV sales by offering lower electricity rates for recharging vehicle batteries.

Most important, PHEVs are not exotic vehicles of the distant future. DaimlerChrysler has already introduced a PHEV prototype, a plug-in hybrid version of the Mercedes-Benz Sprinter Van that has 40 percent lower gasoline consumption than the conventionally powered model. And PHEVs promise to become even more efficient as new technologies improve the energy density of batteries, allowing the vehicles to travel farther on electricity alone.

—D.M.K.

these practices, the results of switching to ethanol range from a 36 percent drop in emissions to a 29 percent increase. Although corn-based ethanol may help the U.S. reduce its reliance on foreign oil, it will probably not do much to slow global warming unless the production of the biofuel becomes cleaner.

But the calculations change substantially when the ethanol is made from cellulosic sources: woody plants such as switchgrass or poplar. Whereas most makers of corn-based ethanol burn fossil fuels to provide the heat for fermentation, the producers of cellulosic ethanol burn lignin—an unfermentable part of the organic material—to heat the plant sugars. Burning lignin does not add any greenhouse gases to the atmosphere, because the emissions are offset by the carbon dioxide absorbed during the growth of the plants used to make the ethanol. As a result, substituting cellulosic ethanol for gasoline can slash greenhouse gas emissions by 90 percent or more.

Another promising biofuel is so-called green diesel. Researchers have produced this fuel by first gasifying biomass—heating organic materials enough that they release hydrogen and carbon monoxide—and then converting these compounds into long-chain hydrocarbons using the Fischer-Tropsch process. (During World War II, German engineers employed these chemical reactions to make synthetic motor

fuels out of coal.) The result would be an economically competitive liquid fuel for motor vehicles that would add virtually no greenhouse gases to the atmosphere. Oil giant Royal Dutch/Shell is currently investigating the technology.

The Need for R&D

Each of these renewable sources is now at or near a tipping point, the crucial stage when investment and innovation, as well as market access, could enable these attractive but generally marginal providers to become major contributors to regional and global energy supplies. At the same time, aggressive policies designed to open markets for renewables are taking hold at city, state and federal levels around the world. Governments have adopted these policies for a wide variety of reasons: to promote market diversity or energy security, to bolster industries and jobs, and to protect the environment on both the local and global scales. In the U.S. more than 20 states have adopted standards setting a minimum for the fraction of electricity that must be supplied with renewable sources. Germany plans to generate 20 percent of its electricity from renewables by 2020, and Sweden intends to give up fossil fuels entirely.

Even President George W. Bush said, in his now famous State of the Union address this past January, that the U.S. is

The Least Bad Fossil Fuel

Although renewable energy sources offer the best way to radically cut greenhouse gas emissions, generating electricity from natural gas instead of coal can significantly reduce the amount of carbon added to the atmosphere. Conventional coal-fired power plants emit 0.25 kilogram of carbon for every kilowatt-hour generated. (More advanced coal-fired plants produce about 20 percent less carbon.) But natural gas (CH_4) has a higher proportion of hydrogen and a lower proportion of carbon than coal does. A combined-cycle power plant that burns natural gas emits only about 0.1 kilogram of carbon per kilowatt-hour.

Unfortunately, dramatic increases in natural gas use in the U.S. and other countries have driven up the cost of the fuel. For the past decade, natural gas has been the fastest-growing source of fossil-fuel energy, and it now supplies almost 20 percent of America's electricity. At the same time, the price of $2.50 to $3 per million Btu in 1997 to more than $7 per million Btu today.

The price increases have been so alarming that in 2003, then Federal Reserve Board Chair Alan Greenspan warned that the U.S. faced a natural gas crisis. The primary solution proposed by the White House and some in Congress was to increase gas production. The 2005 Energy Policy Act included large subsidies to support gas producers, increase exploration and expand imports

of liquefied natural gas (LNG). These measures, however, may not enhance energy security, because most of the imported LNG would come from some of the same OPEC countries that supply petroleum to the U.S. Furthermore, generating electricity from even the cleanest natural gas power plants would still emit too much carbon to achieve the goal of keeping carbon dioxide in the atmosphere below 450 to 550 parts per million by volume. (Higher levels could have disastrous consequences for the global climate.)

Improving energy efficiency and developing renewable sources can be faster, cheaper and cleaner and provide more security than developing new gas supplies. Electricity from a wind farm costs less than that produced by a natural gas power plant if the comparison factors in the full cost of plant construction and forecasted gas prices. Also, wind farms and solar arrays can be built more rapidly than large-scale natural gas plants. Most critically, diversity of supply is America's greatest ally in maintaining a competitive and innovative energy sector. Promoting renewable sources makes sense strictly on economic grounds, even before the environmental benefits are considered.

Jen Christiansen; Source: President's Committee of Advisors on Science and Technology: —D.M.K.

R&D is Key

Spending on research and development in the U.S. energy sector has fallen steadily since its peak in 1980. Studies of patent activity suggest that the drop in funding has slowed the development of renewable energy technologies. For example, the number of successful patent applications in photovoltaics and wind power has plummeted as R&D spending in these fields has declined.

Jen Christiansen; Source: Reversing the incredible shrinking energy R&D Budget;

"addicted to oil." And although Bush did not make the link to global warming, nearly all scientists agree that humanity's addiction to fossil fuels is disrupting the earth's climate. The time for action is now, and at last the tools exist to alter energy production and consumption in ways that simultaneously benefit the economy and the environment. Over the past 25 years, however, the public and private funding of research and development in the energy sector has withered. Between 1980 and 2005 the fraction of all U.S. R&D spending devoted to energy declined from 10 to 2 percent. Annual public R&D funding for energy sank from $8 billion to $3 billion (in 2002 dollars); private R&D plummeted from $4 billion to $1 billion.

To put these declines in perspective, consider that in the early 1980s energy companies were investing more in R&D than were drug companies, whereas today investment by energy firms is an order of magnitude lower. Total private R&D funding for the entire energy sector is less than that of a single large biotech company. (Amgen, for example, had R&D expenses of $2.3 billion in 2005.) And as R&D spending dwindles, so does innovation. For instance, as R&D funding for photovoltaics and wind power has slipped over the past quarter of a century, the number of successful patent applications in these fields has fallen accordingly. The lack of attention to long-term research and planning has significantly weakened our nation's ability to respond to the challenges of climate change and disruptions in energy supplies.

Calls for major new commitments to energy R&D have become common. A 1997 study by the President's Committee of Advisors on Science and Technology and a 2004 report by the bipartisan National Commission on Energy Policy both recommended that the federal government double its R&D spending on energy. But would such an expansion be enough? Probably not. Based on assessments of the cost to stabilize the amount of carbon dioxide in the atmosphere and other studies that estimate the success of energy R&D programs and the resulting savings from the technologies that would emerge, my research group has calculated that public funding of $15 billion to $30 billion a year would be required—a fivefold to 10-fold increase over current levels.

Greg F. Nemet, a doctoral student in my laboratory, and I found that an increase of this magnitude would be roughly comparable to those that occurred during previous federal R&D initiatives such as the Manhattan Project and the Apollo program, each of which produced demonstrable economic benefits in addition to meeting its objectives. American energy companies could also boost their R&D spending by a factor of 10, and it would still be below the average for U.S. industry overall. Although government funding is essential to supporting early-stage technologies, private-sector R&D is the key to winnowing the best ideas and reducing the barriers to commercialization.

Raising R&D spending, though, is not the only way to make clean energy a national priority. Educators at all grade levels, from kindergarten to college, can stimulate public interest and activism by teaching how energy use and production affect the social and natural environment. Nonprofit organizations can establish a series of contests that would reward the first company or private group to achieve a challenging and worthwhile energy goal, such as constructing a building or appliance that can generate its own power or developing a commercial vehicle that can go 200 miles on a single gallon of fuel. The contests could be modeled after the Ashoka awards for pioneers in public policy and the Ansari X Prize for the developers of space vehicles. Scientists and entrepreneurs should also focus on finding clean, affordable ways to meet the energy needs of people in the developing world. My colleagues and I, for instance, recently detailed the environmental benefits of improving cooking stoves in Africa.

But perhaps the most important step toward creating a sustainable energy economy is to institute market-based schemes to make the prices of carbon fuels reflect their social cost. The use of coal, oil and natural gas imposes a huge collective toll on society, in the form of health care expenditures for ailments caused by air pollution, military spending to secure oil supplies, environmental damage from mining operations, and the potentially devastating economic impacts of global warming. A fee on carbon emissions would provide a simple, logical and transparent method to reward renewable, clean energy sources over those that harm the economy and the environment. The tax revenues could pay for some of the social costs of carbon emissions, and a portion could be designated to compensate low-income families who spend a larger share of their income on energy. Furthermore, the carbon fee could be combined with a cap-and-trade program that would set limits on carbon emissions but also allow the cleanest energy suppliers to sell permits to their dirtier competitors. The federal government has used such programs with great success to curb other pollutants, and several northeastern states are already experimenting with greenhouse gas emissions trading.

Best of all, these steps would give energy companies an enormous financial incentive to advance the development and commercialization of renewable energy sources. In essence, the U.S. has the opportunity to foster an entirely new industry. The threat of climate change can be a rallying cry for a clean-technology revolution that would strengthen the country's manufacturing base, create thousands of jobs and alleviate our international trade deficits—instead of importing foreign oil, we can export high-efficiency vehicles, appliances, wind turbines and photovoltaics. This transformation can turn the nation's energy sector into something that was once deemed impossible: a vibrant, environmentally sustainable engine of growth.

DANIEL M. KAMMIEN is Class of 1935 Distinguished Professor of Energy at the University of California, Berkeley, where he holds appointments in the Energy and Resources Group, the Goldman School of Public Policy and the department of nuclear engineering. He is founding director of the Renewable and Appropriate Energy Laboratory and co-director of the Berkeley Institute of the Environment.

The Future of Nuclear Power
An Interdisciplinary MIT Study

STEPHEN ANSOLABEHERE, ET AL.

The generation of electricity from fossil fuels, notably natural gas and coal, is a major and growing contributor to the emission of carbon dioxide—a greenhouse gas that contributes significantly to global warming. We share the scientific consensus that these emissions must be reduced and believe that the U.S. will eventually join with other nations in the effort to do so.

At least for the next few decades, there are only a few realistic options for reducing carbon dioxide emissions from electricity generation:

- increase efficiency in electricity generation and use;
- expand use of renewable energy sources such as wind, solar, biomass, and geothermal;
- capture carbon dioxide emissions at fossil-fueled (especially coal) electric generating plants and permanently sequester the carbon; and
- increase use of nuclear power.

The goal of this interdisciplinary MIT study is not to predict which of these options will prevail or to argue for their comparative advantages. In *our view, it is likely that we shall need all of these options and accordingly it would be a mistake at this time to exclude any of these four options from an overall carbon emissions management strategy.* Rather we seek to explore and evaluate actions that could be taken to maintain nuclear power as one of the significant options for meeting future world energy needs at low cost and in an environmentally acceptable manner.

In our view, it would be a mistake at this time to exclude any of these four options from an overall carbon emissions management strategy.

In 2002, nuclear power supplied 20% of United States and 17% of world electricity consumption. Experts project worldwide electricity consumption will increase substantially in the coming decades, especially in the developing world, accompanying economic growth and social progress. However, official forecasts call for a mere 5% increase in nuclear electricity generating capacity worldwide by 2020 (and even this is questionable), while electricity use could grow by as much as 75%. These projections entail little new nuclear plant construction and reflect both economic considerations and growing anti-nuclear sentiment in key countries. The limited prospects for nuclear power today are attributable, ultimately, to four unresolved problems:

- *Costs: nuclear power has higher overall lifetime costs* compared to natural gas with combined cycle turbine technology (CCGT) and coal, at least in the absence of a carbon tax or an equivalent "cap and trade" mechanism for reducing carbon emissions;
- *Safety: nuclear power has perceived adverse safety, environmental, and health effects,* heightened by the 1979 Three Mile Island and 1986 Chernobyl reactor accidents, but also by accidents at fuel cycle facilities in the United States, Russia, and Japan. There is also growing concern about the safe and secure transportation of nuclear materials and the security of nuclear facilities from terrorist attack;
- *Proliferation: nuclear power entails potential security risks,* notably the possible misuse of commercial or associated nuclear facilities and operations to acquire technology or materials as a precursor to the acquisition of a nuclear weapons capability. Fuel cycles that involve the chemical reprocessing of spent fuel to separate weapons-usable plutonium and uranium enrichment

technologies are of special concern, especially as nuclear power spreads around the world;

- *Waste: nuclear power has unresolved challenges in long-term management of radioactive wastes.* The United States and other countries have yet to implement final disposition of spent fuel or high level radioactive waste streams created at various stages of the nuclear fuel cycle. Since these radioactive wastes present some danger to present and future generations, the public and its elected representatives, as well as prospective investors in nuclear power plants, properly expect continuing and substantial progress towards solution to the waste disposal problem. Successful operation of the planned disposal facility at Yucca Mountain would ease, but not solve, the waste issue for the U.S. and other countries if nuclear power expands substantially.

We believe the nuclear option should be retained, precisely because it is an important carbon-free source of power.

Today, nuclear power is not an economically competitive choice. Moreover, unlike other energy technologies, nuclear power requires significant government involvement because of safety, proliferation, and waste concerns. If in the future carbon dioxide emissions carry a significant "price," however, nuclear energy could be an important—indeed

Global Growth Scenario

Region	Projected 2050 GWe Capacity	Nuclear Electricity Market Share	
		2000	2050
Total World	**1,000**	**17%**	**19%**
Developed world	625	23%	29%
U.S.	300		
Europe & Canada	210		
Developed East Asia	115		
FSU	50	16%	23%
Developing world	325	2%	11%
China, India, Pakistan	200		
Indonesia, Brazil, Mexico	75		
Other developing countries	50		

Projected capacity comes from the global electricity demand scenario in Appendix 2, which entails growth in global electricity consumption from 13.6 to 38.7 trillion kWhrs from 2000 to 2050 (2.1% annual growth). The market share in 2050 is predicated on 85% capacity factor for nuclear power reactors. Note that China, India, and Pakistan are nuclear weapons capable states. Other developing countries includes as leading contributors Iran, South Africa, Egypt, Thailand, Philippines, and Vietnam.

vital—option for generating electricity. We do not know whether this will occur. But *we believe the nuclear option should be retained, precisely because it is an important carbon-free source of power that can potentially make a significant contribution to future electricity supply.*

To preserve the nuclear option for the future requires overcoming the four challenges described above—costs, safety, proliferation, and wastes. These challenges will escalate if a significant number of new nuclear generating plants are built in a growing number of countries. The effort to overcome these challenges, however, is justified only if nuclear power can potentially contribute significantly to reducing global warming, which entails major expansion of nuclear power. In effect, preserving the nuclear option for the future means planning for growth, as well as for a future in which nuclear energy is a competitive, safer, and more secure source of power.

To explore these issues, our study postulates a *global growth scenario* that by mid-century would see 1000 to 1500 reactors of 1000 megawatt-electric (MWe) capacity each deployed worldwide, compared to a capacity equivalent to 366 such reactors now in service. Nuclear power expansion on this scale requires U.S. leadership, continued commitment by Japan, Korea, and Taiwan, a renewal of European activity, and wider deployment of nuclear power around the world. An illustrative deployment of 1000 reactors, each 1000 MWe in size, under this scenario is given in following table.

This scenario would displace a significant amount of carbon-emitting fossil fuel generation. In 2002, carbon equivalent emission from human activity was about 6,500 million tonnes per year; these emissions will probably more than double by 2050. The 1000 GWe of nuclear power postulated here would avoid annually about 800 million tonnes of carbon equivalent if the electricity generation displaced was gas-fired and 1,800 million tonnes if the generation was coal-fired, assuming no capture and sequestration of carbon dioxide from combustion sources.

Fuel Cycle Choices

A critical factor for the future of an expanded nuclear power industry is the choice of the fuel cycle—what type of fuel is used, what types of reactors "burn" the fuel, and the method of disposal of the spent fuel. This choice affects all four key problems that confront nuclear power—costs, safety, proliferation risk, and waste disposal. For this study, we examined three representative nuclear fuel cycle deployments:

We believe that the world-wide supply of uranium ore is sufficient to fuel the deployment of 1,000 reactors over the next half century.

Fuel Cycle Types and Ratings

Economics		Waste	Proliferation	Safety	
				Reactor	Fuel Cycle
Once through	+	× short term − long term	+	×	+
Closed thermal	−	− short term + long term	−	×	−
Closed fast	−	− short term + long term	−	+ to −	−

+ means relatively advantageous; × means relatively neutral; − means relatively disadvantageous

This table indicates broadly the relative advantage and disadvantage among the different type of nuclear fuel cycles. It does not indicate relative standing with respect to other electricity-generating technologies, where the criteria might be quite different (for example, the nonproliferation criterion applies only to nuclear).

- *conventional thermal reactors operating in a "once-through" mode,* in which discharged spent fuel is sent directly to disposal;
- *thermal reactors with reprocessing in a "closed" fuel cycle,* which means that waste products are separated from unused fissionable material that is re-cycled as fuel into reactors. This includes the fuel cycle currently used in some countries in which plutonium is separated from spent fuel, fabricated into a mixed plutonium and uranium oxide fuel, and recycled to reactors for one pass[1];
- *fast reactors[2] with reprocessing in a balanced "closed" fuel cycle,* which means thermal reactors operated world-wide in "once-through" mode and a balanced number of fast reactors that destroy the actinides separated from thermal reactor spent fuel. The fast reactors, reprocessing, and fuel fabrication facilities would be co-located in secure nuclear energy "parks" in industrial countries.

Closed fuel cycles extend fuel supplies. The viability of the once-through alternative in a global growth scenario depends upon the amount of uranium resource that is available at economically attractive prices. *We believe that the world-wide supply of uranium ore is sufficient to fuel the deployment of 1000 reactors over the next half century* and to maintain this level of deployment over a 40 year lifetime of this fleet. This is an important foundation of our study, based upon currently available information and the history of natural resource supply.

The result of our detailed analysis of the relative merits of these representative fuel cycles with respect to key evaluation criteria can be summarized as follows: *The once through cycle has advantages in cost, proliferation, and fuel cycle safety,* and is disadvantageous only in respect to long-term waste disposal; the two closed cycles have clear advantages only in long-term aspects of waste disposal, and disadvantages in cost, short-term waste issues, proliferation

risk, and fuel cycle safety. (See Table.) Cost and waste criteria are likely to be the most crucial for determining nuclear power's future.

We have not found, and based on current knowledge do not believe it is realistic to expect, that there are new reactor and fuel cycle technologies that simultaneously overcome the problems of cost, safety, waste, and proliferation.

Our analysis leads to a significant conclusion: *The once-through fuel cycle best meets the criteria of low costs and proliferation resistance.* Closed fuel cycles may have an advantage from the point of view of long-term waste disposal and, if it ever becomes relevant, resource extension. But closed fuel cycles will be more expensive than once-through cycles, until ore resources become very scarce. This is unlikely to happen, even with significant growth in nuclear power, until at least the second half of this century, and probably considerably later still. Thus our most important recommendation is:

> For the next decades, government and industry in the U.S. and elsewhere should give priority to the deployment of the once-through fuel cycle, rather than the development of more expensive closed fuel cycle technology involving reprocessing and new advanced thermal or fast reactor technologies.

This recommendation implies a major re-ordering of priorities of the U.S. Department of Energy nuclear R&D programs.

Public Attitudes Toward Nuclear Power

Expanded deployment of nuclear power requires public acceptance of this energy source. Our review of survey results shows that a majority of Americans and Europeans oppose building new nuclear power plants to meet future energy needs. To understand why, we surveyed 1350 adults in the U.S. about their attitudes toward energy in general and nuclear power in particular. Three important and unexpected results emerged from that survey:

- The U.S. public's attitudes are informed almost entirely by their perceptions of the technology, rather than by politics or by demographics such as income, education, and gender.
- The U.S. public's views on nuclear waste, safety, and costs are critical to their judgments about the future deployment of this technology. Technological improvements that lower costs and improve safety and waste problems can increase public support substantially.
- In the United States, people do not connect concern about global warming with carbon-free nuclear power. There is no difference in support for building more nuclear power plants between those who are very concerned about global warming and those who are not. Public education may help improve understanding about the link between global warming, fossil fuel usage, and the need for low-carbon energy sources.

There are two implications of these findings for our study: first, the U.S. public is unlikely to support nuclear power expansion without substantial improvements in costs and technology. Second, the carbon-free character of nuclear power, the major motivation for our study, does not appear to motivate the U.S. general public to prefer expansion of the nuclear option.

The U.S. public is unlikely to support nuclear power expansion without substantial improvements in costs and technology.

Comparative Power Costs

Case (Year 2002 $)	Real Levelized Cost Cents/kWe-hr
Nuclear (LWR)	6.7
+ Reduce construction cost 25%	5.5
+ Reduce construction time 5 to 4 years	5.3
+ Further reduce O&M to 13 mills/kWe-hr	5.1
+ Reduce cost of capital to gas/coal	4.2
Pulverized Coal	4.2
CCGT[a] (low gas prices, $3.77/MCF)	3.8
CCGT (moderate gas prices, $4.42/MCF)	4.1
CCGT (high gas prices, $6.72/MCF)	5.6

a. Gas costs reflect real, levelized acquisition cost per thousand cubic feet (MCF) over the economic life of the project.

Power Costs with Carbon Taxes

Carbon Tax Cases Levelized Electricity Cost cents/kWe-hr	$50/ tonne C	$100/ tonne C	$200/ tonne C
Coal	5.4	6.6	9.0
Gas (low)	4.3	4.8	5.9
Gas (moderate)	4.7	5.2	6.2
Gas (high)	6.1	6.7	7.7

Economics

Nuclear power will succeed in the long run only if it has a lower cost than competing technologies. This is especially true as electricity markets become progressively less subject to economic regulation in many parts of the world. We constructed a model to evaluate the real cost of electricity from nuclear power versus pulverized coal plants and natural gas combined cycle plants (at various projected levels of real lifetime prices for natural gas), over their economic lives. These technologies are most widely used today and, absent a carbon tax or its equivalent, are less expensive than many renewable technologies. Our "merchant" cost model uses assumptions that commercial investors would be expected to use today, with parameters based on actual experience rather than engineering estimates of what might be achieved under ideal conditions; it compares the constant or "levelized" price of electricity over the life of a power plant that would be necessary to cover all operating expenses and taxes and provide an acceptable return to investors. The comparative figures given below assume 85% capacity factor and a 40-year economic life for the nuclear plant, reflect economic conditions in the U.S., and consider a range of projected improvements in nuclear cost factors. (See Table.)

We judge the indicated cost improvements for nuclear power to be plausible, but not proven. The model results make clear why electricity produced from new nuclear power plants today is not competitive with electricity produced from coal or natural gas-fueled CCGT plants with low or moderate gas prices, unless *all* cost improvements for nuclear power are realized. The cost comparison becomes worse for nuclear if the capacity factor falls. It is also important to emphasize that the nuclear cost structure is driven by high up-front capital costs, while the natural gas cost driver is the fuel cost; coal lies in between nuclear and natural gas with respect to both fuel and capital costs.

Nuclear does become more competitive by comparison if the social cost of carbon emissions is internalized, for example through a carbon tax or an equivalent "cap and trade" system. Under the assumption that the costs of carbon emissions are imposed, the accompanying table illustrates the impact on the competitive costs for different power sources, for emission costs in the range of $50 to $200/tonne carbon. (See Table.) The ultimate cost will depend on both societal choices (such as how much carbon dioxide emission to permit) and technology developments, such as the cost and feasibility of large-scale carbon capture and long-term sequestration. Clearly, costs in the range of $100 to $200/tonne C would significantly affect the relative cost competitiveness of coal, natural gas, and nuclear electricity generation.

The carbon-free nature of nuclear power argues for government action to encourage maintenance of the nuclear option, particularly in light of the regulatory uncertainties facing the use of nuclear power and the unwillingness of investors to bear the risk of introducing a new generation of nuclear facilities with their high capital costs.

We recommend three actions to improve the economic viability of nuclear power:

> The government should cost share for site banking for a number of plants, certification of new plant designs by the Nuclear Regulatory Commission, and combined construction and operating licenses for plants built immediately or in the future; we support U.S. Department of Energy initiatives on these subjects.

> The government should recognize nuclear as carbon-free and include new nuclear plants as an eligible option in any federal or state mandatory renewable energy portfolio (i.e., a "carbon-free" portfolio) standard.

> The government should provide a modest subsidy for a small set of "first mover" commercial nuclear plants to demonstrate cost and regulatory feasibility in the form of a production tax credit.

We propose a production tax credit of up to $200 per kWe of the construction cost of up to 10 "first mover" plants. This benefit might be paid out at about 1.7 cents per kWe-hr, over a year and a half of full-power plant operation. We prefer the production tax credit mechanism because it offers the greatest incentive for projects to be completed and because it can be extended to other carbon free electricity technologies, for example renewables, (wind currently enjoys a 1.7 cents per kWe-hr tax credit for ten years) and coal with carbon capture and sequestration. The credit of 1.7 cents per kWe-hr is equivalent to a credit of $70 per avoided metric ton of carbon if the electricity were to have come from coal plants (or $160 from natural gas plants). Of course, the carbon emission reduction would then continue without public assistance for the plant life (perhaps 60 years for nuclear). If no new nuclear plant is built, the government will not pay a subsidy.

These actions will be effective in stimulating additional investment in nuclear generating capacity if, and only if, the industry can live up to its own expectations of being able to reduce considerably capital costs for new plants.

Advanced fuel cycles add considerably to the cost of nuclear electricity. We considered reprocessing and one-pass fuel recycle with current technology, and found the fuel cost, including waste storage and disposal charges, to be about 4.5 times the fuel cost of the once-through cycle. Thus use of advanced fuel cycles imposes a significant economic penalty on nuclear power.

Safety

We believe the safety standard for the global growth scenario should maintain today's standard of less than one serious release of radioactivity accident for 50 years from all fuel cycle activity. This standard implies a ten-fold reduction in the expected frequency of serious reactor core accidents, from 10^{-4}/reactor year to 10^{-5}/reactor year. This reactor safety standard should be possible to achieve in new light water reactor plants that make use of advanced safety designs. International adherence to such a standard is important, because an accident in any country will influence public attitudes everywhere. The extent to which nuclear facilities should be hardened to possible terrorist attack has yet to be resolved.

We do not believe there is a nuclear plant design that is totally risk free. In part, this is due to technical possibilities; in part due to workforce issues. Safe operation requires effective regulation, a management committed to safety, and a skilled work force.

The high temperature gas-cooled reactor is an interesting candidate for reactor research and development because there is already some experience with this system, although not all of it is favorable. This reactor design offers safety advantages because the high heat capacity of the core and fuel offers longer response times and precludes excessive temperatures that might lead to release of fission products; it also has an advantage compared to light water reactors in terms of proliferation resistance.

These actions will be effective in stimulating additional investment in nuclear generating capacity if, and only if, the industry can live up to its own expectations of being able to reduce considerably overnight capital costs for new plants.

Because of the accidents at Three Mile Island in 1979 and Chernobyl in 1986, a great deal of attention has focused on reactor safety. However, the safety record of reprocessing plants is not good, and there has been little safety analysis of fuel cycle facilities using, for example, the probabilistic risk assessment method. More work is needed here.

Our principal recommendation on safety is:

> The government should, as part of its near-term R&D program, develop more fully the capabilities to analyze life-cycle health and safety impacts of fuel

cycle facilities and focus reactor development on options that can achieve enhanced safety standards and are deployable within a couple of decades.

Waste Management

The management and disposal of high-level radioactive spent fuel from the nuclear fuel cycle is one of the most intractable problems facing the nuclear power industry throughout the world. No country has yet successfully implemented a system for disposing of this waste. We concur with the many independent expert reviews that have concluded that geologic repositories will be capable of safely isolating the waste from the biosphere. However, implementation of this method is a highly demanding task that will place great stress on operating, regulatory, and political institutions.

We do not believe a convincing case can be made, on the basis of waste management considerations alone, that the benefits of advanced, closed fuel cycles will outweigh the attendant safety, environmental, and security risks and economic costs.

For fifteen years the U.S. high-level waste management program has focused almost exclusively on the proposed repository site at Yucca Mountain in Nevada. Although the successful commissioning of the Yucca Mountain repository would be a significant step towards the secure disposal of nuclear waste, we believe that a broader, strategically balanced nuclear waste program is needed to prepare the way for a possible major expansion of the nuclear power sector in the U.S. and overseas.

The global growth scenario, based on the once-through fuel cycle, would require multiple disposal facilities by the year 2050. To dispose of the spent fuel from a steady state deployment of one thousand 1 GWe reactors of the light water type, new repository capacity equal to the nominal storage capacity of Yucca Mountain would have to be created somewhere in the world every three to four years. This requirement, along with the desire to reduce longterm risks from the waste, prompts interest in advanced, closed fuel cycles.

These schemes would separate or partition plutonium and other actinides—and possibly certain fission products—from the spent fuel and transmute them into shorter-lived and more benign species. The goals would be to reduce the thermal load from radioactive decay of the waste on the repository, thereby increasing its storage capacity, and to shorten the time for which the waste must be isolated from the biosphere.

We have analyzed the waste management implications of both once-through and closed fuel cycles, taking into account each stage of the fuel cycle and the risks of radiation exposure in both the short and long-term. *We do not believe that a convincing case can be made on the basis of waste management considerations alone that the benefits of partitioning and transmutation will outweigh the attendant safety, environmental, and security risks and economic costs.* Future technology developments could change the balance of expected costs, risks, and benefits. For our fundamental conclusion to change, however, not only would the expected long term risks from geologic repositories have to be significantly higher than those indicated in current assessments, but the incremental costs and short-term safety and environmental risks would have to be greatly reduced relative to current expectations and experience.

We further conclude that waste management strategies in the once-through fuel cycle are potentially available that could yield long-term risk reductions at least as great as those claimed for waste partitioning and transmutation, with fewer short-term risks and lower development and deployment costs. These include both incremental improvements to the current mainstream mined repositories approach and more far-reaching innovations such as deep borehole disposal. Finally, replacing the current ad hoc approach to spent fuel storage at reactor sites with an explicit strategy to store spent fuel for a period of several decades will create additional flexibility in the waste management system.

Our principal recommendations on waste management are:

The DOE should augment its current focus on Yucca Mountain with a balanced long-term waste management R&D program.

A research program should be launched to determine the viability of geologic disposal in deep boreholes within a decade.

A network of centralized facilities for storing spent fuel for several decades should be established in the U.S. and internationally.

Nonproliferation

Nuclear power should not expand unless the risk of proliferation from operation of the commercial nuclear fuel cycle is made acceptably small. We believe that nuclear power can expand as envisioned in our global growth scenario with acceptable incremental proliferation risk, provided that reasonable safeguards are adopted and that deployment of reprocessing and enrichment are restricted. The international community must prevent the acquisition of weapons-usable material, either by diversion (in the case of plutonium) or by misuse of fuel cycle facilities (including related facilities, such as research reactors or hot cells). Responsible governments must control, to the extent possible, the know-how relevant to produce and process either highly enriched uranium (enrichment technology) or plutonium.

Three issues are of particular concern: existing stocks of *separated* plutonium around the world that are directly usable for weapons; nuclear facilities, for example in Russia, with inadequate controls; and transfer of technology, especially enrichment and reprocessing technology, that brings nations closer to a nuclear weapons capability. The proliferation risk of the global growth scenario is underlined by the likelihood that use of nuclear power would be introduced and expanded in many countries in different security circumstances.

Nuclear power should not expand unless the risk of proliferation from operation of the commercial nuclear fuel cycle is made acceptably small.

An international response is required to reduce the proliferation risk. The response should:

- re-appraise and strengthen the institutional underpinnings of the IAEA safeguards regime in the near term, including sanctions;
- guide nuclear fuel cycle development in ways that reinforce shared nonproliferation objectives.

Accordingly, we recommend:

The International Atomic Energy Agency (IAEA) should focus overwhelmingly on its safeguards function and should be given the authority to carry out inspections beyond declared facilities to suspected illicit facilities;

Greater attention must be given to the proliferation risks at the front end of the fuel cycle from enrichment technologies;

IAEA safeguards should move to an approach based on continuous materials protection, control and accounting using surveillance and containment systems, both in facilities and during transportation, and should implement safeguards in a risk-based framework keyed to fuel cycle activity;

Fuel cycle analysis, research, development, and demonstration efforts must include explicit analysis of proliferation risks and measures defined to minimize proliferation risks;

International spent fuel storage has significant nonproliferation benefits for the growth scenario and should be negotiated promptly and implemented over the next decade.

Analysis, Research, Development, and Demonstration Program

The U.S. Department of Energy (DOE) analysis, research, development, and demonstration (ARD&D) program should support the technology path leading to the global growth scenario and include diverse activities that balance risk and time scales, in pursuit of the strategic objective of preserving the nuclear option. *For technical, economic, safety, and public acceptance reasons, the highest priority in fuel cycle ARD&D, deserving first call on available funds, lies with efforts that enable robust deployment of the once-through fuel cycle.* The current DOE program does not have this focus.

Every industry in the United States develops basic analytical models and tools such as spreadsheets that allow firms, investors, policy makers, and regulators to understand how changes in the parameters of a process will affect the performance and cost of that process. But we have been struck throughout our study by the absence of such models and simulation tools that permit indepth, quantitative analysis of trade-offs between different reactor and fuel cycle choices, with respect to all key criteria. The analysis we have seen is based on point designs and does not incorporate information about the cost and performance of operating commercial nuclear facilities. Such modeling and analysis under a wide variety of scenarios, for both open and closed fuel cycles, will be useful to the industry and investors, as well as to international discussions about the desirability about different fuel cycle paths.

We call on the Department of Energy, perhaps in collaboration with other countries, to establish a major project for the modeling, analysis, and simulation of commercial nuclear power systems—The Nuclear System Modeling Project. This project should provide a foundation for the accumulation of information about how variations in the operation of plants and other parts of the fuel cycle affect costs, safety, waste, and proliferation resistance characteristics. The models and analysis should be based on real engineering data and, wherever possible, practical experience. This project is technically demanding and will require many years and considerable resources to be carried out successfully.

For technical, economic, safety, and public acceptance reasons, the highest priority in fuel cycle R&D, deserving first call on available funds, lies with efforts that enable robust deployment of the once-through fuel cycle.

We believe that development of advanced nuclear technologies—either fast reactors or advanced fuel cycles employing reprocessing—should await the results of the *Nuclear System Modeling Project* we have proposed above. Our analysis makes clear that there is ample time for the project to compile the necessary engineering and economic analyses and data before undertaking expensive development programs, even if the project should take a decade to complete. Expensive programs that plan for the development or deployment of commercial reprocessing based on any existing advanced fuel cycle technologies are simply not justified on the basis of cost, or the unproven safety, proliferation risk, and waste properties of a closed cycle compared to the once-through cycle. Reactor concept evaluation should be part of the Nuclear System Modeling Project.

On the other hand, we support a modest laboratory scale research and analysis program on *new* separation methods and associated fuel forms, with the objective of learning about approaches that emphasize lower cost and more proliferation resistance. These data can be important inputs to advanced fuel cycle analysis and simulation and thus help prioritize future development programs.

The modeling project's research and analysis effort should only encompass technology pathways that do not produce weapons-usable material during normal operation (for example, by leaving some uranium, fission products, and/or minor actinides with the recycled plutonium). *The closed fuel cycle currently practiced in Western Europe and Japan, known as PUREX/MOX, does not meet this criterion.* There are advanced closed fuel cycle concepts involving combinations of reactor, fuel form, and separations technology that satisfy these conditions and, with appropriate institutional arrangements, can have significantly better proliferation resistance than the PUREX/MOX fuel cycle, and perhaps approach that of the open fuel cycle. Accordingly, the governments of nuclear supplier countries should discourage other nations from developing and deploying the PUREX/MOX fuel cycle.

Government R&D support for advanced design LWRs and for the High Temperature Gas Reactor (HTGR) is justified because these are the two reactor types that are most likely to play a role in any nuclear expansion. R&D support for advanced design LWRs should focus on measures that reduce construction and operating cost. Because the High Temperature Gas Reactor (HTGR) has potential advantages with respect to safety, proliferation resistance, modularity and efficiency, government research and limited development support to resolve key uncertainties, for example, the performance of HTGR fuel forms in reactors and gas power conversion cycle components, is warranted.

Waste management also calls for a significant, and redirected, ARD&D program. The DOE waste program, understandably, has been singularly focused for the past several years on the Yucca Mountain project. We believe DOE must broaden its waste R&D effort or run the risk of being unable to rigorously defend its choices for waste disposal sites. More attention needs to be given to the characterization of waste forms and engineered barriers, followed by development and testing of engineered barrier systems. We believe deep boreholes, as an alternative to mined repositories, should be aggressively pursued. These issues are inherently of international interest in the growth scenario and should be pursued in such a context.

The closed fuel cycle currently practiced in Western Europe and Japan, known as PUREX/MOX, does not meet this nonproliferation criterion.

There is opportunity for international cooperation in this ARD&D program on safety, waste, and the Nuclear System Modeling Project. A particularly pertinent effort is the development, deployment, and operation of a word wide materials protection, control, and accounting tracking system. There is no currently suitable international organization for this development task. A possible approach lies with the G-8 as a guiding body.

Our global growth scenario envisions an open fuel cycle architecture at least until mid-century or so, with the advanced closed fuel cycles possibly deployed later, but only if significant improvements are realized through research. The principal driver of this conclusion is our judgment that natural uranium ore is available at reasonable prices to support the open cycle at least to late in the century in a scenario of substantial expansion. This gives the open cycle clear economic advantage with proliferation resistance an important additional feature. The DOE should undertake a global uranium resource evaluation program to determine with greater confidence the uranium resource base around the world.

Accordingly, we recommend:

The U.S. Department of Energy should focus its R&D program on the once-through fuel cycle;

The U.S. Department of Energy should establish a Nuclear System Modeling project to carryout the analysis, research, simulation, and collection of engineering data needed to evaluate all fuel cycles from the viewpoint of cost, safety, waste management, and proliferation resistance;

The U.S. Department of Energy should undertake an international uranium resource evaluation program;

The U.S. Department of Energy should broaden its waste management R&D program;

The U.S. Department of Energy should support R&D that reduces Light Water Reactor (LWR) costs and for development of the HTGR for electricity application.

We believe that the ARD&D program proposed here is aligned with the strategic objective of enabling a credible growth scenario over the next several decades. Such a ARD&D program requires incremental budgets of almost $400 million per year over the next 5 years, and at least $460 million per year for the 5–10 year period.

Notes

1. This fuel cycle is known as Plutonium Recycle Mixed Oxide, or PUREX/MOX.

2. A fast reactor more readily breeds fissionable isotopes-potential fuel-because it utilizes higher energy neutrons that in turn create more neutrons when absorbed by fertile elements, e.g. fissile Pu^{239} is bred from neutron absorption of U^{238} followed by beta (electron) emission from the nucleus.

Personalized Energy
The Next Paradigm

In the future, energy will be more in the control of neighborhoods and homeowners. But for that to happen, new technologies need to be developed that bring efficiency and reliability up and costs down, says a technology futurist.

STEPHEN M. MILLETT

Consumers want more control over their energy, and they want it cheaper, cleaner, more convenient, and more reliable. They want energy, like other commodities, to be more personalized. Technological improvements will focus on meeting those demands, but they won't happen quickly. Current forecasts for energy supply and demand are limited by a mind-set stuck in the past. But new technologies and new consumer imperatives will spawn new ideas about energy that could get us off the grid and bring power generation into neighborhoods and even into homes.

The primary question we need to ask about the future of energy is whether the old supply-and-demand paradigm of fossil fuels still applies. In that case, the key to solving our energy woes lies in finding ways to increase production of traditional hydrocarbon fuels (such as oil, natural gas, and coal) and promote consumer conservation. But if the old paradigm is out, there may be a whole new paradigm emerging, where new technologies, for instance, could change the whole energy picture.

Right now, we hear too many discussions about drilling more oil, conserving energy, and other actions based on old-paradigm thinking. Indeed, statistics show a big gap between projected energy demand and supplies in the United States: Oil and natural gas consumption are going up and available quantities are going down, so we're going to have a big projected shortfall.

The biggest jump in American energy consumption in the twentieth century was the use of petroleum, and that's almost exclusively transportation. Transportation relies on petroleum to meet 95% of its energy needs, according to the U.S. Bureau of Transportation Statistics. The story on coal is a little bit different. Developed economies such as the United States used to use coal in homes for heating, but that's done almost nowhere anymore. Americans are using more and more coal, but it's to generate electricity in large power plants. Coal is now rarely used at the individual level.

The 2001 report of the president's National Energy Policy Development Group stated, "Renewable and alternative fuels offer hope for America's energy future but they supply only a small fraction of present energy needs. The day they fulfill the bulk of our [energy] needs is still years away. Until that day comes we must continue meeting the nation's energy requirements by the means available to us."

This assertion assumes no changes to the existing energy paradigm: no new technological breakthroughs, no shifts in people's values or consumers' demands, no surprising events—natural or manmade—to alter the energy picture. But this paradigm-blinder limits our thinking—and our forecasts. Paradigms and social systems are rarely permanent, and new technology often drives a transition to other paradigms.

My thesis is that we have just begun the shift away from what I call the "carbon-combustion paradigm" to a new "electro-hydrogen paradigm." The shift is going to be very dramatic in the next 20 years, but the full integration is going to take easily a hundred years. We're going to see a lot of exciting technology innovations in the laboratory and in prototype systems in the next 10 to 20 years, but to go from our current paradigm and all its infrastructure to a new paradigm and all of its infrastructures is going to take a very long time.

Energy and the Consumer

Changes in consumer behavior are driving many trends. In the U.S. market, baby boomers seek convenience, while the elderly put heavy emphasis on the reliability and affordability of power. The question for policy makers and the energy industry is how reliable the electric grid will be in the future. Both baby boomers and Generation X'ers value customization—the personalization of products, especially computers and cell phones. Consumers also want more mobility and longevity in their products. And of course we all want inexpensive energy.

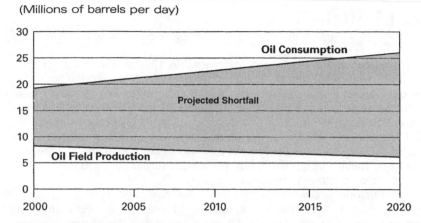

Figure 1 U.S. oil consumption projections.
Sources: National Energy Policy Report: Battelle.

Along with consumer behavior, there are marketplace trends working toward this paradigm shift, including the effects of more-stringent environmental-quality regulations. The bad news is that no energy system will ever be 100% environmentally friendly; the good news is that the next paradigm will be a lot friendlier than the last one was.

U.S. energy policy calls for greater energy self-sufficiency. An additional need, a new element since September 11, 2001, is security of the energy infrastructure. Before then, people didn't even worry about infrastructure security, but now it's a major issue. The electric grid system is absolutely vulnerable to weather and potential security compromises.

Another marketplace issue is the need for energy-cost stability and continued necessity for economic growth. If you take a strictly conservationist approach to this issue and say this cost stability and continued economic growth can be realized by voluntary simplicity, then you are limited by the old energy paradigm. This low-growth scenario has a low probability of occurring. We need more energy for continued economic growth, and that is still what most people in the world want.

Also impacting marketplace trends are emerging technologies such as those behind a gradual shift now beginning from central-station generation to decentralized, or distributed, generation of power from local sources. The paradigm shift to distributed generation, or distributed resources, parallels the paradigm shift from fossil fuels to hydrogen. For example, one of the biggest needs we have is gasification of coal, though surprisingly little work is yet being done in this area. The current gasification technology has been in existence for at least 20 years, it isn't really very good, and it's very expensive. It's just not competitive.

The Real Hydrogen Future

It's easy to speculate about the hydrogen economy's potential and to go off into science-fiction scenarios. But because Battelle deals with the real-world challenges of governments and corporations, we spend a lot of time separating science from science fiction. So here's a little reality check: All forms of energy are going to have some negative environmental consequences. We need to recognize that fact and then try to make the energy system a lot better in the future than it is today. Another reality

check is that no fuels will be free: There will always be costs for both the fuels and their infrastructure of production and distribution. The challenge is to find the new ways that improve the value (benefits/costs) relationship.

The challenge in this transitional period to the hydrogen future is to extract hydrogen from hydrocarbon sources in an affordable way. There are many potential avenues being explored today. The approach at Battelle is to develop a universal reformer for the fuel cell, where we would take methane, methanol, and even gasoline and convert it into hydrogen at the point of burning it. The ability to extract sufficiently pure hydrogen from methane, methanol, or gasoline means that we could continue to use the existing infrastructure (such as all of those gas stations) to distribute safe liquid fuels without the expense and hazards of storing hydrogen today. Avoiding new infrastructure costs in the short run would greatly help the transition to the electrohydrogen paradigm of fuel cells for both transportation and stationary power generation applications.

Economics really favor the current, fading energy paradigm; economics do not yet favor the next one. We're going to have to see a lot of economic and regulatory changes, as well as technological changes. The challenge is cost. We can make fuel cells, we can produce hydrogen, but we can't do it at competitive prices relative to the existing hydrocarbon system, the carbon-combustion system. Electric utilities use a benchmark of $1,500 per kilowatt capacity; anything costing more than that is simply not competitive enough. Researchers at United Technologies, for example, are getting the cost of the fuel cell down below $3,000 per kilowatt—it has been as high as $15,000—but the price is still too high for general commercialization.

And what about solar cells? There's no question that there's a market today for solar cells, but it's largely a vanity technology for people who put it on their houses. If you take a pocket calculator that has solar panels, generating power measured in milliwatts, and normalize that to a kilowatt, the cost might be as high as $17,000. So, clearly, there's a long way to go before solar technology can replace the carbon-combustion system.

Alternative fuels like wind power, which is now growing, have been attractive because of a number of government incentives and subsidies at both the federal and the state levels.

About Battelle

The Battelle Memorial Institute was established in 1929 in Columbus, Ohio, and now manages or co-manages four of the 16 U.S. national labs. We are in the business of technology development, management, and commercialization. We do mostly government work, but we also have industrial clients. We're independent, meaning we have stakeholders but no stockholders. We are technically not for profit but, as we're reminded daily by our CEO, we are not for loss. We do more than $1.7 billion in business a year, and that's a lot of R&D. Battelle scientists have contributed to a wide range of breakthroughs, such as copy machines, optical digital recording, and bar codes; Battelle's R&D yields between 50 and 100 patented inventions a year.

All corporations and organizations face the challenge of keeping up with and anticipating change. At Battelle, we do futuring. I like to use the word *futuring* because the participle adds the action of making or doing something. We do trend analysis, expert focus groups, and expert judgment, and we have our own process of scenario analysis based on cross-impact analysis. We also have our own scenarios software, which we've been using for 20 years for our work with corporations.

Battelle studies "consumer value zones," where marketplace trends, new customer demands, and emerging technologies all converge. The study of energy's consumer value zone leads us to conclude, for example, that the future of energy is personal; that is, energy production will increasingly move from large, centralized power plants to distributed power.

Among its outreach projects, Battelle does an annual technology forecast and maintains a separate Web site for our scenarios and trends: www.dr-futuring.com.

For more information, see Battelle's Web site, www.battelle.org.

—Stephen M. Millett

Technologies to Watch

The technologies to watch include:

- Innovations in materials for batteries and fuel cells, especially PEM (polymer electrolyte membrane) and solid-oxide fuel cells.
- Breakthroughs in reducing diesel emissions.
- Innovations for reconfiguring backup and emergency power generation into distributed generation systems.
- Biofuel development.
- New approaches to the gasification of coal.
- Global warming and carbon-dioxide management.

For fuel cells and batteries, the biggest challenge is in materials development. Exciting new developments in battery technology include the sodium sulfur battery, which the American Electric Power Company is working on in Columbus, Ohio.

For PEM and solid-oxide fuel cells, the name of the game is the materials and getting their costs down. For instance, the current membrane material used in PEM fuel cells now costs as much as $800 per square meter. We need to get the cost down to $8 to make this transition to the next paradigm. So we need technology breakthroughs that bring costs down. In addition, the current use of platinum as the catalyst is obviously very expensive and needs to be changed.

Reducing diesel emissions is another significant area of research at Battelle and other institutions. Biofuel blending with diesel and other fuels is a very exciting growth area.

Bringing diesel emissions down will promote the transition of current backup generation to distributed resources coordinated with the power grid. Most utilities now dismiss customer-driven backup generation as simply being irrelevant to the grid, but if you can make all of those diesel generators environmentally compliant, and if you can coordinate them with the grid, then you've got a prototype distributed-generation system already in place.

Biofuel development, not just bioblending, is another breakthrough area. The DNA revolution in agriculture is very exciting because we could design plants—not just corn, but also chickweed or garbage grass, for instance—that could be engineered for high-starch content to be more easily converted into methanol.

Gasification of coal is a huge area of research and development. Affordable and efficient coal gasification would enable us to break down the constituent parts of coal and get the hydrogen atoms out of it. In an ideal process, we would be able to separate sulfur and other undesirable constituents out of the coal and extract pure hydrogen. We could also separate out the carbon content that produces carbon-dioxide emission from stacks. An innovative coal gasification technology would be a tremendous boon for the American economy—and the economies of Germany, Russia, China, and India, to mention just a few others. Hydrogen from coal would be a major step in the transition to fuel cells.

The new energy paradigm is also about the environment, and we at Battelle are concerned about global climate change and carbon-dioxide management. To that end, Battelle is actively pursuing approaches like carbon sequestration. We are currently working with the U.S. Department of Energy to evaluate how to capture carbon dioxide and store it underground so that it cannot escape into the atmosphere.

Toward a Distributed Power System

We're not suddenly going to do away with our coal-burning plants, but there are emerging opportunities to use large fuel cells and batteries in conjunction with central generation. This could produce emergency and peak power at the generation site as well as provide supplementary power at distributed sites. By "distributed," I don't mean we're going from the big power plant to the home all in one jump. Energy will be distributed first at the level of neighborhoods and districts, and then we'll work it on down to homes generating their own power. It'll go step by step, but the trend favors personalized energy.

We're going to see some exciting technologies developed in the next 10 years, but it's going to be a slow process toward full-blown commercialization. If we're on a low technology-development trajectory, it will take more time. If we get a couple of breakthroughs in technology or some regulatory changes, then we can be on a faster track, but no sooner than 2008 or even 2010 at the earliest unless a desperate need for power drives the trends

faster. Slow progress favors the "tracker" and "adapter" companies and organizations, while fast progress favors the early innovators. Many companies are now agonizing over whether to be the progress leaders or the followers (or fast followers).

Who's going to lead this energy paradigm shift? Who really is going to provide the thought leadership and the breakthroughs? The Japanese are clearly ahead of the Americans in fuel cells. Honda and Toyota are ahead of the Big Three auto manufacturers in Detroit on energy breakthroughs for transportation. Honda in particular is the world leader in thinking through distributed power generation.

As for regulatory leadership, the question is who is going to provide the standards. There's a dearth of leadership for the new energy paradigm in the United States. Neither the federal government nor the states are showing signs of leadership, and there are very few progressive electric and gas utilities out there in the United States. Wherever the leadership comes from for the new energy paradigm, that's who will likely succeed and capture the largest market share.

But for now, those who should lead seem to be saying, "Change is good. You go first."

STEPHEN M. MILLETT is a thought leader at Battelle and co-author of *A Manager's Guide to Technology Forecasting and Strategy Analysis Methods.* His address is Battelle, 505 King Avenue, Columbus, Ohio 43201. E-mail milletts@battelle.org; Web site www.battelle.org.

Originally published in the July/August 2004 issue of *The Futurist,* pp. 44–48. Copyright © 2004 by World Future Society, 7910 Woodmont Avenue, Suite 450, Bethesda, MD 20814. Telephone: 301/656-8274; Fax: 301/951-0394; http://www.wfs.org. Used with permission from the World Future Society.

Hydrogen: Waiting for the Revolution

Everybody agrees it's the future fuel of choice. Why hasn't the future arrived?

BILL KEENAN

Here's how you'll live in the Hydrogen Age: Your car, powered by hydrogen fuel cells and electric motors, quietly drives along smog-free highways. At night, when you return your vehicle to the garage, you hook up its fuel cell to a worldwide distributed-energy network; the central power grid automatically purchases your battery's leftover energy, offsetting your overall energy costs.

In the garage, you also have a suitcase-sized electrolyzer, or other conversion device, plugged into the electrical system to pump a fresh batch of hydrogen into your car. (The fuel cell uses hydrogen to produce electricity, which powers the motor.) If you need a refill as you're driving along one of the nation's highways, you pull up to a clean, quiet hydrogen fueling station to top off in less time than it takes today to fill a car with gasoline.

The electricity in your home will also come from hydrogen, either via small local fuel-cell power plants or residential fuel cells in your basement. "Moreover," says Jeremy Rifkin, president of the Foundation on Economic Trends and author of *The Hydrogen Economy,* "sensors attached to every appliance or machine powered by electricity—refrigerators, air-conditioners, washing machines, security alarms—will provide up-to-the-minute information on energy prices, as well as on temperature, light, and other environmental conditions, so that factories, offices, homes, neighborhoods, and whole communities can continuously and automatically adjust their energy consumption to one another's needs and to the energy load flowing through the system."

The U.S. Department of Energy is only slightly less enthusiastic, maintaining in a report that in the hydrogen economy, "America will enjoy a secure, clean, and prosperous energy sector that will continue for generations to come. It will be produced cleanly, with near-zero net carbon emissions, and it will be transported and used safely. [Hydrogen] will be the fuel of choice for American businesses and consumers."

The new energy regime will have economic and political ramifications as well. Oil companies and utility companies will merge and morph into "energy companies" with a focus on generating renewable energy and local power distribution, including purchasing power from residential customers. Distributed energy production will also result in a worldwide "democratization of energy," bringing low-cost power to underdeveloped areas.

Oil and Hydrogen Don't Mix

Driving the interest in a hydrogen-based energy system: threats to the economy, the environment, and national security. Oil production, by current estimates, will likely peak sometime between 2020 and 2040. At this point, the world's economies will have consumed half of the known oil reserves, with two-thirds of the remaining oil in the volatile Middle East. As a result, prices will rise dramatically, and global consumers will experience increasingly frequent shortages.

Global warming is another significant threat that a shift to hydrogen might ameliorate. The release of carbon dioxide into the atmosphere from the burning of fossil fuels such as coal, oil, and natural gas makes up about 85 percent of greenhouse-gas emissions in the United States. This increase has resulted in an unprecedented rate of global warming, according to most scientific experts. The thinning of the polar ice caps, the retreat of glaciers around the world, the spread of tropical diseases to more temperate climates, and the rising of global sea levels are all evidence of global warming. Says Rifkin: "Weaning the world away from a fossil-fuel energy regime will limit carbon-dioxide emissions to only twice their pre-industrial levels and mitigate the effects of global warming on the Earth's already beleaguered biosphere."

The Mechanics of Hydrogen

While still untested on a large scale, the promise of a hydrogen economy is based on a number of undeniable realities. Hydrogen can be burned or converted into electricity in a way that creates virtually no pollution. It is also Earth's most abundant element, available everywhere in the world. While hydrogen is scarce naturally in pure form, it can be generated easily by reforming gasoline, methanol, natural gas, and other readily available resources. It can also be created by electrolysis, a process by which electricity is run through water to separate the oxygen and hydrogen molecules.

The fuel cell, which combines oxygen in the air with hydrogen to create electricity and water, is the vital link in the hydrogen vision. It closes the energy loop and allows electricity to be stored and transported via hydrogen and then reconverted back into electricity.

In an ideal future, renewable-energy sources such as wind, solar, or water power will be used to create hydrogen through electrolysis. The hydrogen can be converted again to electricity locally by means of a fuel cell to power a car, provide energy for a home, power a laptop, or operate any number of other products.

—B.K.

Add to these threats the burden of growing world populations, an increasingly unstable political situation in the Middle East, and the likelihood of longer and more frequent blackouts and brown-outs resulting from an aging and vulnerable power grid in the United States, and the promise of a safe, pollution-free, and distributed power system based on hydrogen becomes increasingly attractive.

Pathways and Roadblocks

Does all this sound too good to be true? It is: The hydrogen economy faces serious obstacles. More than 90 percent of the hydrogen produced today comes from reformulated natural gas generated through a process that creates a significant amount of carbon dioxide. Energy for this process, or for electrolysis, a more expensive way of generating hydrogen would also come from power plants fueled by oil or natural gas. So in the near term, a shift to hydrogen will not greatly reduce the world's dependence on fossil fuels and, in fact, may well hasten the greenhouse effect and global warming by increasing carbon-dioxide emissions.

Consequently, a lot of discussion about the hydrogen economy revolves around the various "pathways," or means of producing hydrogen. Atakan Ozbek, director of energy research at ABI Research, a technology-research think tank, points out that while hydrogen can come from virtually any fuel, energy from oil and gas is currently cheaper and more efficient than energy from renewable resources such as wind, sun, or water. Then, too, in the event of an oil crisis and resultant electricity shortage, coal will likely be pressed into service, regardless of the environmental cost. Nuclear power plants can also provide electricity to create hydrogen, but nuclear energy's high cost—plus the still-hot controversy over waste disposal—make such a pathway less than certain.

"What we're trying to find out right now," Ozbek says, "is how to get hydrogen to the fuel cell in a way that is economically feasible and makes sense engineering-wise."

Environmental considerations are paramount: If coal is reintroduced in a large way into our "energy portfolio"—whether to produce hydrogen or as part of our existing energy plan to replace oil—carbon-dioxide emissions will rise significantly.

The Department of Energy roadmap anticipates this, and the DOE is funding research into the "sequestration" of carbon-dioxide gases created by coal processing and natural-gas reformation. This would involve capturing these gases at some point in the energy process and permanently storing them underground or in the ocean.

To many, this is unrealistic. Jon Ebacher, vice president of power-systems technology for GE Energy, won't say that sequestration is impossible, but his comments fall short of an endorsement. Even a fairly efficient coal plant, Ebacher says, produces millions of tons of carbon dioxide each year. "So if you're going to sequester carbon dioxide from all of the plants that use hydrocarbon fuels," he says, "that's a pretty massive undertaking."

Only a hydrogen economy based 100 percent on renewable power would result in zero emissions—the vision that has captured so many imaginations. And that vision remains decades away. In the meantime, Ebacher says, "natural gas can see us through a transition period until we get solar and other renewable-energy efficiencies up to a much higher level." That transition period, he suggests, might last twenty-five to fifty years.

Another potential roadblock: transport and storage of hydrogen. Less dense than other fuels, the gas must be compressed or liquefied to be stored or moved efficiently, adding to costs and inconvenience. While the existing natural-gas infrastructure would seem to offer a convenient pathway to hydrogen delivery, this can't be done without a major retrofit. Indeed, Ebacher says, almost all of the country's existing natural-gas pipeline would have to be modified to handle hydrogen.

Finally, fuel-cell researchers must make significant advances. The power produced by a fuel cell is significantly more expensive per unit than that produced by an internal-combustion engine. Fuel-cell vehicle development is also beset by problems and costs related to type of fuel, storage, and performance. A number of prototype and "concept car" fuel-cell vehicles have been produced and displayed at auto shows and fuel-cell conferences around the world—but at a development cost of about $250,000 or more per vehicle. GM estimates that it spent between $1 million and $2 million to develop its Hy-wire fuel-cell concept car. A consumer version would cost far less, obviously, but likely would still take sticker shock to a whole new dimension.

Putting a Brake on Hydrogen Cars

Linking the hydrogen age to cars could be a critical policy mistake, according to Joseph Romm, former acting assistant secretary for the DOE's Office of Energy Efficiency and Renewable Energy and author of *The Hype About Hydrogen*. Despite car-company promises to have fuel-cell vehicles in dealer showrooms by 2010, if not sooner, Romm argues that the cost of fuel cells, problems with onboard storage of hydrogen in vehicles, and the issues related to creating a hydrogen delivery infrastructure are likely to push the market for hydrogen fuel-cell vehicles well into the future.

The focus on hydrogen as an immediate goal in the transportation sector amounts to confusing a means (hydrogen) with an end (greenhouse gas reduction), Romm explains. This could have harmful consequences, since, he estimates, it will take thirty to fifty years for hydrogen vehicles to have a significant impact on greenhouse gases. A recent National Academy of Sciences study seconds this point, stating, "In the best-case scenario, the transition to a hydrogen economy would take many decades, and any reductions in oil imports and carbon-dioxide emissions are likely to be minor during the next twenty-five years."

"If the goal is to reduce greenhouse gases," Romm argues, "then there are technologies available right now that can have a more immediate effect"—hybrid vehicles, for instance. And diverting existing (and limited) natural-gas supplies to create hydrogen for vehicles "would make that fuel less available where its use could result in a more immediate reduction in greenhouse-gas emissions—in replacing existing oil and coal-burning electric-power plants in the nation's energy grid with cleaner natural-gas power plants."

As a fuel, hydrogen is "simply a better mousetrap."

In fact, some hydrogen-technology companies have back-burnered research and development on transportation applications. "The horizons for fuel-cell vehicles keep getting pushed out further and further, and it's unlikely that somebody's going to license and commit to a uniform, standardized hydrogen technology for at least ten to fifteen years," says Stephen Tang, an industry consultant and former president and CEO of Eatontown, NJ.-based Millennium Cell, which

As GE Goes, So Goes the Nation?

Fuel cells will probably not be a viable market until a company like General Electric gets into the business in a big way, say critics of the hydrogen economy.

GE is indeed researching fuel cells, albeit cautiously, in keeping with its approach to most other energy markets. "We do have an investment in fuel cells," says Jon Ebacher, the company's vice president of power-systems technology. "I don't know if it will ever get to the dimensions where it will work at the huge volumes that were once forecast, but I think it's quite viable in niche markets." Right now, he sees a possible market in "industrial facilities that have isolated power needs, where you have a maintenance crew that deals with heating, ventilating, and air-conditioning." But, he says, "there's a distance between where they are today and the huge potential in consumer markets that was forecast at one time."

GE has also invested substantially in researching a type of fuel-cell system that would employ a gas turbine and hydrogen system working as a combined cycle. Right now, Ebacher says, GE is considering creating a power plant based on this system by 2013. It could be sooner, depending on external factors, including political developments around both fuel and the environment, the price of fuel, and the types of fuel that are available. But there are still unresolved technical challenges that could push that back.

Natural-gas prices in particular are an important barometer. "During the California energy crisis, the price of gas spiked up to $7 per million BTUs," Ebacher says. Higher gas prices, he says, "could spark some other research efforts that may come in front of fuel cells—coal, for instance. It's possible to run a combined-cycle system on coal. You put a chemical plant beside a combined-cycle power plant to process the coal into a gaseous fuel to run the electrical plant." But right now, Ebacher says, "the capital cost of doing that doesn't cross the goal line. However, if the price of natural gas or its availability gets in a bad place, all of a sudden the capital cost of doing that might not look so bad.

"It all revolves around availability, economics, and the environment—where the pressures are, what are the levers. But if you talk about running out of hydrocarbon fuels, then you would have to say that hydrogen had better be in the cards."

—B. K.

makes a system called Hydrogen on Demand that supplies hydrogen to fuel cells.

To pay the bills in the meantime, Tang says, "Millennium Cell has targeted markets that it believes can tolerate the price of hydrogen and fuel cells, such as consumer electronic devices, standby power, and military portables. In all of those markets, you're competing with an incumbent technology that is rather expensive in its own right and also has some limitations in performance. In these markets, then, we can focus on hydrogen as a performance fuel and not focus so much on the environmental benefits or the energy-independence benefits—attributes that buyers have difficulty valuing. It's simply a better mousetrap: Hydrogen allows you to run your cell phone much longer, or your laptop much longer, without being a slave to the energy grid or inferior batteries."

Who Will Lead?

Despite the limitations, there is growing momentum for hydrogen vehicles. Hybrid vehicles may be a "bridging technology" toward the hydrogen age, but it's one that "doesn't at all curb the nation's appetite for oil," says Chris Borroni-Bird, GM's director of design-technology fusion. Therefore, the automaker directs about a third of its R&D—over $1 billion thus far and involving more than six hundred people—toward fuel cells. The company insists that it will have a commercially viable fuel-cell vehicle available by the end of the decade.

In other business sectors, investment in hydrogen technology is slowly returning after the boom in hydrogen technology stocks in 1999–2000 and the subsequent bust that lasted until last year. "Behind a lot of the hype, there was tremendous capital inflow in the mid-1990s going into 2000," Tang says. Unfortunately, the number of commercial products—and the resulting revenue—in the industry have been "underwhelming" relative to investment dollars. That has made the investment community more cautious so far, but things are changing. "Right now there is a much more realistic view of the possibilities," Tang says. "The investor today is looking more toward interesting niche strategies and early market penetration rather than the hope of the mass market, the home run where fifty million cars are going to be sold with your product in it."

With the investment community poised and the technology issues coming together, says ABI Research's Ozbek, "Everything is feeding into this giant equation—you can consider it a giant chemical reaction—and once everything has been fed in and the equation solved, it's going to change the whole energy infrastructure." Federal support and direction will be especially important. While Ozbek considers President Bush's $1.7 billion State of the Union pledge for energy research a good start, he would like the government to provide such research incentives as Japan and the European Union have in recent years.

And though the president disappointed many hydrogen proponents by making no specific mention of hydrogen-energy R&D in his 2004 address, his proposed 2005 budget did increase funding for hydrogen research. The federal government, Ozbek argues, should provide enhanced tax credits for buyers of fuel-cell vehicles and fuel credits for energy companies and other investing in building a hydrogen infrastructure. Jeremy Rifkin agrees, urging the federal government to take the lead by establishing benchmarks—mandating tougher fuel-efficiency standards and requiring a greater use of renewable energy sources by power companies—as the European Union currently does.

California's Hydrogen Highway

One state isn't waiting for action from companies or the federal government. In California, the new Schwarzenegger administration has committed to an energy plan that aims to create a "hydrogen highway" in the state by 2010. The ambitious plan proposes the construction of hydrogen fueling stations every twenty miles along the state's twenty-one major interstate highways. By taking this step to break the chicken-and-egg dilemma (which comes first, the vehicle or the fueling infrastructure?) and by continuing to impose strict mandates on automakers for fuel efficiencies, California could jump-start the hydrogen economy.

"The pieces are all on the table," says Terry Tamminen, secretary of California's Environmental Protection Agency, "and there have been demonstration projects, but they have not been pulled together into any kind of unified vision, something that average people can

Is GM's Hy-Wire the Car of the Future?

General Motors has had a reputation for being rather conservative when it comes to both new technological developments and vehicle design, but it seems to have leapt ahead of other carmakers with its concept car, the Hy-wire.

The idea, says Chris Borroni-Bird, director of design-technology fusion for GM, is that "if you design a vehicle around the fuel cell and hydrogen tanks, you might be able to create a better vehicle than if you just put those same systems in a car designed for an internal-combustion engine."

The Hy-wire design puts the fuel cell and hydrogen storage tanks into a skateboard-like chassis that allows for greater flexibility and interchangeability of body types. Customized car bodies are then effectively "docked into" the uniform chassis.

And because the fuel cell can provide much greater electrical output than today's batteries, GM's designers have replaced mechanical and hydraulic systems for steering and braking with an electronically controlled one. "This system provides more design freedom, because those electrical wires can be routed in numerous ways, replacing a fixed steering column," Borroni-Bird says.

The Hy-wire prototype has no gas engine, no brake pedals, and no instrument panel. The fuel cell enables you to operate everything by wire. The electronic controls are included in a compact handgrip console that extends from the floor from between the front seats of the vehicle. Drivers can steer, brake, or accelerate with the controls built into the handgrips.

Because GM puts the hydrogen directly on board the vehicle, there is no need for the car to convert fossil fuels or other renewable sources into hydrogen. As a result, it can claim to offer a zero-emission vehicle and market the car to be compatible with a network of hydrogen fueling stations.

To that end, GM "applauds any hydrogen infrastructure projects, anywhere in the world," says Tim Vail, GM's director of business development for fuel-cell activities. Yet it will take a lot of applause to get the government to invest the estimated $11 billion to get a sufficient mass of hydrogen refueling stations to support 1 million vehicles, in proximity to 70 percent of the nation's population. "But," says the optimistic Vail, "$11 billion is nothing compared to past infrastructure projects such as the highways or the railroads. So it's not that big an issue to overcome. You just have to have the will to do it."

—B. K.

use and where we can more fully commercialize the technology. So we're taking a lot of this work that's already been done, bringing it together, adding some timetables and leadership, and then of course asking for some federal money to help with the pieces that aren't paid for by private industry or other investments."

California could jumpstart the hydrogen economy.

California already has several hydrogen fueling stations, serving research projects and some municipal fleets, and about a dozen more are in the works. For instance, SunLine Transit Agency, a local public-transit company, now operates a hydrogen fueling station that it uses to test its hydrogen-powered buses. And AC Transit, which provides public transportation in the San Francisco Bay area, expects to have three fuel-cell-powered buses later this year.

The state's goal is to provide an infrastructure of fueling stations to support a consumer market for fuel-cell vehicles. "If we can deliver such a network by a certain date," Tamminen explains, "we can then ask car companies to deliver on their promises to start delivering cars to showrooms."

One of the things driving California's plan is a California Energy Commission report that, Tamminen says, "includes credible evidence that in three to five years we are going to have serious shortages of refined fuels in the state. Not because there's not enough petroleum under the sands of Iraq but, rather, because we don't have enough refinery capacity in the state—or in the country—to keep up with the demand created by longer commutes, poorer fuel economy, and a growing population. The report predicts a likelihood of $3 to $5 per gallon gasoline prices and periodic shortages.

"During the oil embargo of the mid-1970s, we had twenty-four thousand retail gasoline outlets in the state, compared to ten thousand today. If there are shortages, not only will there be gas lines—they will be twice as long."

Consequently, it's not a question of if but when we move toward a hydrogen economy. Even Romm, who is dubious about short-term prospects for hydrogen, concludes: "The longer we wait to act, and the more inefficient, carbon-emitting infrastructure that we lock into place, the more expensive and the more onerous will be the burden on all segments of society when we finally do act."

BILL KEENAN is a freelance business writer and former editor of *Selling* magazine.

UNIT 4

Biosphere: Endangered Species

Unit Selections

Key Points to Consider

- Are there ways to assess the value or worth of living organisms other than those from whom we derive direct benefits (our domesticated plant and animals species)? What are the relationships between economic assessments of the biosphere and moral or value judgments on the preservation of species?

- Why is the spread of invasive species through the global trading network so difficult to control and what kinds of damage do invasive species produce? What suggestions would you make to remedy the problem?

- What kinds of changes are taking place in the chemistry, physics, and biology of the ocean—the world's largest ecosystem? Explain some of the interconnections between such things as changing ocean temperatures, and the population of phytoplankton and the marine species that feed upon them.

Student Web Site

www.mhcls.com/online

Internet References

Further information regarding these Web sites may be found in this book's preface or online.

Endangered Species
http://www.endangeredspecie.com/

Friends of the Earth
http://www.foe.co.uk/index.html

Natural Resources Defense Council
http://nrdc.org

Smithsonian Institution Web Site
http://www.si.edu

World Wildlife Federation (WWF)
http://www.wwf.org

There have been five great extinction events over the last 550 million years. In each more than 50% of all species were lost. We are now in the middle of the sixth. Some of the past extinctions have been attributed to specific causes, such as a meteorite impact, while the explanations for others are less clear. Our sixth extinction event is tied closely to human interaction. Early hunters effectively reduced the number of large species. Agrarian practices and human settlement further reduced species diversity by modifying the local environment with animal sanctuaries becoming increasingly isolated and fenced off. And now the future of many species is threatened by climate change associated with global warming. In the past, animals could adapt to environmental warming or cooling by migrating to more habitable climates. The isolation of animal habitats that exist today make that natural transition impossible.

We are changing the genetic makeup of our planet in other ways as well. While species are being lost due to extinction, new ones are being 'created' by the process of genetic modification, where DNA of different organisms are mixed to create something never before seen. These creations are called genetically modified organisms, or GMOs. Scientists have successfully inserted the gene of a flounder into a tomato. Other modifications include rice that synthesizes vitamin A, corn that is resistant to herbicide, and soybeans that synthesize their own pesticide. While there are clear benefits to these genetic modifications, there are also fears that they could escape into the wild, with unforeseen consequences for the delicate ecological balances that have taken millions of years to develop.

The oceans are not immune to environmental degradation. Covering 70% of the Earth's surface and supplying much of the world with food, our oceans are being polluted and overfished. Coral reefs, with their incredible biodiversity, are dying due to increasing ocean temperatures and acidity caused by fossil fuel burning. We are working ever harder to catch ever-dwindling supplies of fish. In the Pacific Ocean, there exists the 'Great Pacific Garbage Patch,' a soupy concentration of discarded plastic twice the size of Texas.

In this unit, problems concerning the biosphere are addressed in three separate articles. The article "Strangers in Our Midst," by Jeffrey A. McNeely, recognizes the new-found pressure of invasive species on our ecosystems. In the past, isolation of individual ecosystems could exists for millennia. Now, species are accidentally or intentionally introduced into a region where they have never existed.

Tundi Agardy's "America's Coral Reefs: Awash with problems" discusses the vulnerability of the most diverse biome in the Ocean, and the final article, "Markets for Biodiversity Services," examines the potential for private foundations to take the lead in protecting species diversity.

Strangers in Our Midst
The Problem of Invasive Alien Species

JEFFREY A. MCNEELY

Invasive alien species—non-native species that become established in a new environment then proliferate and spread in ways that damage human interests—are now recognized as one of the greatest biological threats to our planet's environmental and economic well-being.[1]

Most nations are already grappling with complex and costly invasive-species problems: Zebra mussels (*Dreissena polymorpha*) from the Caspian and Black Sea region affect fisheries, mollusk diversity, and electric-power generation in Canada and the United States; water hyacinth (*Eichornia crassipes*) from the Amazon chokes African and Asian waterways; rats originally carried by the first Polynesians exterminate native birds on Pacific islands; and deadly new disease organisms (such as the viruses causing SARS, HIV/AIDS, and West Nile fever) attack human, animal, and plant populations in temperate and tropical countries. For all animal extinctions where the cause is known, invasive alien species are the leading culprits, contributing to the demise of 39 percent of species that have become extinct since 1600.[2] The 2000 IUCN Red List of Threatened Species reported that invasive alien species harmed 30 percent of threatened birds and 15 percent of threatened plants.[3] Addressing the problem of these invasive alien species is urgent because the threat is growing daily and the economic and environmental impacts are severe.

A key question is whether the global reach of modern human society can be matched by an appropriate sense of responsibility. One critical element of this question is the definition of "native," a concept with challenging spatial and temporal dimensions. While every species is native to a particular geographic area, this is just a snapshot in time, because species are constantly expanding and contracting their ranges, sometimes with human help. For example, Britain has nearly 40 more species of birds today than were recorded 200 years ago. About a third of these are deliberate introductions, such as the Little Owl (*Athene noctua*), while the others are natural colonizations that may be taking advantage of climate change.[4]

An invasive alien species is not a "bad" species but rather one "behaving badly" in a particular context.

According to one view, local biological "enrichment" by non-native species always harms native species at some level, so any introduction should be regarded, at least in principle, as undesirable. An opposing view is that because species are constantly expanding or contracting their range, new species—especially those that are beneficial to people, such as crops, ornamental plants, and pets—should be welcomed as "increasing biodiversity" unless they are clearly harmful. According to this perspective, in the case of British birds noted above, only those introduced by people and that are causing ecological or economic damage, such as pigeons, are considered to be invasive.

All continental areas have suffered from invasions of alien species, losing biological diversity as a result, but the problem is especially acute on islands in general and for small islands in particular. The physical isolation of islands over millions of years has favored the evolution of unique species and ecosystems, so islands often have a high proportion of endemic species. The evolutionary processes associated with isolation have also meant that island species are especially vulnerable to predators, pathogens, and parasites from other areas. More than 90 percent of the 115 birds known to have become extinct over the past 400 years were endemic to islands.[5] Most of these evolved in the absence of mammalian predators, so the arrival of rats and cats carried by people has had a devastating impact.

Island plants are also affected. For example, the tree *Miconia calvescens* replaced the forest canopy on more than 70 percent of the island of Tahiti over a 50-year time span, starting with a few trees in two botanical gardens. Some 40–50 of the 107 plant species endemic to the island of Tahiti are believed to be on the verge of extinction primarily due to this invasion.[6] Introduced animals also can affect plants. For example, goats introduced on St. Clemente Island, California, have caused the extinction of eight endemic species of plants and have endangered eight others.[7]

An invasive alien species is not a "bad" species but rather one "behaving badly" in a particular context, usually due to inappropriate human agency or intervention. A species may be so threatened in its natural range that it is given legal protection, yet it may generate massive ecological and other damage elsewhere.

The degradation of natural habitats, ecosystems, and agricultural lands (through loss of vegetation and soil and pollution of land and waterways) that has occurred throughout the world has made it easier for non-native species to become invasive, opening up new possibilities for them. For all of these reasons, and others that will become apparent below, the issue of invasive alien species is receiving growing international attention.

The Vectors: How Species Move Around the World

The natural barriers of oceans, mountains, rivers, and deserts have provided the isolation that has enabled unique species and ecosystems to evolve. But in just a few hundred years, these barriers have been overcome by technological changes that helped people move species vast distances to new habitats, where some of them became invasive. The growth in the volume of international trade, from US$192 billion in 1960 to almost $6 trillion in 2003,[8] provides more opportunities than ever for species to be spread either accidentally or deliberately.

Some movement seems accidental, or at least incidental, in that transporting the species was not the purpose of the transporter. For example, ballast water is now regarded as the most important vector for transoceanic movements of shallow-water coastal organisms, dispersing fish, crabs, worms, mollusks, and microorganisms from one ocean to another. Enclosed water bodies like San Francisco Bay are especially vulnerable. The bay already has at least 234 invasive alien species, causing significant economic damage. California has one of the toughest ballast water laws in the nation, requiring ships from foreign ports to exchange their ballast water 200 miles from the California coastline, but enforcement remains spotty at best.

Ballast water may also be important in the epidemiology of waterborne diseases affecting plants and animals. One study measured the concentration of the bacteria *Vibrio cholerae*—which cause human epidemic cholera—in the ballast water of vessels arriving to the Chesapeake Bay from foreign ports, finding the bacteria in plankton samples from all ships.[9]

Other invasives are hitchhikers on global trade. For example, the Asian long-horned beetle (*Anoplophora glabripennis*) is one of the newest and most harmful invasive species in the United States. Originating in northeastern Asia, it finds its way to the United States through packing crates made of low-quality timber (that which is too infested for other uses). The number of insects found in materials imported from China increased from 1 percent of all interceptions in 1987 to 20 percent in 1996.[10] Outbreaks were reported in and around Chicago as early as 1992, in Brooklyn in August 1996, and in California in 1997. The beetle finds a congenial home among native maples, elders, elms, horse chestnuts, and others. The U.S. Department of Agriculture predicted that if the beetle becomes established, it could denude Main Street, USA, of shade trees, affect lumber and maple sugar production, threaten tourism in infested areas, and reduce biological diversity in forests.[11]

Another dangerous trade-related species for North America is the Asian gypsy moth (*Lymantria dispar*), which was first reported in the United States in 1991, entering as egg masses attached to ships or cargo from eastern Siberia. The caterpillars of this species are known to feed on more than 600 species of trees, and as moths, the females can disperse themselves over long distances. Scientists fear that this species could cause vastly more damage than the European gypsy moth, which already defoliates 1.5 million hectares of forest per year in North America.

With almost 700 million people crossing international borders as tourists each year, the opportunities for them to carry potential invasive species, either knowingly or unknowingly, are profound and increasing. Many tourists return with living plants that may become invasive, or carry exotic fruits that may be infested with invasive insects that can plague agriculture back home. Travelers may also carry diseases between countries, as apparently happened with the SARS virus. Tourism is considered an especially efficient pathway for invasive alien species on subAntarctic islands such as South Georgia. Visitors to the island reached 15,000 in 1999. Part of the problem is that many tourists are visiting similar islands on the same trip, increasing the chances of a seed, fruit, or insect being carried, more than would be expected from a single landing of a few people who spend an extended time on one island.[12]

Many species are introduced on purpose but have unintended consequences. One example of purposeful introduction gone wrong is the extensive stocking program that introduced African tilapia *Oreochromis* into Lake Nicaragua in the 1980s, resulting in the decline of native populations of fish and the imminent collapse of one of the world's most distinctive freshwater ecosystems. The alteration of Lake Nicaragua's ecosystem is likely to have effects on the planktonic community and primary productivity of the entire lake—Central America's largest—destroying native fish populations and likely leading to unanticipated consequences.[13]

Sport fishers have also had an influence, importing their favorite game fish into new river systems, where they can have significant negative impacts on native species. For example, the northern pike (*Esox lucius*) has invaded rivers in Alaska and is replacing native species of salmon. While the northern pike occurs naturally in some parts of Alaska, it was introduced to the salmon-rich south-central area in the 1950s, probably by a fisherman who brought it to Bulchitna Lake. Flooding in the 1980s subsequently spread the pike into the streams of the Susitna and Matanuska river basins. Pike have now occupied at least a dozen lakes and four rivers in some of the richest salmon and trout habitat in the Pacific Northwest. Rainbow trout are an even greater threat. Originating in western North America, they have been introduced into 80 new countries, often with devastating impacts on native fish.

Pets are also a problem. Domestic cats can plunder ecosystems that they did not previously inhabit. On Marion Island in the sub-Antarctic Indian Ocean, cats were estimated to kill about 450,000 seabirds annually.[14] Exotic pets may escape—or be released when they have outlived their novelty—and become established in their new home. Stories of crocodiles in the Manhattan's sewer system are probably fanciful, but many former pets are becoming established in the wild. For example, Monk parakeets (*Myiopsitta monachus*), descended from former pets that were released possibly in the 1960s, have invaded some 76 localities in 15 U.S. states.[15] Native to southern South America, they are the only parrots that build their own nests, some of which support several hundred individuals and have separate families living in different chambers. Some believe that they soon will become widespread throughout the lower 48 states, posing a significant threat to at least some agricultural lands by feeding on ripening crops. And Burmese pythons (*Python molurus*) have become established in Everglades National Park, where they reach a very large size and prey on many native species, even alligators.

Pet stores often advertise invasive species that are legally controlled. For example, the July 2000 issue of the magazine *Tropical Fish Hobbyist* recommended several species of the genus *Salvinia* as aquarium plants, even though they are considered noxious weeds in the United States and prohibited by Australian quarantine laws.

The globalization of trade and the power of the Internet offer new challenges, as sales of seeds and other organisms by mail order or over the Internet pose new and very serious risks to the ecological security of all nations. Controls on harvest and export of species are required as part of a more responsible attitude of governments toward the potential of spreading genetic pollution through invasive species. Further, all receiving countries want to ensure that they are able to control what is being imported. Virtually all countries in the world have serious problems in this regard, an issue that some countries are calling "biosecurity."

The Science of Understanding Invasions

Biodiversity is dynamic, and the movement of species around the world is a continuing process that is accelerating through expanding global trade. By trying to identify which species are especially likely to become invasive, and hence harmful to people, ecologists are improving the quality of invasion biology as a predictive science so that people can continue to benefit from global biodiversity without paying the costs resulting from species that later become harmful.

Previous examples indicate the characteristics that can make a species invasive. For instance, coastal ecosystems are frequently invaded by microorganisms from ballast water for three main reasons. First, concentrations of bacteria and viruses exceed those reported for other taxonomic groups in ballast water by 6 to 8 orders of magnitude, and the probability of successful invasion increases with inoculation concentration. Second, the biology of many microorganisms combines a high capacity for increase, asexual reproduction, and the ability to form dormant resting stages. Such flexibility in life history can broaden the opportunity for successful colonization, allowing rapid population growth when suitable environmental conditions occur. And third, many microorganisms can tolerate a broad range of environmental conditions, such as salinity or temperature, so many sites may be suitable for colonization.[16] Insects are a major problem because they can lay dormant or travel as egg masses and are difficult to detect. The African tilapia introduced to Lake Nicaragua adapted well, because they are able to grow rapidly; feed on a wide range of plants, fish, and other organisms; and

form large schools that can migrate long distances. Further, they are maternal mouth brooders, so a single female can colonize a new environment by carrying her young in her mouth.[17] Rapid growth, generalized diet, ability to move large distances, and prolific breeding are all characteristics of successful invaders.

It is not always simple, however, to distinguish a beneficial non-native species from one at significant risk of becoming invasive. A non-native species that is useful in one part of a landscape may invade other parts of the landscape where its presence is undesirable, and some species may behave well for decades before suddenly erupting into invasive status. The Nile Perch (*Lates niloticus*), for example, was introduced to Lake Victoria in the 1950s but did not become a problem until the 1980s, when it was a key factor in the extinction of as many as half of the lake's 500 species of endemic fish, attractive prey for the perch.[18] That said, ecologists over the past several decades have agreed on some broad principles for guiding risk assessment. First, the probability of a successful invasion increases with the initial population size and with the number of attempts at introduction. While it is possible for a species to invade with a single gravid female or fertile spore, the odds of doing so are very low. Second, among plants, the longer a non-native plant has been recorded in a country and the greater the number of seeds or other propagules that it produces, the more likely it will become invasive. Third, species that are successful invaders in one situation are likely to be successful in other situations; rats, water hyacinth, microorganisms, and many others fall into this category. Fourth, intentionally introduced species may be more likely to become established than are unintentionally introduced species, at least partly because the vast majority of these have been selected for their ability to survive in the environment where they are introduced. Fifth, plant invaders of croplands and other highly disturbed areas are concentrated in herbaceous families with rapid growth and a wide range of environmental tolerances, while invaders of undisturbed natural areas are usually from woody families, especially nitrogen-fixing species that can live in nitrogen-poor soils.[19] And sixth, fire, like disturbance in general, increases invasion by introduced species. So ecosystems that are naturally prone to fire, such as the fynbos of South Africa, coastal chaparral in California, and maquis in the Mediterranean,[20] can be heavily invaded if fire-liberated seeds of invasive species are available. (These are all shrub communities adapted to cool, wet winters and hot, dry summers, where fire is a regular phenomenon. They are also rich in species: Fynbos have about 8,500 species that include many endemic *Proteaceae*; chaparral have about 5,000 species; and maquis have 25,000—of which about 60 percent are endemic to the Mediterranean region.[21])

Other ecological factors that may favor nonindigenous species include a lack of controlling natural enemies, the ability of an alien parasite to switch to a new host, an ability to be an effective predator in the new ecosystem, the availability of artificial or disturbed habitats that provide an ecosystem the aliens can easily invade, and high adaptability to novel conditions.[22]

It is sometimes argued that systems with great species diversity are more resistant to new species invading. However, a

Invasive Alien Species and Protected Areas

Protected areas are widely perceived as being devoted to conserving natural ecosystems. Ironically, protected areas are in fact heavily damaged by invasive alien species, and many protected-area managers consider this their biggest problem. Some examples:

- Galapagos National Park, a World Heritage site, is being affected by numerous invasive alien species, including pigs, goats, feral cats, fire ants, and mosquitoes.
- Kruger National Park, South Africa's largest, has recorded 363 alien plant species, including water weeds that pose a serious threat to the park's rivers.
- In the Wadden Sea, a biosphere reserve and Ramsar site protected by the Netherlands, Germany, and Denmark, the Pacific oyster has invaded, having escaped captive management. It is disrupting tourism because of its sharp shells. It has also carried with it numerous other invasive alien species.
- The Wet Tropic World Heritage Area of North Queensland, Australia, is infested by numerous invasive alien species, of which the worst is the pond apple from

Florida, which has invaded creeks and riverbanks, wetlands, melaleuca swamps, and mangrove communities. Feral pigs, another invasive species, help to spread the species. The pond apple is now rare in its native range in the Florida Everglades.

- Everglades National Park in Florida, another World Heritage site, is threatened by the invasion of melaleuca from Queensland, demonstrating that species that may behave well in their natural habitat can be a serious problem when they invade somewhere else.
- Tongariro National Park, New Zealand, is also a World Heritage site, but a third of its territory has been infested by heather, a European plant deliberately introduced into New Zealand by an early park warden in 1912 in an attempt to reproduce the moors of Scotland.

These are just a few examples among many that could be cited that demonstrate that even the most strictly protected areas can be extremely vulnerable to invasion by non-native species.

---

study in a California riparian system found that the most diverse natural assemblages are in fact the most invaded by non-native plants, and protected areas worldwide are heavily invaded by non-native plants and animals.[23] Dalmatian toadflax (*Linaria dalmatica*) is invading relatively undisturbed shrub-steppe habitat in the Pacific Northwest, wetland nightshade (*Solanum tampicense*) is invading cypress wetlands in central and south Florida, and garlic mustard (*Allilaria officinalis*) is often found in relatively undisturbed systems in the northern parts of North America.

This work helps resolve the controversy over the relationship between biodiversity and invasions, suggesting that the scale of investigation its a critical factor. Theory suggests that non-native species should have a more difficult time invading a diverse ecosystem, because the web of species interactions should be more efficient in using resources such as nutrients, light, and water than would fewer species, leaving fewer resources available for the nonnative species. But even in well-protected landscapes such as national parks, invaders often seem to be more successful in diverse ecosystems. Even though diversity does matter in fending off invasives, its effects are negated by other factors at larger scales. The most diverse ecosystems might be at the greatest risk of invasion, while losses of species, if they affect community-scale diversity, may erode invasion resistance.[24]

The Economic Impacts of Invasion

One reason invasive alien species are attracting more attention is that they are having substantial negative impacts on numerous economic sectors, even beyond the obvious impacts on agriculture

(weeds), forestry (pests), and health (diseases or disease vectors). The probability that any one introduced species will become invasive may be low, but the damage costs and costs of control of the species that do become invasive can be extremely high (such as the recent invasion of eastern Canada by the European brown spruce longhorn beetle (*Tetropium fuscum*), which threatens the Canadian timber industry).

Estimates of the economic costs of invasive alien species include considerable uncertainty, but the costs are profound—and growing (see Table 1).

Most of these examples come from the industrialized world, but developing countries are experiencing similar, and perhaps proportionally greater, damage. Invasive alien insect pests—such as the white cassava mealybug (*Phenacoccus herreni*) and larger grain borer (*Prostephanus truncates*) in Africa—pose direct threats to food security. Alien weeds constrain efforts to restore degraded land, regenerate forests, and improve utilization of water for irrigation and fisheries. Water hyacinth and other alien water weeds that choke waterways currently cost developing countries in Africa and Asia more than US$100 million annually. Invasive alien species pose a threat to more than $13 billion of current and planned World Bank funding to projects in the irrigation, drainage, water supply, sanitation, and power sectors.[25] And a study of three developing nations (South Africa, India, and Brazil) found annual losses to introduced pests of $138 billion per year.[26]

In addition to the direct costs of managing invasives, the economic costs also include their indirect environmental consequences and other nonmarket values. For example, invasives may cause changes in ecological services by disturbing the operation of the hydrological cycle, including flood control

Table 1 Indicative Costs of Some Invasive Alien Species (in U.S. Dollars)

Species	Economic Variable	Economic Impact
Introduced disease organisms	Annual cost to human, plant, and animal health in the United States	$41 billion per year[a]
A sample of alien species of plants and animals	Economic costs of damage in the United States	$137 billion per year[b]
Salt cedar	Value of ecosystem services lost in western United States	$7–16 billion over 55 years[c]
Knapweed and leafy spurge	Impact on economy in three U.S. states	Direct costs of $40.5 million per year; indirect costs of $89 million[d]
Zebra mussel	Damages to U.S. industry	Damage of more than $2.5 billion to the Great Lakes fishery between 1998–2000;[e] $5 billion to U.S. industry by 2000[f]
Most serious invasive alien plant species	Costs 1983–1992 of herbicide control in England	$344 million per year for 12 species[g]
Six weed species	Costs in Australia agroecosystems	$105 million per year[h]
Pinus, Hakeas, and Acacia	Costs on South African floral kingdom to restore to pristine state	$2 billion total for impacts felt over several decades[i]
Water hyacinth	Costs in seven African countries	$20–50 million per year[j]
Rabbits	Costs in Australia	$373 million per year (agricultural losses)[k]
Varroa mite	Economic cost to beekeeping in New Zealand	An estimated $267–602 million over the next 35 years[l]

[a] P. Daszak, A. Cunningham, and A. D. Hyatt, "Emerging Infectious Diseases of Wildlife: Threats to Biodiversity and Human Health," *Science,* 21 January 2000, 443–49.

[b] D. Pimentel, L. Lach, R. Zuniga, and D. Morrison, "Environmental and Economic Costs of Non-indigenous Species in the United States," *BioScience* 50 (2000): 53–65.

[c] E. Zavaleta, "Valuing Ecosystem Services Lost to Tamarix Invasion in the United States," in H. A. Mooney and R. J. Hobbs, eds., *Invasive Species in a Changing World* (Washington DC: Island Press, 2000).

[d] D. A. Bangsund, F. L. Leistritz, and J. A. Leitch, "Assessing Economic Impacts of Biological Control of Weeds: The Case of Leafy Spurge in the Northern Great Plains of the United States," *Journal of Environmental Management* 56 (1999): 35–43; and D. A. Bangsund, S. A. Hirsch, and J. A. Leitch, *The Impact of Knapweed on Montana's Economy* (Fargo, ND: Department of Agricultural Economics, North Dakota State University, 1996).

[e] P. C. Focazio, "Coordinated Issue Area: Aquatic Nuisance, Non-Indigenous, and Invasive Species," *Coastlines* 30, no.1 (2001): 4–5.

[f] "Coatings to Repel Zebra Mussels," U.S. Army Construction Engineering Research Laboratory fact sheet, http://www.cecer.army.mil/facts/sheets/FL10.html.

[g] M. Williamson, "Measuring the Impact of Plant Invaders in Britain," in S. Starfinger, K. Edwards, I. Kowarik, and M. Williamson, eds., *Plant Invasions: Ecological Mechanisms and Human Responses* (Leiden, Netherlands: Backhuys, 2000), 57–70.

[h] A. Watkinson, R. Freckleton, and P. Dowling, "Weed Invasions of Australian Farming Systems: From Ecology to Economics," in C. Perrings, M. Williamson, and S. Dalmazzone, eds., *The Economics of Biological Invasions* (Cheltenham, UK: Edward Elgar, 2000), 94–116.

[i] J. Turpie and B. Heydenrych, "Economic Consequences of Alien Infestation of the Cape Floral Kingdom's Fynbos Vegetation," in Perrings, Williamson, and Dalmazzone, ibid., pages 152–82.

[j] S. Joffe and S. Cook, *Management of the Water Hyacinth and Other Aquatic Weeds: Issues for the World Bank* (Cambridge, UK: Commonwealth Agriculture Bureau International (CABI) Bioscience, 1997).

[k] P. White and G. Newton-Cross, "An Introduced Disease in an Invasive Host: The Ecology and Economics of Rabbit Carcivirus Disease (RCD) in Rabbits in Australia," in Perrings, Williamson, and Dalmazzone, note h above, pages 117–37.

[l] R. Wittenberg and M. J. W. Cock, eds., *Invasive Alien Species: A Tool Kit of Best Prevention and Management Practices* (Wallingford, UK: Global Invasive Species Programme and CABI, 2001).

Source: J. A. McNeely.

and water supply, waste assimilation, recycling of nutrients, conservation and regeneration of soils, pollination of crops, and seed dispersal. Such services have current-use value and option value (the potential value of such services in the future). In the South African fynbos, for example, the establishment of invasive tree species—which use more water than do native species—has decreased water supplies for nearby communities and increased fire hazards, justifying government expenditures equivalent to US$40 million per year for both manual and chemical control.[27]

Customs and quarantine practices, developed in an earlier time, are inadequate safeguards against the rising tide of species that threaten native biodiversity.

Many people in today's globalized economy are driven especially by economic motivations. Those who are importing non-native species are usually doing so with a profit motive and often seek to avoid paying for possible associated negative impacts if those species become invasive. The fact that these negative impacts might take several decades to appear make it all the easier for the negative economic impacts to be ignored. Similarly, those who are ultimately responsible for such "accidental" introductions (for example, through infestation of packing materials or organisms carried in ballast water) seek to avoid paying the economic costs that would be required to prevent these "accidental," but predictable, invasions. In both cases, the potential costs are externalized to the larger society, and to future generations.

Responses

Customs and quarantine practices, developed in an earlier time to guard against diseases and pests of economic importance, are inadequate safeguards against the rising tide of species that threaten native biodiversity. Globally, about 165 million 6-meter-long, sealed containers are being shipped around the world at any given time. This number is far larger than custom officers can reasonably be expected to examine in detail. In the United States, some 1,300 quarantine officers are responsible for inspecting 410,000 planes and more than 50,000 ships, with each ship carrying hundreds of containers. While they intercept alien species nearly 50,000 times a year, it is highly likely that at least tens of thousands more enter the country uninspected each year. In Europe, inspection at the port of entry is also desperately overextended, and once a container enters the European Union, no further border inspections are done. This is a recipe for disaster.

Instead, a different set of strategies is now needed to deal with invasive species. These include prevention (certainly the most preferable), early eradication, special containment, or integrated management (often based on biological control). Mechanical, biological, and chemical means are available for controlling invasive species of plants and animals once they have arrived. Early warning, quarantine, and various other health measures are involved to halt the spread of pathogens.[28]

The international community has responded to the problem of invasive alien species through more than 40 conventions or programs, and many more are awaiting finalization or ratification.[29] The most comprehensive is the 1992 Convention on Biological Diversity, which calls on its 188 parties to "prevent the introduction of, control, or eradicate those alien species which threaten ecosystems, habitats, or species" (Article 8h).[30] A much older instrument, one that is virtually universally applied, is the 1952 International Plant Protection Convention, which applies primarily to plant pests, based on a system of phytosanitary certificates. Regional agreements further strengthen this convention. Other instruments deal with invasive alien species in specific regions (such as Antarctica), sectors (such as fishing in the Danube River), or vectors (such as invasive species in ballast water, through the International Maritime Organization). The fact that the problem continues to worsen indicates that the international response to date has been inadequate.

On the national level, some legal measures can offer very straightforward methods of preventing or managing invasions.

For example, to deal with the problem of Asian beetle invasions, the United States now requires that all solidwood packing material from China must be certified free of bark (under which insects may lurk) and heat-treated, fumigated, or treated with preservatives. China might reasonably issue a reciprocal regulation, as North American beetles are a hazard there.

The nursery industry is by far the largest intentional importer of new plant taxa. Issuing permits for imported species is a good way for the agencies responsible for managing such invasions to keep track of what is being traded and moved around the country. Some people believe that it is impossible to issue a regulation containing a list of permitted and prohibited species, at least partly because the ornamental horticulture industry is always seeking new species. But the Florida Nurserymen and Growers Association recently identified 24 marketed species on a black list drawn up by Florida's Exotic Pest Plant Council and decided to discourage trade in 11 of the species (the least promising sellers in any case).[31]

Sometimes nature itself can fight back against invasive alien species, at least when they reach plague proportions. For example, the zebra mussels that have invaded the North American Great Lakes with disastrous effects are now declining because a native sponge (*Eunapius fragilis*) is growing on the mussels, preventing them from opening their shells to feed or to breathe. The sponge has become abundant in some areas, while the zebra mussel population has fallen by up to 40 percent, although it is not yet clear whether the sponges will be effective in controlling the invasive mussels in the long term.[32]

Biological control—the intentional use of natural enemies to control an invasive species—is an important tool for managers. Some early efforts at biological control agents had disastrous effects, such as South American cane toads (*Bufo marinus*) in Australia, Indian common mynahs (*Acridotheres tristis*) in Hawaii, and Asian mongooses (*Herpestes javanicus*) in the Caribbean. Not only did these species not deal with the problem species upon which they were expected to prey, but they ended up causing havoc to native species and ecosystems. On the other hand, biological control programs are now much more carefully considered and in many cases are the most efficient, most effective, cheapest, and least damaging to the environment of any of the options for dealing with invasives that have already arrived.[33] Examples include the use of a weevil (*Cyrtobagous salviniae*) to control salvinia fern (*Salvinia molesta*), another weevil (*Neohydronomus affinis*) to control water lettuce (*Pistia stratiotes*), and a predatory beetle (*Hyperaspis pantherina*) to control orthezia scale (*Orthezia insignis*) that threatened the endemic national tree of Saint Helena (*Commidendrum robustum*).[34]

Those seeking to use viruses or other disease organisms to control an invasive species need to understand ecological links. When millions of rabbits died after the intentional introduction of the myxomatosis virus in the United Kingdom, for example, populations of their predators, including stoats, buzzards, and owls, declined sharply. The impact affected other species indirectly, leading to local extinction of the endangered large blue butterfly (*Maculina arion*) because of reduced grazing by rabbits on heathlands, which removed the habitat for an ant species that assists developing butterfly larvae.[35] But the use of the myxoma virus in conjunction with 1080 poison on the Phillip Island in the South Pacific successfully eradicated invasive rabbits, allowing the recovery of the island's vegetation (including the endemic *Hibiscus insularis*).[36]

At small scales of less than one hectare, it appears possible with current technology to eradicate invasive species of plants through use of herbicides, fire, physical removal, or a combination of these, but the costs of eradication rise quickly as the area covered increases. With the right approach and technology, invasive alien mammals can be eradicated from islands of thousands of hectares in size. Rat eradication from islands of larger than 2,000 hectares has been successful, and large mammals have been removed from much bigger ones than that, primarily by hunting and trapping.

Environmentally sensitive eradication also requires the restoration of the community or ecosystem following the removal of the invasive. For example, the eradication of Norway rats from Mokoia Island in New Zealand was followed by greatly increased densities of mice, also alien species. Similarly, the removal of Pacific rats (*Rattus exulans*) from Motupao Island, New Zealand, to protect a native snail led to increases of an exotic snail to the detriment of the natives. And on Motunau Island, New Zealand, the exotic box-thorn (*Lycium ferocissimum*) increased after the control of rabbits. On Santa Cruz Island, off the west coast of California, removing goats led to dramatic increases in the abundance of fennel (*Foeniculum vulgare*) and other alien species of weeds. Thus reversing the changes to native communities caused by non-native species will often require a sophisticated understanding of ecological relationships. It is now well recognized that eradication programs are only the first step in a long process of restoration.[37] Sometimes native species become dependent on invasive ones, causing dilemmas for managers. For example, giant kangaroo rats (*Dipodomys ingens*) in the American West continually modify their burrow precincts by digging tunnels, clipping plants, and other activities. This chronic disturbance to soil and vegetation sometimes promotes the establishment of invasive species of plants that were originally imported as ornamentals from the Mediterranean so that they constitute a very large proportion of the vegetation on giant kangaroo rat territories. They have significantly larger seeds than do native species so are favored by the grain-eating kangaroo rats.[38] Because the kangaroo rats depend on non-native plant species for food and the non-native plant species depend upon the kangaroo rats to disturb their habitat continually, the relationship is mutualistic. This strong relationship may also inhibit population growth of native grassland plants that occupy disturbed habitats but have difficulty competing with nonnative weeds for resources. This mutualism presents an intractable conservation management dilemma, suggesting that it may be impossible to restore valley grasslands occupied by endangered kangaroo rats to conditions where native species dominate.

High-tech management measures are also being tried. For example, Australian scientists are planning to insert a gene known as "daughterless" into invasive male carp (*Cyprinus carpio*) in the Murray-Darling River, the country's longest, thereby ensuring that their offspring are male. The objective is to release them into the wild, sending wild carp populations into a decline and making room for the native species that are being threatened by the invasive European carp.[39] Using genetic modification can help eradicate an invasive alien species, but if the detrimental gene is released into nature and starts to flourish, many other species could be negatively affected. Thus the precautionary approach needs to be applied to control techniques as well as to introductions.

The problems of invasive alien species are so serious that actions must be taken even before we can be "certain" of all of their effects. However, mechanical removal, biocontrol, chemical control, shooting, or any other approach to controlling alien invasive species needs to be carefully considered prior to use to ensure that the implications have been fully and carefully considered, including impacts on human health, other species, and so forth. A public information program is also needed to ensure that the proposed measures are likely to be effective as well as socially and politically acceptable. Many animal-rights groups oppose the killing of any species of wildlife, for instance, even if they are causing harm to native species of plants and animals. The recent controversy surrounding the population of mute swans in the Chesapeake Bay is a good example.[40]

Conclusions

Ecosystems have been significantly influenced by people in virtually all parts of the world; some have even called these "engineered ecologies." Thus, a much more conscious and better-informed management of ecosystems—one that deals with non-native species—is critical.

In just a few hundred years, major global forces have rendered natural barriers ineffective, allowing non-native species to travel vast distances to new habitats and become invasive alien species. The globalization and growth in the volume of trade and tourism, coupled with the emphasis on free trade, provide more opportunities than ever for species to be spread accidentally or deliberately. This inadvertent ending of millions of years of biological isolation has created major ongoing environmental problems that affect developed and developing countries, with profound economic and ecological implications.

Because of the potential for economic and ecological damage when an alien species becomes invasive, every alien species needs to be treated for management purposes as if it is potentially invasive, unless and until convincing evidence indicates that it is harmless in the new range. This view calls for urgent action by a wide range of governmental, intergovernmental, private sector, and civil institutions.

A comprehensive solution for dealing with invasive alien species has been developed by the Global Invasive Species Programme.[41] It includes 10 key elements:

- *An effective national capacity to deal with invasive alien species.* Building national capacity could include designing and establishing a "rapid response mechanism" to detect and respond immediately to the presence of potentially invasive species as soon as they appear, with sufficient funding and regulatory support; as well as implementing appropriate training and education programs to enhance individual capacity, including customs officials, field staff, managers, and policymakers. It could also include developing institutions at national or regional levels that bring together biodiversity specialists with agricultural quarantine specialists. Building basic border control and quarantine capacity and ensuring that agricultural quarantine, customs, and food inspection officers are aware of the elements of the Biosafety Protocol are other ways to deal with invasive alien species on a national level.

- *Fundamental and applied research at local, national, and global levels.* Research is required on taxonomy, invasion pathways, management measures, and effective monitoring. Further understanding on how and why species become established can lead to improved prediction on which species have the potential to become invasive; improved understanding of lag times between first introduction and establishment of invasive alien species; and better methods for excluding or removing alien species from traded goods, packaging material, ballast water, personal luggage, and other methods of transport.

The problems of invasive alien species are so serious that actions must be taken even before we can be "certain" of all their effects.

- *Effective technical communications.* An accessible knowledge base, a planned system for review of proposed introductions, and an informed public are needed within countries and between countries. Already, numerous major sources of information on invasive species are accessible electronically and more could also be developed and promoted, along with other forms of media.
- *Appropriate economic policies.* While prevention, eradication, control, mitigation, and adaptation all yield economic benefits, they are likely to be undersupplied, because it is difficult for policymakers to identify specific beneficiaries who should pay for the benefits received. New or adapted economic instruments can help ensure that the costs of addressing invasive alien species are better reflected in market prices. Economic principles relevant to national strategies include ensuring that those responsible for the introduction of economically harmful invasive species are liable for the costs they impose; ensuring that use rights to natural or environmental resources include an obligation to prevent the spread of potential invasive alien species; and requiring importers of such potential species to have liability insurance to cover the unanticipated costs of introductions.

- *Effective national, regional, and international legal and institutional frameworks.* Coordination and cooperation between the relevant institutions are necessary to address possible gaps, weaknesses, and inconsistencies and to promote greater mutual support among the many international instruments dealing with invasive alien species.
- *A system of environmental risk analysis.* Such a system could be based on existing environmental impact assessment procedures that have been developed in many countries. Risk analysis measures should be used to identify and evaluate the relevant risks of a proposed activity regarding alien species and determine the appropriate measures that should be adopted to manage the risks. This would also include developing criteria to measure and classify impacts of alien species on natural ecosystems, including detailed protocols for assessing the likelihood of invasion in specific habitats or ecosystems.
- *Public awareness and engagement.* If management of invasive species is to be successful, the general public must be involved. A vigorous public awareness program would involve the key stakeholders who are actively engaged in issues relevant to invasive alien species, including botanic gardens, nurseries, agricultural suppliers, and others. The public can also be involved as volunteers in eradication programs of certain nonnative species, such as woody invasives of national parks for suggested actions that individuals can take.
- *National strategies and plans.* The many elements of controlling invasive alien species need to be well coordinated, ensuring that they are not simply passed on to the Ministry of Environment or a natural resource management department. A national strategy should promote cooperation among the many sectors whose activities have the greatest potential to introduce them, including military, forestry, agriculture, aquaculture, transport, tourism, health, and water-supply sectors. The government agencies with responsibility for human health, animal health, plant health, and other relevant fields need to ensure that they are all working toward the same broad objective of sustainable development in accordance to national and international legislation. Such national strategies and plans can also encourage collaboration between different scientific disciplines and approaches that can seek new approaches to dealing with problems caused by invasive alien species.
- *Invasive alien species issues built into global change initiatives.* Global change issues relevant to invasives begin with climate change but also include changes in nitrogen cycles, economic development, land use, and other fundamental changes that might enhance the possibilities of these species becoming established. Further, responses to global change issues, such as sequestering carbon, generating biomass energy, and recovering degraded lands, should be designed in ways that use native species and do not increase the risk of the spread of non-native invasives.

What Can an Individual Do?

While the problem of invasive alien species seems daunting, an individual can make an important contribution to the problem, and if thousands of individuals work toward reducing the spread of invasive aliens, real progress can be made. Here are some steps that can be taken:

- Become informed about the issue.
- Grow native plants, keep native pets, and avoid releasing non-natives into the wild.
- Avoid carrying any living materials when traveling.
- Never release plants, fish, or other animals into a body of water unless they came out of that body of water.
- Clean boats before moving them from one body of water to another, and avoid using non-native species as bait.
- Support the work of organizations that are addressing the problem of invasive alien species.

- *Promotion of international cooperation.* The problem of invasive alien species is fundamentally international, so international cooperation is essential to develop the necessary range of approaches, strategies, models, tools, and potential partners to ensure that the problems of such species are effectively addressed. Elements that would foster better international cooperation could include developing an international vocabulary, widely agreed upon and adopted; cross-sector collaboration among international organizations involved in agriculture, trade, tourism, health, and transport; and improved linkages among the international institutions dealing with phytosanitary, biosafety, and biodiversity issues and supporting these by strong linkages to coordinated national programs.

Because the diverse ecosystems of our planet have become connected through numerous trade routes, the problems caused by invasive alien species are certain to continue. As with maintaining and enhancing health, education, and security, perpetual investments will be required to manage the challenge they present. These 10 elements will ensure that the clear and present danger of invasive species is addressed in ways that build the capacity to address any future problems arising from expanding international trade.

Notes

1. H. A. Mooney, J. A. McNeely, L. E. Neville, P. J. Schei, and J. K. Waage, eds., *Invasive Alien Species: Searching for Solutions* (Washington, DC: Island Press, 2004).

2. B. Groombridge, ed., *Global Biodiversity: Status of the Earth's Living Resources* (Cambridge, UK: World Conservation Monitoring Centre, 1992).

3. C. Hilton-Taylor, IUCN *Red List of Threatened Species* (Gland, Switzerland: IUCN–The World Conservation Union (IUCN). 2000).

4. R. May, "British Birds by Number," *Nature,* 6 April 2000, 559–60.

5. Ibid., and Groombridge, note 2 above.

6. J-Y. Meyer, "Tahiti's Native Flora Endangered by the Invasion of *Miconia calvesens,*" *Journal of Geography* 23 (1997): 775–81.

7. D. Pimentel, L. Lach, R. Zuniga, and D. Morrison, "Environmental and Economic Costs of Nonindigenous Species in the United States," *BioScience* 50 (2000): 53–65.

8. World Trade Organization (WTO), *International Trade Statistics 2003* (Geneva: WTO, 2004).

9. G. M. Ruiz et al., "Global Spread of Microorganisms by Ships," *Nature,* 2 November 2000, 49–50.

10. J. E. Pasek, "Assessing Risk of Foreign Pest Introduction via the Solid Wood Packing Material Pathway," presentation made at the North American Plant Protection Organization Symposium on Pet Risk Analysis, Puerto Vallerta, Mexico, 18–21 March 2002.

11. U.S. Department of Agriculture, *Agricultural Research Service Research to Combat Invasive Species,* www.invasivespecies. gov/toolkit/arsisresearch.doc (accessed 27 April 2004).

12. S. L. Chown and K. J. Gaston, "Island-Hopping Invaders Hitch a Ride with Tourists in South Georgia," *Nature,* 7 December 2000, 637.

13. K. R. McKaye et al., "African Tilapia in Lake Nicaragua," *BioScience* 45 (1995): 406–11.

14. L. Winter, "Cats Indoors!" *Earth Island Journal,* Summer 1999, 25–26.

15. G. Zorpette, "Parrots and Plunder," *Scientific American,* July 1997, 15–17.

16. Ruiz et al., note 9 above.

17. McKaye et al., note 13 above.

18. A. J. Ribbink, "African Lakes and Their Fishes: Conservation Scenarios and Suggestions," *Environmental Biology of Fishes* 19 (1987): 3–26; T. Goldschmit, F. Witte, and J. Wanink, "Cascading Effects of the Introduced Nile Perch on the Detritivorousphytoplanktivorous Species in the Sublittoral Areas of Lake Victoria," *Conservation Biology* 7 (1993): 686–700; and R. Ogutu-Ohwayo, "Nile Perch in Lake Victoria: The Balance between Benefits and Negative Impacts of Aliens," in O. T. Sandlund, P. J. Schei, and A. Viken, eds., *Invasive Species and Biodiversity Management* (Dordrecht, Netherlands: Kluwer Academic Publishers, 1999), 47–64.

19. M. L. McKinney and J. L. Lockwood, "Biotic Homogenization: A Few Winners Replacing Many Losers in the Next Mass Extinction," *Tree* 14 (1999): 450–53.

20. C. M. D'Antonio, T. L. Dudley, and M. Mack, "Disturbance and Biological Invasions: Direct Effects and Feedbacks," in L. Locker, ed., *Ecosystems of Disturbed Ground* (Amsterdam: Elziveer, 1999).

21. B. Groombridge and M. D. Jenkins, *World Atlas of Biodiversity: Earth's Living Resources in the 21st Century* (Berkeley, CA: University of California Press, 2002).

22. Pimentel, Lach, Zuniga, and Morrison, note 7 above.

23. N. L. Larson, P. J. Anderson, and W. Newton, "Alien Plant Invasion in Mixed-Grass Prairie: Effects of Vegetation Type and Anthropogenic Disturbance," *Ecological Applications* 11 (2001): 128–41; and J. M. Levine, "Species Diversity and Biological Invasions: Relating Local Process to Community Pattern," *Science,* 5 May 2000, 852–54.

24. Ibid.

25. S. Noemdoe, "Putting People First in an Invasive Alien Clearing Programme: Working for Water Programme," in J. A. McNeely, ed., *The Great Reshuffling: Human Dimensions of Invasive Alien Species,* (Gland, Switzerland: IUCN, 2001), 121–26.

26. D. Pimentel et al., "Economic and Environmental Threats of Alien Plant, Animal, and Microbe Invasions," *Agriculture, Ecosystems and Environment* 84 (2001): 1–20.

27. Ibid.

28. J. Kaiser, "Stemming the Tide of Invading Species," *Science,* 17 September 2000, 836–841.

29. C. Shine, N. Williams, and L. Gündling, *A Guide to Designing Legal and Institutional Frameworks on Alien Invasive Species* (Bonn, Germany: IUCN, 2000).

30. L. Glowka, F. Burhenne-Guilmin, and H. Synge, *A Guide to the Convention on Biological Diversity* (Gland, Switzerland: IUCN, 1994). See also K. Raustiala and D. G. Victor, "Biodiversity Since Rio: The Future of the Convention on Biological Diversity," *Environment,* May 1996, 16–20, 37–45.

31. Kaiser, note 28 above.

32. A. Ricciardi, F. Sneider, D. Kelch, and H. Reiswig, "Lethal and Sub-Lethal Effects of Sponge Overgrowth on Introduced Dreissenide Mussels in the Great Lakes—St. Lawrence River System," *Canadian Journal of Fisheries and Aquatic Sciences* 52:2695–703.

33. M. S. Hoddle, "Restoring Balance: Using Exotic Species to Control Invasive Exotic Species," *Conservation Biology* 18 (2004): 38–49; S. M. Louda and P. Stiling, "The Double-Edged Sword of Biological Control in Conservation and Restoration," *Conservation Biology* 18 (2004): 50–53; and R. Wittenberg and M. J. W. Cock, eds., *Invasive Alien Species: A Tool Kit of Best Prevention and Management Practices* (Wallingford, UK: Global Invasive Species Programme and Commonwealth Agricultural Bureau International, 2001).

34. Wittenberg and Cock, ibid.

35. P. Daszak, A. Cunningham, and A. D. Hyatt. "Emerging Infectious Diseases of Wildlife: Threats to Biodiversity and Human Health," *Science,* 21 January 2000, 443–49.

36. P. Coyne, "Rabbit Eradication on Phillip Island," in Wittenberg and Cock, note 33 above, page 176.

37. R. C. Klinger, P. Schuyler, and J. D. Sterner, "The Response of Herbaceous Vegetation and Endemic Plant Species to the Removal of Feral Sheep from Santa Cruz Island, California," in C. R. Veitch and M. N. Klout, eds., *Turning the Tide: the Eradication of Invasive Species* (Gland, Switzerland, and Cambridge, UK: IUCN, 2002), 141–54.

38. P. Schiffman, "Promotion of Exotic Weed Establishment by Endangered Giant Kangaroo Rats (*Dipodomys ingens*) in a California Grassland," *Biodiversity and Conservation* 3 (1994): 524–37.

39. R. Nowak, "Gene Warfare: One Small Tweak and a Whole Species Will Be Wiped Out," *New Scientist,* 11 May 2002, 6.

40. See B. Engle, "No Swansong in the Chesapeake Bay," *Environment,* December 2003, 7.

41. The Global Invasive Species Programme (GISP) was established in 1997 as a consortium of the Scientific Committee on Problems of the Environment (SCOPE), CABI, and IUCN, in partnership with the United Nation Environment Programme and with funding from the Global Environment Facility (GEF). See J. A. McNeely, H. A. Mooney, L. E. Neville, P. J. Schei, and J. K. Waage, eds., *Global Strategy on Invasive Alien Species* (Gland, Switzerland: IUCN, 2001).

JEFFREY A. MCNEELY is chief scientist at IUCN-The World Conservation Union in Gland, Switzerland. His research focuses on a broad range of topics relating to conservation and sustainable use of biodiversity, with a particular focus in recent years on the relationship between agriculture and wild biodiversity, the relationship between biodiversity and human health, and the impacts of war on biodiversity. McNeely has written or edited more than 30 books, from *Mammals of Thailand* (Association for the Conservation of Wildlife, 1975), his first, to *Eco-agriculture: Strategies to Feed the World and Save Wild Biodiversity* (Island Press, 2003). He has also published extensively on biodiversity, protected areas, and cultural aspects of conservation. He may be reached at jam@iucn.org.

From *Environment,* July/August 2004, pp. 17–31. Reprinted by permission of the Helen Dwight Reid Educational Foundation. Published by Heldref Publications, 1319 Eighteenth St., NW, Washington, DC 20036-1802. Copyright © 2004. www.heldref.org

America's Coral Reefs: Awash with Problems

Government must acknowledge the magnitude of the crisis and fully engage the scientific and conservation communities in efforts to solve it.

TUNDI AGARDY

America's coral reefs are in trouble. From the disease-ridden dying reefs of the Florida Keys, to the over-fished and denuded reefs of Hawaii and the Virgin Islands, this country's richest and most valued marine environment continues to decline in size, health, and productivity.

How can this be happening to one of our greatest natural treasures? Reefs are important recreational areas for many and are loved even by large portions of the public who have never had the opportunity to see their splendor firsthand. Coral reefs are sometimes referred to as the "rainforests of the sea," because they teem with life and abound in diversity. But although only a small number of Americans have ever had rainforest experiences, many more have had the opportunity to dive and snorkel in nearshore reef areas. And in contrast to the obscured diversity of the forests, the gaudily colored fish and invertebrates of the reef are there for anyone to see. Once they have seen these treasures, the public becomes transformed from casual observers to strong advocates for their protection. This appeal explains why many zoos have rushed in recent years to display coral reef fishes and habitats, even in inland areas far from the coasts (such as Indianapolis, site of one of the largest of the country's public aquaria). Coral reefs have local, national, and even global significance.

Even when one looks below the surface (pun intended) of the aesthetic appeal of reefs, it is easy to see why these biological communities command such respect. Coral reefs house the bulk of known marine biological diversity on the planet, yet they occur in relatively nutrient-poor waters of the tropics. Nutrient cycling is very efficient on reefs, and complicated predator-prey interactions maintain diversity and productivity. But the fine-tuned and complex nature of reefs may spell their doom: Remove some elements of this interconnected ecosystem, and things begin to unravel. Coral reefs are one of the few marine habitats that undergo disturbance-induced phase shifts: an almost irreversible phenomenon in which diverse reef ecosystems dominated by stony corals dramatically turn into biologically impoverished wastelands overgrown with algae. Worldwide, some 30 percent of reefs have been destroyed in the past few decades, and another 30 to 50 percent are expected to be destroyed in 20 years' time if current trends continue. In the Caribbean region, where many of the reefs under U.S. jurisdiction can be found, coral cover has been reduced by 80 percent during the past three decades.

The U.S. government is fully aware of the value of these marine ecosystems and the fact that they are in trouble. In 1998, the Clinton administration established the U.S. Coral Reef Task Force (USCRTF), a high-level interagency group charged with examining reef problems and finding solutions. Executive Order 13089 stipulated that a task force be established to oversee that "all Federal agencies whose actions may affect U.S. coral reef ecosystems shall: (a) identify their actions that may affect U.S. coral reef ecosystems; (b) utilize their programs and authorities to protect and enhance the conditions of such ecosystems; and (c) to the extent permitted by law, ensure that any actions they authorize, fund, or carry out will not degrade the conditions of such ecosystems." The task force comprises 11 federal agencies, plus corresponding state, territorial, and tribal authorities.

The USCRTF has looked for ways to better monitor the condition of reefs, share information, and coordinate management. Among the key government players are the National Oceanic and Atmospheric Administration (NOAA), Department of the Interior, Environmental Protection Agency (EPA), Department of Defense, Department of Agriculture, Department of Justice, Department of State, National Science Foundation (NSF), and NASA. Yet however well-intentioned this move on the part of government, coral reef health has continued to decline, and the USCRTF, while elevating the profile of the issue, has not been able to stem the degradation. The reasons for this ineffectiveness are complex and go beyond the "too little, too late" offered as the standard criticism. Although the response of the government may have indeed come too late for many of America's reefs, the shortcomings of the task force have more to do with its reluctance to fully engage with the scientific community, take advantage of emerging technologies, and raise awareness about the consequences of reef degradation. If this is happening to our most treasured marine environments, what can the future be for our less-well-loved, less charismatic marine areas?

Threats to U.S. Reefs

Even as we are becoming more fully aware of their enormous ecological and economic value, coral reefs are being lost in the United States, just as they are being destroyed in other parts of the world. Some 37 percent of all corals in Florida have died since 1996, and the incidence of coral disease at sampling sites there went up by 446 percent in the same short period. The U.S. has jurisdiction over a surprisingly large proportion of extant coral reefs, including the world's third largest barrier reef in Florida; a vast tract of reef systems throughout the Hawaiian Islands; and extensive reefs in U.S. territories such as Puerto Rico, the U.S. Virgin Islands, Guam, American Samoa, and the Northern Mariana Islands. These reef resources contribute an estimated $375 billion to the U.S. economy annually, yet virtually all of these reef ecosystems are under threat, and many may be destroyed altogether in the coming decades.

Although in many parts of the world coral reefs are deliberately destroyed in the process of coastal development or to obtain construction materials, in the United States coral reefs suffer the classic death of a thousand cuts. They are strongly affected by eutrophication: the overfertilization of waters caused by the inflow of nutrients from fertilizer, sewage, and animal wastes. The overabundance of nutrients causes algae to overgrow and smother coral polyps; in extreme cases, leading to totally altered and biologically impoverished alternate ecosystems. Reefs are also sensitive to sediments that increase turbidity and reduce the sunlight reaching the coral colonies. (Though corals are animals, they have symbiotic dinoflagellates called zooxanthellae living within their tissues. The photosynthesis undertaken by these plant symbionts provides corals with the extra energy needed to create the calcium carbonate that forms their skeletons and thus the reef structure.) Sedimentation is a common threat to U.S. coral reefs, especially in areas where unregulated coastal development or deforestation causes soil runoff into nearshore waters.

Because energy flows in coral reef ecosystems are largely channeled into ecosystem maintenance and little surplus is available for harvest, reefs are highly sensitive to overfishing. The removal of grazing fishes, for instance, increases the likelihood that algae will dominate the reef, causing a subsequent decline in productivity and diversity. Reef communities denuded of even relatively small numbers of fishes are also less likely to recover from episodic bleaching events, because recruitment is inhibited by the lack of grazing fishes to create settlement space. Similarly, declines in sea turtle species such as hawksbill and green turtles

negatively affect reef ecology. The removal of top predators such as reef sharks, jacks, and barracudas can also cause cascading effects resulting in reduced overall diversity and declines in productivity. Despite these impacts, very few coral reef areas of the United States have fishing regulations expressly designed to prevent these ecological cascading effects from occurring. In fact, most people would be surprised to find out that even in seemingly protected reefs, such as those that occur within the Virgin Islands Biosphere Reserve around St. John, U.S.V.I., almost all forms of recreational and commercial fishing are allowed.

Coral reefs are also extremely vulnerable to changes in their ambient environment, having narrow tolerance ranges in temperature and salinity. Warming affects both coral polyp physiology and the pH of seawater, which in turn affects the calcification rates of hard corals and their ability to create reef structure. For this reason, even a slight warming of sea temperatures has dramatic effects, especially when coupled with other negative impacts such as eutrophication and overfishing. There is some indication that warming sea temperatures may render coral colonies vulnerable to the spread of disease or to increased mortality in response to normally nonpathogenic viruses and bacteria. The spread of known coral diseases and the emergence of new, even more debilitating diseases are alarming phenomena in the Florida Keys reefs and underlie many of the die-back episodes there in the past decade.

The effects of warming are most clearly manifested in coral bleaching. Bleaching is an event in which the zooxanthellae of the corals, which give corals their beautiful colors, are expelled from the coral polyps, leaving the colonies white. Bleached corals cannot lay down calcium carbonate skeletons and thus enter a period of stasis. A bleached coral is not necessarily a dead coral, however, and corals have been known to recover from bleaching events (we also know from paleoarcheology that bleaching is a natural event that preceded greenhouse gas-related warming of the atmosphere). Because some reefs do fully recover after bleaching, it is difficult to predict what consequences warming events such as periodic El Niños will have on the long-term health of any reef. This uncertainty has been seized on by both doomsayers and naysayers in the debate about the future of reefs: The doomsayers declare that the majority of reefs face certain death from bleaching, while the naysayers claim that bleaching is not only natural but adaptive. However, one thing is absolutely clear: Stressed reefs have a heightened sensitivity to temperature changes and are far less likely to recover from bleaching events. And with a few exceptions (some parts of the northwest Hawaiian Islands and Palmyra Atoll, for instance), all of the coral reefs in the United States are highly stressed by a combination of land-based sources of pollution, overfishing, and the destruction of habitats that are ecologically critical to reef communities, such as seagrass beds and mangrove forests. This does not bode well for a future in which sea temperatures will undoubtedly continue to rise.

These losses affect more than our personal environmental sensibilities. Reefs support some of the most important industries in the United States and the rest of the world: 5 percent of world commercial fisheries are reef-based, and over 50 percent of U.S. federally managed fishery species depend on reefs during some part of their life cycle. Herman Cesar, Lauretta Burke, and Lida Pet-Soede argue in a recent monograph on the economics of coral reef degradation that the costs of better managing reefs are far outweighed by the net benefits provided by reefs. In the Florida Keys, for example, they claim that a proposed wastewater treatment plant that would mitigate many of the threats to the Florida Keys reef tract would cost $60 to $70 million in capital costs and about $4 million in annual maintenance costs. At the same time, the benefits to the local population (estimated to be greater than the net present value of $700 million) would far eclipse the outlays. In Hawaii and the reef-fringed territories, coastal tourism is tightly coupled to intact reefs. Reefs in these regions not only provide tourist destinations, they also play important roles in controlling beach erosion and buffering land from storms. In such places, it is easy to see how an investment in better reef protection would be a small cost in contrast to the great benefits provided by sustained tourism revenues.

Inadequate Responses

The failure to respond to the coral reef crisis in this country has to do with many factors:

Incomplete understanding of the problem and communication failures. Although there is an appreciation of the crisis worldwide, there is still reluctance on the part of some U.S. managers to consider the crisis "our problem." Everyone is quick to lament the destruction of Southeast Asian and Indian Ocean reefs by dynamite fishers, or the use of cyanide in collecting coral reef fish in the Philippines, but the reefs under U.S. jurisdiction have hardly fared better. In the past decade, we have seen a slow awakening to the problems facing U.S. reefs, but the response has been to collect more data, slowly and painstakingly. At first independently, and then in a more coordinated fashion with the establishment of the USCRTF, government agencies have made greater efforts to monitor reefs in certain regions, but the massive amounts of data collected often create problems in data interpretation and management. Too little emphasis has been placed on either synthesizing the data collected or collecting data in new ways to make it more relevant for conservation. Lacking a synthesis or periodic syntheses, we end up burying our heads in the sand about what is happening to our coral reefs.

The USCRTF and the government agencies it represents have not actively looked for ways to partner with academic, scientific, and nongovernmental organizations to take advantage of information being collected and disseminated by them. Instead, the government has relied almost solely on the efforts of its own scientists. Many of its scientists, such as Charles Birkeland of the U.S. Geological Survey, are indeed world leaders in coral reef ecology and management, but collectively the research being undertaken by government agencies is either substandard, too conservative, or both. Virtually every new advance in coral reef ecosystem understanding has been made not by government scientists but by academics or researchers in the private sector. U.S. government scientists have not explored the potential of

new technologies such as biochemical markers that indicate reef stress (pioneered by the private sector), nor have they properly harnessed the remote sensing technologies they have deployed in order to improve reef surveillance.

Even the knowledge that has been gained is inadequately communicated to the public and to decisionmakers. Part of the problem has been the rush to oversimplify what is actually a very complex set of issues, in the hopes that decisionmakers higher up will take both notice and action. In the Florida Keys, for instance, advocates for improving the water quality of the nearshore environment have fought against the restoration of the Florida Everglades, arguing that the increased water flows into Florida Bay would bring higher concentrations of pollutants to the reef tract. In casting the reef problems in such a simplistic light, proponents of singular solutions actually impede responsible government agencies from tackling reef problems head-on and in the comprehensive manner that is required.

The U.S. government has serious shortcomings when it comes to communicating and raising awareness about complicated environmental issues. For this reason, it would behoove the USCRTF to partner with organizations that have good outreach mechanisms in place, such as environmental groups. Such public-private partnerships would also ease the financial burden of the cash-strapped government agencies, allowing them to spend funds in short supply on management and on measuring management efficacy.

Poor use of cutting-edge science and the at-large scientific community. Although the United States is one of the most technologically advanced countries in the world, it has not adequately harnessed science to address the coral reef crisis. In a 1999 article in the journal Marine and Freshwater Research, Michael Risk compares the response of the scientific community to the coral reef crisis with its response to two other crises affecting the United States: acid rain in the Northern Hemisphere and eutrophication of the Great Lakes. Risk argues that whereas there was effective engagement of the scientific community in tackling the latter two issues, neither U.S. nor international scientists have helped craft an effective response to the large-scale death of reefs.

Risk is right to ask why science has failed coral reefs, but I take issue with his assessment of the nation's inadequate response to the crisis. It is not the fault of the scientific community that the government has been slow to act to save reefs, but rather the fault of government in not knowing how to use science and scientists effectively. Decisionmakers have not engaged the scientific community and have failed to heed what scientific advice has been put forward. For instance, the government did not fully mobilize nongovernmental academic institutions and conservation organizations to help draft its National Action Plan to Conserve Coral Reefs, and as a result the plan has been criticized as lacking in rigor and ambition. It is telling that a World Bank project to undertake global coral reef-targeted research, which assembles international teams of leading researchers to address critical issues of bleaching, disease, connectivity, remote sensing, modeling, and restoration, has a paucity of U.S.

government scientists in all six of the working groups. This targeted research project is crucial: It intends to identify the key questions that managers need to have answered in order to better protect reefs, and it aims to do intensive applied research to answer those questions.

The National Action Plan to Conserve Coral Reefs was produced by the USCRTF and published on March 2, 2000. It is a general document describing why coral reefs are important and what needs to be done to protect them. There are two main sections: understanding coral reef ecosystems and reducing the adverse impacts of human activities. The first section discusses four action items: (1) create comprehensive maps, (2) conduct long-term monitoring and assessment, (3) support strategic research, and (4) incorporate the human dimension (undertake economic valuation, etc.). The second section is a bit more ambitious: (1) create and expand a network of marine protected areas (MPAs), (2) reduce impacts of extractive uses, (3) reduce habitat destruction, (4) reduce pollution, (5) restore damaged reefs, (6) reduce global threats to coral reefs, (7) reduce impacts from international trade in coral reef species, (8) improve federal accountability and coordination, and (9) create an informed public. All well and good, but despite its moniker the action plan provides almost no guidance on how to do these things. It called for each federal agency to develop implementation plans (required by Executive Order 13089) by June 2000. However, those plans were only to cover fiscal years 2001 and 2002, and the plans were never formalized or made public. The USCRTF recognized that a greater investment needed to be made to figure out how each agency was going to contribute to carrying out the action plan and pushed agencies to develop post-2002 strategies. To date, only the Department of Defense and NOAA have completed such strategies. NOAA's plan is embodied in its National Strategy for Conserving Coral Reefs document published in September 2002. Both the action plan and the NOAA strategy are available on the USCRTF Web site (www.coralreef.gov).

The plans put forward by the USCRTF, however, place far too much emphasis on monitoring and mapping and far too little emphasis on abating threats and effectively managing reefs. The focus of research has been to monitor existing conditions rather than to set up applied experiments that would tell us which threats are most critical to tackle. This is not to say that all government research has been worthless. Regular monitoring in the Florida Keys allowed NOAA to understand the alarming "blackwater" event in January 2003 (in which fishermen noticed black water, later found to be a combination of a plankton bloom and tannins, moving from the Everglades toward the reefs) and reassure the public that it was a natural event, because they had several years of monitoring information with which they could hindcast. Similarly, the mapping investment, although too high a priority, has led to some interesting revelations: There are newly discovered reefs in the northeastern portion of the Gulf of Mexico that are now on the public's radar screen, for instance.

Although the USCRTF has recognized the importance of MPAs in conserving reefs, it has not given the government agencies that have responsibility for implementation guidance on how to optimally design these protected areas. The action plan thus codifies a dangerous tendency to use simplistic formulae for designing protected areas. The plan states these as its goals: "establish additional no-take ecological reserves to provide needed protection to a balanced suite of representative U.S. coral reefs and associated habitats, with a goal to protect at least 5% of all coral reefs . . . by 2002; at least 10% by 2005, and at least 20% by 2010." By adopting a policy of conserving 20 percent of reef areas within no-take reserves, without requiring planners to fully understand the threats to a particular reef and without guiding planners to locate such protected areas in the most ecologically critical areas, the plan pushes decisionmakers to implement ineffective MPAs, thus squandering opportunities for real conservation. In some jurisdictions, these area targets have already been reached, with 20 percent of reef areas set aside as no-take zones, but because these areas were chosen more for their ease of establishment and less for their ecological importance, little conservation has been accomplished. In a true display of lack of ambition and creativity, the USCRTF and its agencies have not considered using ocean zoning outside of MPAs to conserve reefs, and the MPA directives remain an old-school, one-size-fits-all approach.

Poor governmental coordination and lots of infighting. Since its formation in June of 1998, the USCRTF has made some strides toward better monitoring, information sharing, and management coordination for reefs under U.S. jurisdiction. This is no minor feat, because until the task force was established, no effort had been made to promote communication and cooperation between the multitude of agencies and bureaus that each have a role to play in coral reef management. NOAA's National Ocean Service and National Marine Fisheries Service, the Department of the Interior's U.S. Fish and Wildlife Service, and the EPA are the key players in the USCRTF, but also important are the National Parks Service, DOD, Department of Agriculture, Department of Justice, Department of State, NSF, and NASA. Although the major players (in particular, NOAA and the Department of the Interior) are engaged in internecine warfare over territorial claims and access to funding, some of the more minor players have taken their charge very seriously. DOD, for example, has developed its own plan for conserving the reefs under its jurisdiction, which include some of the most pristine reefs in the nation, such as the reefs of Johnston Atoll in the central Pacific.

Unlike many terrestrial habitats, coral reefs suffer both from human activity that directly affects the marine environment (such as dredging, fishing, and marine tourism) and from activity on land that has an indirect but highly insidious effect on reef health and productivity. Thus, in order to better understand and manage reefs, it is imperative that the United States continues, and now strengthens, coordinating mechanisms between the various government entities that control the wide array of human activities that damage reefs.

The USCRTF now has a roadmap to increase understanding about coral reefs and better protect them from further destruction, embodied in the National Action Plan to Conserve Coral Reefs. A subsequent report prepared by NOAA, in cooperation with the USCRTF, was submitted to Congress in 2002. The 156-page National Coral Reef Action Strategy provides a nationwide status report on implementation of the National Action Plan to Conserve Coral Reefs and the Coral Reef Conservation Act of 2000.

Will the USCRTF now be able to do what it could not in the first five years of its existence: stem the tide of degradation affecting U.S. coral reefs? Or is the U.S. government merely creating a façade of improved management, while government researchers and managers continue to work in isolation from cutting-edge researchers in U.S. academe, nongovernmental organizations, and international institutions? Will new policy developments, such as the administration's support for broad environmental exemptions for DOD's military training and anti-terrorism operations, act to wholly undermine any substantive progress made by the USCRTF and the government agencies it represents?

Only time will answer these questions with certainty, but the initial impressions are not promising. The National Action Plan to Conserve Coral Reefs is too heavily invested in relatively easily accomplished activities such as mapping the nation's coral reefs, and its formulaic and simplistic approach to creating MPAs will not likely result in meaningful protection. Already overburdened and underfunded agencies are not getting the political mentoring they need to ensure that appropriations will be sufficient to allow them to carry out their mandates under these plans. Without public-private partnerships and private-sector financial support, too many elements of the plan will fall by the wayside. Neither the Action Plan nor the NOAA strategy provide adequate information on the true choices and tradeoffs that decisionmakers will have to consider and act on in order to create a revolution in the way we manage coral reefs. And clearly a revolution is needed; business as usual will only continue to put U.S. coral reef ecosystems in harm's way. In the end, the United States may fall far short of its goal of demonstrating how to effectively manage coral reefs in a way that all the world can see. Instead, it may well win the race to destroy the inventory of one of the world's most diverse and precious environments.

Global forces are at play: The United States is not an island. Were the United States suddenly to act more effectively to protect reefs under its jurisdiction, our reef ecosystems would still be in some peril, for many reasons. First, many damaging activities occur out of sight, especially in remote reef areas with little or no surveillance. Second, the open nature of marine systems means that reefs are affected by the condition of the environment far from the reef tracts themselves. Sometimes larval propagules travel long distances, and the origin of recruits is tens or hundreds of kilometers away, in areas that could be entirely outside U.S. jurisdiction. Similarly, pollution from outside the U.S. can easily find its way to reefs within America's borders. Finally, some threats to reefs are global in nature, such as rising temperatures caused by global warming. These threats will not diminish unless meaningful international agreements succeed in tackling the root causes of the threats. For all these reasons, protection of U.S. reefs will require more than administering the reefs within our borders; it will also require international negotiation, cooperation, and capacity-building.

Promising Sign

Is there hope for U.S. coral reefs? Yes, as long as we can more fully engage the private sector and the scientific community in the struggle to save reefs, and at the same time convince decisionmakers of the need to take significant steps to protect these fragile ecosystems. It is a promising sign that in June 2003, NOAA, EPA, and the Department of the Interior convened a meeting in Hawaii to discuss coral bleaching and ways to gain better collaboration between the scientific community at large and the government agencies charged with managing reefs. The USCRTF is beginning to reach out to scientists involved in coral reef research and management outside the United States, such as the coral reef-targeted research working groups formed under the recent World Bank initiative. In this way, the U.S. government can begin to take advantage of the significant strides in scientific understanding that have been made by nongovernmental researchers, both in the United States and abroad.

New advances in technology may help coral reefs as well, and just in time. For instance, the Planetary Coral Reef Foundation has teamed up with the Massachusetts Institute of Technology and other academic institutions to attempt to launch a satellite that will provide real-time information about the condition of reefs worldwide. Such a satellite mission would make it possible to know the extent of coral bleaching and the presence of fishing operations anywhere in the world at any time. With such a system in place, traditional surveillance could be cut back, allowing money to be redirected toward conservation. At the same time, donors could get a better sense of where their investments are paying off in terms of real conservation of reefs and could identify trouble spots quickly enough to get funds flowing to places where emergency measures are needed.

With the full engagement of the scientific community and with partnering to remove some of the burden from beleaguered government agencies, managers will be able to tailor responses to the given threats at any reef location. Where fishing is deemed to be a major stressor, the United States will have to find the political will to manage reef-based fisheries more effectively. Where pollution (whether nutrients, toxics, debris, or alien species) is undermining reef health and resilience, coastal zone and agricultural agencies will have to work to find ways to reduce pollutant loading. Where visitor overuse and diver damage are issues, managers will have to look for ways to prevent people from loving reefs to death. And in all areas, managers will have to resist oversimplifying the situation and begin to better inform the public and decisionmakers about the hard choices to be made.

The coral reef crisis is indeed our problem. It affects our natural heritage and the livelihoods of a great number of our citizens. Only when the people in power recognize the magnitude of the problem will effective steps be taken to engage the wider scientific and conservation community in safeguarding reefs. When future generations look back at the dawn of the millennium and the environmental choices that were made, they will either curse us for letting one of nature's most wondrous ecosystems be extinguished or praise us for recognizing the great value of reefs and moving to protect them. I hope it is the latter.

TUNDI AGARDY (tundiagardy@earthlink.net) is the executive director of Sound Seas in Bethesda, Maryland.

Markets for Biodiversity Services
Potential Roles and Challenges

Michael Jenkins, Sara J. Scherr, and Mira Inbar

Historically, it has been the responsibility of governments to ensure biodiversity protection and provision of ecosystem services. The main instruments to achieve such objectives have been

- direct resource ownership and management by government agencies;
- public regulation of private resource use;
- technical assistance programs to encourage improved private management; and
- targeted taxes and subsidies to modify private incentives.

But in recent decades, several factors have stimulated those concerned with biodiversity conservation services to begin exploring new market-based instruments. The model of public finance for forest and biodiversity conservation is facing a crisis as the main sources of finance have stagnated, despite the recognition that much larger areas require protection. At the same time, increasing recognition of the roles that ecosystem services play in poverty reduction and rural development is highlighting the importance of conservation in the 90 percent of land outside protected areas. It is thus urgent to find new means to finance the provision of ecosystem services, yet under current conditions private actors lack financial incentives to do so.

Crisis in Biodiversity Conservation Finance

Financing and management of natural protected areas has historically been perceived as the responsibility of the public sector. According to the United Nations Environment Programme, there are presently 102,102 protected areas worldwide, covering an area of 18.8 million square kilometers. Seventeen million square kilometers of these areas—11.5 percent of the Earth's terrestrial surface—are forests. Two-thirds of these have been assigned to one of the six protected-area management categories designated by the World Conservation Union (IUCN).

However, over the last few decades, severe cutbacks in the availability of public resources have undermined the effectiveness of such strategies. Protected areas in the tropics are increasingly dependent on international public or private donors for financing. Yet budgets for government protection and management of forest ecosystem services are declining, as are international sources from overseas development assistance. Land acquisition for protected areas and compensation for lost resource-based livelihoods are often prohibitively expensive. For example, it has been estimated that $1.3 billion would be required to fully compensate inhabitants in just nine central African parks.[1] The donation-driven model is often unsustainable, both economically and environmentally. Sovereignty is also an issue: About 30 percent of private forest concessions in Latin America and the Caribbean and 23 percent in Africa are already foreign owned. At the same time, public responsibility for nature protection is shifting with processes of devolution and decentralization, and new sources of financing for local governments to take on biodiversity and ecosystem service protection have not been forthcoming.

Moreover, scientific studies increasingly indicate that biodiversity cannot be conserved by a small number of strictly protected areas.[2] Conservation must be conceived in a landscape or ecosystem strategy that links protected areas within a broader matrix of land uses that are compatible with and support biodiversity conservation in situ. To achieve such outcomes, it will be essential to engage private actors in conservation finance on a large scale. Yet the markets for products from natural areas and forests face at least three serious challenges: declining commodity prices for traditionally important products, such as timber; competition from illegal sources; and poorly functioning, overregulated markets. Thus, private forest owners and landowners need to find new revenue streams to justify retaining forests on the landscapes and to manage them well in the context of declining commodity prices and competition in natural forests from illegal sources of timber.

Rural Development, Poverty Reduction, and Biodiversity

The vast majority of biodiversity resources in the world are found in populated landscapes, and it can be argued that the biodiversity that underpins ecosystem services critical to human health and livelihoods should have high priority in conservation efforts. An estimated 240 million rural people live in the world's high-canopy forest landscapes. In Latin America, for example, 80 percent of all forests are located in areas of medium to high human population density.[3] Population growth in the world's remaining "tropical wilderness areas" is twice the global average. More than a billion people live in the 25 biodiversity "hotspots" identified by Conservation

Table 1 Estimated Financial Flows for Forest Conservation (in millions, U.S. dollars)

Sources of Finance	SFM (early 1990s)	SFM (early 2000)	PAS (early 1990s)	PAS (early 2000)
Official development assistance	$2,000–$2,200	$1,000–$1,200	$700–$770	$350–$420
Public expenditure	NA	$1,600	NA	$598
Philanthropy[a]	$85.6	$150	NA	NA
Communities[b]	$365–$730	$1,300–$2,600	NA	NA
Private companies	NA	NA	NA	NA

[a]Underestimates self-financing and in-kind nongovernmental organization contributions.
[b]Self-financing and in-kind contributions from indigenous and other local communities.
Note: In 1990, there were an estimated 100 million hectares of community-managed forests worldwide. SFM is "sustainable forest management." PAS stands for "protected area system."
Source: A. Molnar, S. J. Scherr, and A. Khare, *Current Status and Future Potential of Markets for Ecosystem Services of Tropical Forests: An Overview* (Washington, DC: Forest Trends, 2004).

International; in 16 of these hotspots, population growth is higher than the world average.[4] While species richness is lower in drylands and other ecosystems not represented among the "hot spots," the species that play functional ecosystem roles are all the more important and difficult to replace.

Poor rural communities are especially dependent upon natural biodiversity. Low-income rural people rely heavily on the direct consumption of wild foods, medicines, and fuels, especially for meeting micronutrient and protein needs, and during "hungry" periods. An estimated 350 million poor people rely on forests as safety nets or for supplemental income. Farmers earn as much as 10 to 25 percent of household income from nontimber forest products. Bushmeat is the main source of animal protein in West Africa. The poor often harvest, process, and sell wild plants and animals to buy food. Sixty million poor people depend on herding in semiarid rangelands that they share with large mammals and other wildlife. Thirty million low-income people earn their livelihoods primarily as fishers, twice the number of 30 years ago. The depletion of fisheries has serious impacts on food security. Wild plants are used in farming systems for fodder, fertilizer, packaging, fencing, and genetic materials. Farmers rely on soil microorganisms to maintain soil fertility and structure for crop production, and they also rely on wild species in natural ecological communities for crop pollination and pest and predator control. Wild relatives of domesticated crop species provide the genetic diversity used in crop improvement. The rural poor rely directly on ecosystem services for clean and reliable local water supplies. Ecosystem degradation results in less water for people, crops, and livestock; lower crop, livestock, and tree yields; and higher risks of natural disasters.

More than a billion people live in the 25 biodiversity "hotspots" identified by Conservation International; in 16 of these hotspots, population growth is higher than the world average.

Three-quarters of the world's people living on less than $1 per day are rural. Strategies to meet the United Nations Millennium Development Goals in rural areas—to reduce hunger and poverty

and to conserve biodiversity—must find ways to do so in the same landscapes. Crop and planted pasture production—mostly in low-productivity systems—dominate at least half the world's temperate, subtropical, and tropical forest areas; a far larger area is used for grazing livestock.[5] Food insecurity threatens biodiversity when it leads to overexploitation of wild plants and animals. Low farm productivity leads to depletion of soil and water resources and increases the pressure to clear additional land that serves as wildlife habitat. Some 40 percent of cropland in developing countries is degraded. Of more than 17,000 major protected areas, 45 percent (accounting for one-fifth of total protected areas) are heavily used for agriculture, while many of the rest are islands in a sea of farms, pastures, and production forests that are managed in ways incompatible for long-term species and ecosystem survival.[6]

Despite this high level of dependence by the poor on biodiversity, the dominant model of conservation seeks to exclude people from natural habitats. In India, for example, 30 million people are targeted for resettlement from protected areas.[7] From the perspective of poverty reduction and rural development, it is thus urgent to identify alternative conservation systems that respect the rights of forest dwellers and owners and address conservation objectives in the 90 percent of forests outside public protected areas. Markets for ecosystem services potentially offer a more efficient and lower-cost approach to forest conservation.[8]

Need for Financial Incentives to Provide Ecosystem Services

There is growing recognition that regulatory and protected area approaches, while critical, are insufficient to adequately conserve biodiversity. A fundamental problem is financial, especially for resources that lie outside protected areas. For these to be conserved, they need to be more valuable than the alternative uses of the land. And for such resources to be well managed, good stewardship needs to be more profitable than bad stewardship. The failure of forest owners and producers to capture financial benefits from conserving ecosystem benefits leads to overexploitation of forest resources and undersupply of ecosystem services.

This reality is hard for many people to accept, because most ecosystem services are considered "public goods." The "polluter pays" principle has argued that the right of the public to these services trumps the private rights of the landowner or manager. Yet good management has a cost. While the individual who manages his or her resources to protect biodiversity produces public benefits, the costs incurred are private. Under current institutions, those who benefit from these services have no incentive to compensate suppliers for these services. In most of the world, forest ecosystem services are not traded and have no "price." Thus, where the opportunity costs of forest land for agricultural enterprises, infrastructure, and human settlements are higher than the use or income value of timber and nontimber forest products (NTFPs), habitats will be cleared and wild species will be allowed to disappear. Because they receive little or no direct benefit from them, resource owners and producers ignore the real economic and noneconomic values of ecosystem services in making decisions about land use and management.

A lower-cost approach to securing conservation is to pay only for the biodiversity services themselves, by paying landowners to manage their assets so as to achieve biodiversity or species conservation.

Mechanisms are needed by which resource owners are rewarded for their role as stewards in providing biodiversity and ecosystem services. Anticipation of such income flows would enhance the value of natural assets and thus encourage their conservation. Compared to previous approaches to forest conservation, market-based mechanisms promise increased efficiency and effectiveness, at least in some situations. Experience with market-based instruments in other sectors has shown that such mechanisms, if carefully designed and implemented, can achieve environmental goals at significantly less cost than conventional "command-and-control" approaches, while creating positive incentives for continual innovation and improvement. Markets for ecosystem services could potentially contribute to rural development and poverty reduction by providing financial benefits from the sale of ecosystem services, improving human capital through associated training and education, and strengthening social capital through investment in local cooperative institutions.

New Market Solutions to Conserve Biodiversity

The market for biodiversity protection can be characterized as a nascent market. Many approaches are emerging to financially remunerate the owners and managers of land and resources for their good stewardship of biodiversity (see Table 2). Market mechanisms to pay for other ecosystem services—watershed services, carbon sequestration or storage, landscape beauty, and salinity control, for example—can be designed to conserve biodiversity as well. However, in general, biodiversity services are the most demanding to protect because of the need to conserve many different elements essential for diverse, interdependent species to thrive. Figure 1 illustrates potential market solutions and some of the complexities involved.

Land Markets for High-Biodiversity-Value Habitat

National governments (in the form of public parks and protected areas), NGO conservation organizations (for example, The Nature Conservancy), and individual conservationists have long paid for the purchase of high-biodiversity-value forest habitats. Direct acquisition can be expensive, as underlying land and use values are also included. Local sovereignty concerns arise when buyers are from outside the country—or even the local area—or where extending the area of noncommercial real estate reduces the local tax base. New commercial approaches are being developed to encourage the establishment of privately owned conservation areas, such as conservation communities (the purchase of a plot of land by a group of people mainly for recreation or conservation purposes), ecotourism-based land protection projects, and ecologically sound real estate projects being organized in Chile.[9] These build on growing consumer demand for housing and vacation in biodiverse environments.

Payments for Use or Management

A lower-cost approach to securing conservation is to pay only for the biodiversity services themselves, by paying landowners to manage their assets so as to achieve biodiversity or species conservation. It is likely that the largest-scale payments for land-use or management agreements belong to one of two categories. One encompasses government agroenvironmental payments made to farmers in North America and Europe for reforesting conservation easements. The other category describes management contracts aiming to conserve aquatic and terrestrial wildlife habitat. In Switzerland, "ecological compensation areas," which use farming systems compatible with biodiversity conservation, have expanded to include more than 8 percent of total agricultural land. In the tropics, diverse approaches include nationwide public payments in Costa Rica for forest conservation and in Mexico for forested watershed protection.

Conservation agencies are organizing direct payments systems, such as conservation concessions being negotiated by Conservation International, and forest conservation easements negotiated by the *Cordão de Mata* ("linked forest") project with dairy farmers in Brazil's Atlantic Forest. The dairy farmers in the latter example receive, in exchange, technical assistance and investment resources to raise crop and livestock productivity. Some countries that use land taxes are using tax policies in innovative ways to encourage the expansion of private and public protected areas.

Payment for Private Access to Species or Habitat

Private sector demand for biodiversity has tended to take the form of payments for access to particular species or habitats that function as "private goods" but in practice serve to cover some or all of the costs of providing broader ecosystem services. Pharmaceutical companies have contracted for bioprospecting rights in tropical forests. Ecotourism companies have paid forest owners for the right to bring tourists into their lands to observe wildlife, while private individuals are willing to pay forest owners for the right to hunt, fish, or gather nontimber forest products.

Table 2   Types of Payments for Biodiversity Protection

Purchase of high-value habitat

Type	Mechanism
Private land acquisition	Purchase by private buyers or nongovernmental organizations explicitly for biodiversity conservation
Public land acquisition	Purchase by government agency explicitly for biodiversity conservation
Payment for access to species or habitat	
Bioprospecting rights	Rights to collect, test, and use genetic material from a designated area
Research permits	Right to collect specimens, take measurements in area
Hunting, fishing, or gathering permits for wild species	Right to hunt, fish, and gather
Ecotourism use	Rights to enter area, observe wildlife, camp, or hike
Payment for biodiversity-conserving management	
Conservation easements	Owner paid to use and manage defined piece of land only for conservation purposes; restrictions are usually in perpetuity and transferable upon sale of the land
Conservation land lease	Owner paid to use and manage defined piece of land for conservation purposes for defined period of time
Conservation concession	Public forest agency is paid to maintain a defined area under conservation uses only; comparable to a forest logging concession
Community concession in public protected areas	Individuals or communities are allocated use rights to a defined area of forest or grassland in return for commitment to protect the area from practices that harm biodiversity
Management contracts for habitat or species conservation on private farms, forests, or grazing lands	Contract that details biodiversity management activities and payments linked to the achievement of specified objectives
Tradable rights under cap-and-trade regulations	
Tradable wetland mitigation credits	Credits from wetland conservation or restoration that can be used to offset obligations of developers to maintain a minimum area of natural wetlands in a defined region
Tradable development rights	Rights allocated to develop only a limited total area of natural habitat within a defined region
Tradable biodiversity credits	Credits representing areas of biodiversity protection or enhancement that can be purchased by developers to ensure they meet a minimum standard of biodiversity protection
Support biodiversity-conserving businesses	
Biodiversity-friendly businesses	Business shares in enterprises that manage for biodiversity conservation
Biodiversity-friendly products	Eco-labeling

Source: S. J. Scherr, A. White, and A. Khare, *Current Status and Future Potential of Markets for Ecosystem Services in Tropical Forests: An Overview* (Washington, DC. Forest Trends, 2003).

Tradable Rights and Credits within a Regulatory Framework

Multifactor markets for ecosystem services have been successfully established, notably for sulfur dioxide emissions, farm nutrient pollutants, and carbon emissions. These create rights or obligations within a broad regulatory framework and allow those with obligations to "buy" compliance from other landowners or users. Developing such markets for biodiversity is more complicated, because specific site conditions matter so much. The United States has operated a wetlands mitigation program since the early 1980s in which developers seeking to destroy a wetland must offset that by buying wetland banks conserved or developed elsewhere. A similar

approach is used for "conservation banking," described in the box on the next page.

A variant of this approach is being designed for conserving forest biodiversity in Brazil by permitting flexible enforcement of that country's "50 percent rule," which requires landholders in Amazon forest areas to maintain half of their land in forest. This rule is also applied in other regions in Brazil, where lesser proportional areas are set aside for forest use. Careful designation of comparable sites is required.

Another approach, biodiversity credits, is under development in Australia. In this system, legislation creates new property rights for private landholders who conserve biodiversity values on their land. These landholders can then sell resulting "credits" to a

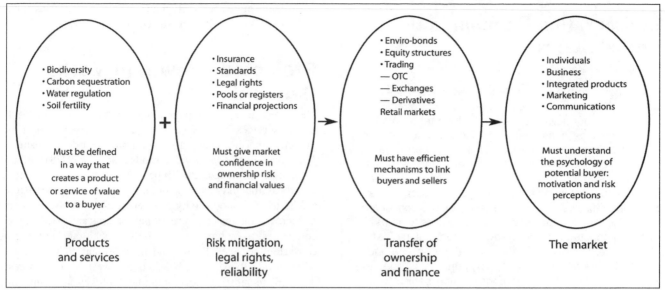

Figure 1 New Market Solutions to Conserve Biodiversity.

Note: OTC ("over-the-counter") trading involves direct negotiation with buyers and sellers rather than an official stock market.

Source: D. Brand, "Emerging Markets for Forest Services and Implications for Rural Development, Forest Industry, and Government," presentation to the Katoomba Group Meeting, "Developing Markets for Ecosystem Services," Vancouver, October 2000.

A New Fund to Finance Forest Ecosystem Services

The Mexican government recently announced the creation of a new fund to pay indigenous and other communities for the forest ecosystem services produced by their land.[1] Indigenous and other communities own approximately 80 percent of all forests in Mexico—totaling some 44 million hectares—as collectively held, private land. The Mexican Forestry Fund has been under design since 2002, guided by a consultative group with government, nongovernmental organization, and industry representatives. The purpose of the US$20 million fund is to promote the conservation and sustainable management of natural forests, leverage additional financing, contribute to the competitiveness of the forest sector, and catalyze the development of mechanisms to finance forest ecosystem services. Operational manuals are being prepared, and priority conservation sites have already been identified. The fund proposes to pay $40 per hectare (ha) per year to owners of deciduous forests in critical mountain areas and $30 per ha per year to other forest types.

Notes

1. Comisión Nacional Forestal (CONAFOR), presentation given at the Mexican Forestry Expo, Guadalajara, Mexico, 8 August 2003.

common pool. The law also creates obligations for land developers and others to purchase those credits. The approach requires that the "value" of the biodiversity unit can be translated into a dollar value.

Biodiversity-Conserving Businesses

Conservation values are beginning to inform consumer and investor decisions. Eco-labeling schemes are being developed that advertise or certify that products were produced in ways consistent with biodiversity conservation. The global trade in certified organic agriculture was worth $21 billion worldwide in 2000.[10] International organic standards are expanding to landscape-scale biodiversity impacts. The Rainforest Alliance and the Sustainable Agriculture Network certify coffee, bananas, oranges, and other products grown in and around high-biodiversity-value areas. The Sustainable Agriculture Initiative is a coalition of multinational commercial food producers (Nestle, Dannon, Unilever, and others) who are seeking to ensure that all of the products they purchase along the supply chain come from producers who are protecting biodiversity. In 2002, more than 100 million hectares of forest were certified (a fourfold increase over 1996), although only 8 percent of the total certified area is in developing countries, and most of that is in temperate forests.

Current Market Demand

Available information suggests that biodiversity protection services are presently the largest market for ecosystem services. A team from McKinsey & Company, the World Resources Institute, and The Nature Conservancy estimated the annual international finance for the conservation market (conservation defined as protecting land from development) at $2 billion, with the forest component a large share of that.[11] Buyers are predominantly development banks and foundations in the United States and Europe.

A study by the International Institute for Environment and Development (IIED) of 72 cases of markets for forest biodiversity protection services in 33 countries found that the main buyers of biodiversity services (in declining order of prevalence) were private corporations, international NGOs and research institutes, donors, governments, and private individuals.[12] Communities, public agencies, and private individuals predominate as sellers.

Conservation Banking in the United States

Amendments to the United States Endangered Species Act in 1982 provided for an "incidental take" of enlisted species, if "a landowner provides a long-term commitment to species conservation through development of a Habitat Conservation Plan (HCP)." These amendments have opened the door to a series of market-based transactions, described as conservation banking, which permits land containing a natural resource (such as wetlands, forests, rivers, or watersheds) that is conserved and maintained for specified enlisted species to be used to offset impacts occurring elsewhere to the same natural resource.[1] A private landowner may request an "incidental take" permit and mitigate it by purchasing "species credits" from preestablished conservation banks. Credits are administered according to individuals, breeding pairs, acres, nesting sites, and family units. Conservation banking has maximized the value of underutilized commercial real estate and given private landowners incentive to conserve habitat.

California was the first state to authorize the use of conservation banking and has established 50 conservation banks since 1995. Other states, including Alabama, Colorado, and Indiana, have followed suit. In April 2002, the Indiana Department of Transportation, the Federal Highway Association Indiana Division, and four local government agencies finalized an HCP for the endangered Indiana bat as part of the improvement of transportation facilities around Indianapolis International Airport. These highway improvements will occur in an area of known Indiana bat habitat that is predicted to experience nearly $1.5 billion in economic development during the next ten years. Under the HCP, approximately 3,600 acres will be protected, including 373 acres of existing bat habitat.

Notes

1. A. Davis, "Conservation Banking," presentation to the Katoomba Group-Lucarno Workshop, Lucarno, Switzerland, November 2003.

Local actors more commonly focus on protecting species or habitats of particular economic, subsistence, or cultural value.

Projected Growth in Market Demand

The fastest-growing component of future market demand for biodiversity services is likely to be in eco-labeling of crop, livestock, timber, and fish products for export and for urban consumers. In 1999, the value of the organic foods market was US$14.2 billion. Its value is growing at 20–30 percent a year in the industrialized world, as the international organic movement is strengthening standards for biodiversity conservation.[13] Pressures continue to increase on major international trading and food processing companies to source from suppliers who are not degrading ecosystem services. Donor and international NGO conservation will continue to expand as NGOs begin to establish entire research departments aimed at developing new market-based instruments. Voluntary biodiversity offsets are also a promising source of future demand, as many large companies are seeking ways to maintain their "license to operate" in environmentally sensitive areas, and offsets are of increasing interest to them.

The costs of and political resistance to land acquisition are rising. Construction of biological corridors in and around production areas is an increasingly important conservation objective. At the same time, however, many of the most important sites for biodiversity conservation are in more densely populated areas with high opportunity costs for land. Thus we are likely to see a major shift from land acquisition to various types of direct payments for easements, land leases, and management contracts.

The fastest-growing component of future market demand for biodiversity services is likely to be in eco-labeling of crop, livestock, timber, and fish products for export and for urban consumers.

A rough back-of-the-envelope estimate suggests that the current value of international, national, and local direct payments and trading markets for ecosystem services from tropical forests alone could be worth several hundred million dollars per year, while the value of certified forest and tropical tree crop products may reach as much as a billion dollars. While this is a large and significant amount, it represents a small fraction of the value of conventional tropical timber and other forest product markets. For example, by comparison, the total value of tropical timber exports is $8 billion (including only logs, sawnwood, veneer, and plywood), which is a small fraction of the total exports and domestic timber, pulpwood, and fuelwood markets in tropical countries. NTFP markets are far larger still.[14] The total value of international trade for NTFPs is $7.5 billion–$9 billion per year, with another $108 billion in processed medicines and medicinal plants.[15] Domestic markets for NTFPs are many times larger (for example, domestic consumption accounted for 94 percent of the global output of fresh tropical fruits 1995–2000.)[16] Nonetheless, these rough figures are quite interesting

Most of these cases took place in Latin America and in Asia and the Pacific. Only four cases were found in Europe and Russia and one was found in the United States.

Three-quarters of the cases in the IIED study were international markets, and the rest were distributed among regional, national, and local buyers. International actors—as well as many on the national level—who demand biodiversity protection services tend to focus on the most biodiverse habitats (in terms of species richness) or those perceived to be under the greatest threat globally (for example, places like the Amazon, where there are a high number of endemic species and where habitat area has greatly declined). Most of the private corporations were interested in eco-labeling schemes for crops or timber, investment in biodiversity-friendly companies, horticultural companies concerned with ecosystem services, or pharmaceutical bioprospecting. Such private payments are usually site-specific.

when compared with the scale of public and donor forest conservation finance summarized in Table 1.

Scaling Up Payments for Biodiversity: Next Steps

Markets for ecosystem services are steadily growing and can be expected to grow even more rapidly in the next decade. Yet they predominate as pilot projects. What will it take to transform these markets to impact ecosystem conservation on the global scale? The four most strategic and catalytic areas for policy and action are to

- structure emerging markets to support community-driven conservation;
- mobilize and organize buyers for ecosystem services;
- connect global and national action on climate change to biodiversity conservation; and
- invest in the policy frameworks and institutions required for functioning ecosystem service payment systems.

Supporting Community-Driven Conservation

The benefits of investments in ecosystem services will be maximized over the long term if markets reward local participation and utilize local knowledge. In community forests and agroforestry landscapes, communities have already established sophisticated conservation strategies. Studies of indigenous timber enterprises document conservation investments on the order of $2 per hectare per year apart from other management activities and investments of community time and labor; this is equal to the average available budget per hectare for protected areas worldwide. Conservation policies must recognize the role that local people are playing in the conservation of forest ecosystems worldwide and support them (either with cash or in-kind support) to continue to be good environmental stewards.

To enable conservation-oriented management to remain or become economically viable, it is important that ecosystem service payments and markets are designed so that they strategically channel financial payments to rural communities. Such payments can be used to develop and invest in new production systems that increase productivity and rural incomes, and enhance biodiversity at a landscape scale—an approach referred to as "ecoagriculture."[17] Ecosystem service payments to poor rural communities that are providing stewardship services of national or international value can help to meet multiple Millennium Development Goals. For any semblance of a sustainable future to be realized, it is crucial that our long-term vision includes biodiversity and natural ecosystems as part of the "natural infrastructure" of a healthy economy and society.

Mobilizing and Organizing Buyers for Ecosystem Services

Turning beneficiaries into buyers is the driving force of ecosystem service markets. Because beneficiaries are often hesitant to pay for goods previously considered free, "willingness to pay" for ecosystem services must be organized on a greater scale. The private sector must be called upon to engage in responsible corporate behavior in conserving biodiversity. For example, Insight Investment, a major financial firm, has developed a biodiversity policy that uses

conservation as a screen for investment. Voluntary payments by consumers, retail firms, and other actors can be encouraged through social advertising. This approach is growing rapidly now for eco-labeling programs (labeling of some personal care products and foods) and voluntary carbon emission offset programs involving investment in reforestation. Stockholder pressure is beginning to influence some firms to avoid investments and activities that harm biodiversity, and this is evolving to positive action. Civil society campaigns can also mobilize willingness to pay for biodiversity offsets and payments to local partners for conservation.

Connecting Climate Action with Biodiversity Conservation

Far more aggressive action must and will be taken to mitigate and adapt to climate change. Land use and land-use change currently contribute more than 20 percent of carbon emissions and other greenhouse gases. Action to reduce these emissions must be a central part of our response, and it is critical that action to sequester carbon through improved land uses accompanies strategies to reduce industrial emissions. There is thus an unprecedented opportunity at this time to structure our responses to climate change so that actions related to land use are also designed to protect and restore biodiversity. Moreover, such actions can be designed in ways that enhance and protect livelihoods, especially for those most vulnerable to the impacts of climate change. Indeed, it is imperative that they do so.

As a result of the deliberations at the Conference of the Parties of the United Nations Framework Convention on Climate Change last year, payments for forest carbon through the Clean Development Mechanism (CDM) of the Kyoto Protocol can be used to finance forest restoration and regeneration projects that conserve biodiversity while providing an alternative income source for local people.[18] But the scale of forest carbon under CDM is very small—too small to have a major impact on climate, biodiversity, or livelihoods. It is critical that we aim for a much larger program in the second commitment period, and it is crucial that nations affiliated with the Organisation for Economic Co-operation and Development (OECD) create initiatives to utilize carbon markets for biodiversity conservation in their own internal trading programs. It is imperative to develop a new principle of international agreements on climate response and carbon trading, one that builds a system that encourages overlap of the major international environmental agreements and the Millennium Development Goals. This could mobilize demand by creating an international framework for investing in good ecosystem service markets. It is also important that emerging private voluntary markets for carbon (that is, with actors who do not have a regulatory obligation) are encouraged to pursue such biodiversity goals as well. The Climate, Community and Biodiversity Alliance, for example, is seeking to develop guidelines and indicators for private investments in carbon projects that will achieve these multiple goals. The Forest Climate Alliance of The Katoomba Group is seeking to mobilize the international rural development community to advocate for such approaches.[19]

Investing in Policy Frameworks and Institutions for Biodiversity Markets

Ecosystem service markets are genuinely new—and biodiversity markets are the newest and most challenging. Every market requires basic rules and institutions in order to function, and this is equally

179

Protecting Brazil's Atlantic Forest: The Guaraqueçaba Climate Action Project

Due to excessive deforestation, the Atlantic Forest of Brazil has been reduced to less than 10 percent of its original size. The Guaraqueçaba Climate Action Project has sought to regenerate and restore natural forest and pastureland.[1] Companies such as American Electric Power Company, General Motors, and Chevron-Texaco have invested US$18.4 million to buy carbon emission offset credits from the approximately 8.4 million metric tons of carbon dioxide that the project is expected to sequester during its lifespan. The project has initiated sustainable development activities both within and outside the project boundary, including ecotourism, organic agriculture, medicinal plant production, and a community craft network. The project has made significant contributions toward enhancing biodiversity in the area, creating economic opportunities for local people (such as jobs), restoring the local watershed, and substantially mitigating climate change.

Notes

1. The Nature Conservancy (TNC), *Climate Action: The Atlantic Forest in Brazil* (Arlington, VA: TNC, 1999).

true of biodiversity markets. The biodiversity conservation community needs to act quickly and strategically to ensure that as these markets develop, they are effective, equitable, and operational and are used sensibly to complement other conservation approaches.

Policymakers and public agencies play a vital role in creating the legal and legislative frameworks necessary for market tools to operate effectively. This includes establishing regulatory rules, systems of rights over ecosystem services, and mechanisms to enforce contracts and settle ownership disputes. Ecosystem service markets pose profound equity implications, as new rules may fundamentally change the distribution of rights and responsibilities for essential ecosystem services. Forest producers and civil society will need to take a proactive role to ensure that rules support the public interest and create development opportunities.

New institutions will also be needed to provide the business services required in ecosystem service markets. For example, in order for beneficiaries of biodiversity services to become willing to pay for them, better methods of measuring and assessing biodiversity in working landscapes must be developed, as well as the institutional capacity to do so. New institutions must be created to encourage transactions and reduce transaction costs. Such institutions could include "bundling" biodiversity services provided by large numbers of local producers, as well as investment vehicles that have a diverse portfolio of projects to manage risks. Registers must be established and maintained, to record payments and trades. For example, The Katoomba Group is developing a Web-based "Marketplace" to slash the information and transaction costs for buyers, sellers, and intermediaries in ecosystem service markets.[20]

Conclusion

Conservation of biodiversity and of the services biodiversity provides to humans and to the ecological health of the planet requires financing on a scale many times larger than is feasible from public and philanthropic sources. It is essential to find new mechanisms by which resource owners and managers can realize the economic values created by good stewardship of biodiversity. Moreover, private consumers, producers, and investors can financially reward that stewardship. New markets and payment systems, strategically shaped to deliver critical public benefits, are showing tremendous potential to move biodiversity conservation objectives to greater scale and significance.

Notes

1. M. Cernea and K. Schmidt-Soltau, 2003. "Biodiversity Conservation versus Population Resettlement, Risks to Nature and Risks to People," paper presented at CIFOR (Center for International Forestry Research) Rural Livelihoods, Forests and Biodiversity Conference, Bonn, Germany, 19–23 May 2003.

2. See S. Wood, K. Sebastian, and S. J. Scherr, *Pilot Analysis of Global Ecosystems: Agroecosystems* (Washington, DC: International Food Policy Research Institute and the World Resources Institute, 2000), 64; and E. W. Sanderson et al., "The Human Footprint and the Last of the Wild," *Bioscience* 52, no. 10 (2002): 891–904.

3. K. Chomitz, *Forest Cover and Population Density in Latin America,* research notes to the World Bank (Washington, DC: World Bank, 2003).

4. R. P. Cincotta and R. Engelman, *Nature's Place: Human Population and the Future of Biological Diversity* (Washington, DC: Population Action International, 2000).

5. Wood, Sebastian, and Scherr, note 2 above.

6. J. McNeely and S. J. Scherr, *Ecoagriculture: Strategies to Feed the World and Conserve Wild Biodiversity* (Washington, DC: Island Press, 2003).

7. A. Khare et al., *Joint Forest Management: Policy Practice and Prospects* (London: International Institute for Environment and Development, 2000).

8. S. J. Scherr, A. White, and D. Kaimowitz, *A New Agenda for Forest Conservation and Poverty Reduction: Making Markets Work for Low-Income Communities* (Washington, DC: Forest Trends, CIFOR, and IUCN-The World Conservation Union, 2004).

9. E. Corcuera, C. Sepulveda, and G. Geisse, "Conserving Land Privately: Spontaneous Markets for Land Conservation in Chile," in S. Pagiola et al., eds., *Selling Forest Environmental Services: Market-Based Mechanisms for Conservation and Development* (London: Earthscan Publications, 2002).

10. J. W. Clay, *Community-Based Natural Resource Management within the New Global Economy: Challenges and Opportunities,* a report prepared by the Ford Foundation (Washington, DC: World Wildlife Fund, 2002).

11. M. Arnold and M. Jenkins, "The Business Development Facility: A Strategy to Move Sustainable Forest Management and Conservation to Scale," proposal to the International Finance Corporation (IFC) Environmental Opportunity Facility from Forest Trends, Washington, DC, 2003.

12. N. Landell-Mills, and I. Porras. 2002. *Markets for Forest Environmental Services: Silver Bullet or Fool's Gold? Markets for Forest Environmental Services and the Poor, Emerging Issues* (London: International Institute for Environment and Development, 2002).

13. International Federation of Organic Agriculture Movements (IFOAM), "Cultivating Communities," 14th IFOAM Organic World Congress, Victoria, BC, 21–28 August 2002.

14. S. J. Scherr, A. White, and A. Khare, *Current Status and Future Potential of Markets for Ecosystem Services of Tropical Forests: A Report for the International Tropical Timber Organization* (Washington, DC: Forest Trends, 2003).

15. M. Simula, *Trade and Environment Issues in Forest Protection,* Environment Division working paper (Washington, DC: Inter-American Development Bank, 1999).

16. Food and Agricultural Organization of the United Nations (FAO), *FAOSTAT* database for 2000, accessible via http://www.fao.org.

17. For more information on ecoagriculture, see the Ecoagriculture Partners' Web site at http://www.ecoagriculturepartners.org.

18. S. J. Scherr and M. Inbar, *Clean Development Mechanism Forestry for Poverty Reduction and Biodiversity Conservation: Making the CDM Work for Rural Communities* (Washington, DC: Forest Trends, 2003).

19. For more information on this project, see http://www.katoombagroup.org/Katoomba/forestcarbon.

20. The Katoomba Group is a unique network of experts in forestry and finance companies, environmental policy and research organizations, governmental agencies and influential private, community, and nonprofit groups. It is dedicated to advancing markets for some of the ecosystem services provided by forests, such as watershed protection, biodiversity habitat, and carbon storage. For more information on the Katoomba Group, see http://www.katoombagroup.org. Forest Trends serves as the secretariat for the group. More information on Forest Trends can be found at http://www.forest-trends.org.

MICHAEL JENKINS is the founding president of Forest Trends, a nonprofit organization based in Washington, D.C., and created in 1999. Its mission is to maintain and restore forest ecosystems by promoting incentives that diversify trade in the forest sector, moving beyond exclusive focus on lumber and fiber to a broader range of products and services. Previously he worked as a senior forestry advisor to the World Bank (1998), as associate director for the Global Security and Sustainability Program of the MacArthur Foundation (1988–1998), as an agroforester in Haiti with the U.S. Agency for International Development (1983–1986), and as technical advisor with Appropriate Technology International (1981–1982). He has also worked in forestry projects in Brazil and the Dominican Republic and was a Peace Corps volunteer in Paraguay. He speaks Spanish, French, Portuguese, Creole, and Guaraní, and can be contacted by telephone at (202) 298-3000 or via e-mail at mjenkins@forest-trends.org. **SARA J. SCHERR** is an agricultural and natural resource economist who specializes in the economics and policy of land and forest management in tropical developing countries. She is presently director of the Ecosystem Services program at Forest Trends, and also director of Ecoagriculture Partners, the secretariat of which is based at Forest Trends. She previously worked as principal researcher at the International Center for Research in Agroforestry, in Nairobi, Kenya, as (senior) research fellow at the International Food Policy Research Institute in Washington, D.C., and as adjunct professor at the Agricultural and Resource Economics Department of the University of Maryland, College Park. Her current work focuses on policies to reduce poverty and restore ecosystems through markets for sustainably grown products and environmental services and on policies to promote ecoagriculture—the joint production of food and environmental services in agricultural landscapes. She also serves as a member of the Board of the World Agroforestry Centre, and as a member of the United Nations Millennium Project Task Force on Hunger. Scherr can be reached by telephone at (202) 298-3000 or via e-mail at sscherr@forest-trends.org. **MIRA INBAR** is program associate with Forest Trends. She works in the Ecosystem Services program, supporting efforts to establish frameworks and instruments for emerging transactions in environmental services worldwide. Before joining Forest Trends, she worked with communities in the Urubamba River Valley of Peru to initiate a forest conservation plan. She has worked with the National Fishery Department in Western Samoa, the Marie Selby Botanical Gardens, and Environmental Defense. Inbar may be reached by telephone at (202) 298-3000 or via e-mail at minbar@forest-trends.org. This article is © The Aspen Institute and is published with permission.

From *Environment*, July/August 2004, pp. 32–42. Copyright © 2004 by The Aspen Institute. Reprinted by permission.

UNIT 5

Resources: Land and Water

Unit Selections

Key Points to Consider

- How are groundwater reserves in the United States being monitored in terms of the differences between the amount of water extracted and the amount of water returned to groundwater reserves? Does current scientific analysis hold the idea that many of our groundwater supplies are being depleted to the point where they will no longer yield usable water?

- Is there a relationship between water pollution and the cost of water? How can conservation methods alter the price of water for commercial, domestic, industrial, and agricultural needs?

- Is it possible to reach a balance between the demands of commercial timber industries for tropical hardwoods and the utility of the forest environment for agriculture? Has the development of a transportation system in the Amazon improved or worsened the environmental prospects for the world's largest tropical forest area?

Student Web Site
www.mhcls.com/online

Internet References
Further information regarding these Web sites may be found in this book's preface or online.

Global Climate Change
http://www.puc.state.oh.us/consumer/gcc/index.html

National Oceanic and Atmospheric Administration (NOAA)
http://www.noaa.gov

National Operational Hydrologic Remote Sensing Center (NOHRSC)
http://www.nohrsc.nws.gov

Terrestrial Sciences
http://www.cgd.ucar.edu/tss/

A confluence of events and conditions has led to extreme hardship for a significant portion of the World's population. Over 20% of people live in extreme poverty. There is no simple explanation or solution for this condition, but a number of environmental factors weigh heavily in the overall equation. Increased population, loss of arable land, expansion of farming into marginal regions, drought, degradation in the quality and availability of drinking water, spread of disease, and civil/political conflict are all partly responsible. Each of these factors is interrelated, each conspiring to make the others worse. As populations grow, there is a greater effort to extract more from the land. But more intense farming practices can cause significant loss of topsoil. As a result, people are forced to inhabit and try to cultivate more marginal farmlands. Increased populations mean higher water demands, draining aquifers and consuming (as well as polluting) available surface water sources. Coupled with overgrazing, removal of wood for fuels and higher temperatures due to global warming, much of the world is in the midst of a crisis, and there is no hope for relief anytime soon.

The tragic situation described above violates most of the fundamental tenets of sustainable development. We utilize numerous resources provided to us by the Earth. Some are free, such as the air we breathe, while others we must pay for. Some are renewable, such as trees, wind, water and sunshine. Others are limited, and once gone, will be gone for all practical purposes, forever. For future generations to benefit from the same resources we have today, we must follow sustainable development. We must utilize technologies that do not damage the environment, and do not deplete the natural resource base. At the same time, people must have access to economic progress, and we must invest in human resources. And finally, we must, at some point, reach some stable population growth. The concept of sustainable development is scalable, so that it applies communities, nations and continents. If a community, nation, or this entire planet squanders the available resources, the next generation will lose the benefits enjoyed by its predecessor.

What the ultimate carrying capacity of the planet is, no one knows. But what is clear is that many resources that we have taken for granted are clearly limited, in some cases critically. What the articles in this unit of *Annual Editions: Environment 08/09* highlight is that possibilities exist for us to turn around our 'business as usual' model and transform our practices to those that follow the ideas of sustainable development.

In the first article, "Tracking U.S. Groundwater: Reserves for the Future?" William M. Alley discusses the problems of

Creatas/PunchStock

depleted groundwater reserves. Groundwater makes up a far larger proportion of fresh water than all lakes and rivers combined. Groundwater is generally clean, and pretty much available at all times. We have used groundwater to turn semi-arid lands into vast farms, and to turn deserts into sprawling cosmopolitan regions. But when the question 'How much groundwater is available?' is posed, the answers can be quite disturbing. Desert communities are, in some cases, beginning to come to terms with the limitations of groundwater. Alley holds out hope that with proper monitoring, we will be able to make the smart choices to insure that groundwater supplies are not depleted. The economic sense of conserving and respecting our water supplies is illustrated in the following short article, "How Much Is Clean Water Worth," by Jim Morrison. Here, we see that a simple cost-benefit analysis made it clear to New York City planners that repairing the health of the watershed in the Catskill Mountains was far cheaper than the alternative: building a multi-billion dollar water treatment facility.

Few things are as depressing as reading about the rapid rate of destruction of the equatorial rain forests. Because of some unique properties of these ecosystems, they are very susceptible to irreparable damage. Once a forest is clear cut, it is pretty much gone. Eirivelthon Lima and others describe how tropical forests can be turned into sustainable ecosystems in "Searching for Sustainability."

Through flexibility, concessions, and decentralization, both sustainability and economic growth are possible.

It is easy for us to look around and see that things are bad and getting worse. But as these essays point out, environmentalists, economists, and policymakers can make a difference, and there is still cause for hope. The easy way forward is the 'hard path,' the business-as-usual model where practices of the past are continued into the future. The 'soft path' is one where we look at the world in a new way, recognizing the worth of our environment and work needed to preserve it. It takes imagination, creativity, and courage to accomplish the necessary tasks.

Tracking U.S. Groundwater: Reserves for the Future?

Wiliam M. Alley

During the past 50 years, groundwater depletion has spread from isolated pockets to large areas in many countries throughout the world. Groundwater occurs almost everywhere beneath the land surface. Its widespread occurrence is a major reason it is used as a source of water supply worldwide. Moreover, groundwater plays a crucial role in sustaining streamflow between precipitation events and especially during protracted dry periods. In addition to human uses, many plants and aquatic animals are dependent upon groundwater discharge to streams, lakes, and wetlands.

A growing awareness of groundwater as a critical natural resource leads to some basic questions. How much groundwater do we have left? Are we running out? Where are groundwater resources most stressed? Where are they most available for future supply? To address these basic and seemingly simple questions requires consideration of several complexities of defining groundwater availability and a review of how one monitors groundwater reserves.

The term "groundwater reserves" is used to emphasize the fact that groundwater, like other limited natural resources, can be depleted. This potential for depletion is a key concept, despite the fact that unlike nonrenewable resources such as mineral deposits, most groundwater resources are replenished. On the other hand, some "fossil" groundwaters in arid and semiarid areas have accumulated over tens of thousand of years (often under cooler, wetter climatic conditions) and are effectively nonrenewable except by artificial recharge of surface water or treated wastewater.

Groundwater management decisions in the United States are made at a local level, which may be a state, municipality, or special district formed for groundwater management. Thus, monitoring of groundwater reserves should be designed to provide the information needed by these entities as a primary consideration. The issues to be addressed are varied and occur at many scales from preservation of a small spring fed by a nearby water source to the management of groundwater development throughout a large aquifer system or river basin.

The nation's groundwater reserve is not a single vast pool of underground water, but rather is contained within a variety of aquifer systems.[1] In general, the locations of the nation's aquifers are known, so much of the current research focuses on characterizing aquifer systems and how they respond to human activities.[2]

Many aquifers cross political divides, including county, state, and international boundaries. This characteristic (as well as the specialized nature of the science of groundwater hydrology) drives the need for a federal role and multijurisdictional collaboration in groundwater monitoring. Concerns about groundwater reserves have become more regional, national, and even global in scale in recent years, as exemplified by interstate and international conflicts over the salinization, contamination, or overexploitation of groundwater.[3] The effects of groundwater development may require many years to become evident. Thus, there is an unfortunate tendency to forgo the data collection and analysis that is needed to support informed decisionmaking until after problems materialize.

How Much Groundwater Is Available?

The volume of water stored as groundwater is often compared to other major global pools of water within the Earth's hydrological cycle. For example, if one ignores water frozen in glaciers and polar ice, groundwater comprises more than 95 percent of the world's freshwater resources. This statistic illustrates the considerable value of groundwater, but it is also misleading, as it misses the variation in quantity, quality, and availability from location to location. The volume of groundwater in storage, its quality, and the yield to wells vary greatly across the planet. Typically, groundwater is used locally, so the effects of localized pumping on a given region are the primary concern of hydrogeologists.

Estimates of the volume of groundwater are poorly known relative to other pools of water. For example, the volume of the Earth's oceans has been well known for many years, whereas global estimates for groundwater storage vary by orders of magnitude (see Table 1). In part, this variability is due to different considerations of depth and salinity in defining the global groundwater pool. In addition, the variability reflects less knowledge about groundwater than other global pools of water. Early

Table 1 Volume of Water Attributed to Oceans and Groundwater Over Time

Date	Cubic kilometers of water (in thousands) Oceans	Groundwater
1945	1,372,000	250
1967	1,320,000	8,350
1978	1,338,000	10,530–23,400
1979	1,370,000	4,000–60,000
1997	1,350,000	15,300

Note. These data come from different studies of the world water balance. Significant figures are largely retained from original sources.

Source: W. M. Alley, J. W. LaBaugh, and T. E. Reilly, "Groundwater as an Element in the Hydrological Cycle," in M. Anderson, ed., *The Encyclopedia of Hydrological Sciences* (Chichester, UK: John Wiley and Sons Ltd., 2005), 2215–28.

estimates of the global groundwater pool greatly underestimated its volume. It was not until after development began in earnest in the mid-twentieth century that an appreciation of the large storage volume of groundwater emerged universally. More recently, scientists have viewed this resource as an important component of the world water cycle and have expressed increasing interest in quantifying its role.[4]

As a practical matter, it is virtually impossible to remove all water from storage with pumping wells. However, the volume of recoverable groundwater in storage for a particular area or aquifer can be estimated as the product of the area, saturated thickness, and specific yield (accounting as appropriate for differences in the estimates of saturated thickness and specific yield among multiple layers or zones).[5] To assess the value and limitations of estimates of groundwater in storage, it is helpful to first consider how aquifers are drained and then look at their dynamic links to the surface environment.

Aquifer Drainage

The mechanism of aquifer drainage depends on whether an aquifer is unconfined or confined (see Figure 1). In an unconfined aquifer, the upper surface of the saturated zone (water table) is free to rise and decline. The principal source of water from pumping an unconfined aquifer is the dewatering of the aquifer material by gravity drainage. The volume of water that is usable in practice is limited by the aquifer's permeability (how easily water moves through a rock unit), water quality, cost of drilling wells, and design of the well and pump.

Consider as an example the unconfined High Plains aquifer, which underlies an area stretching from southern South Dakota to the end of the panhandle of Texas and is the most heavily pumped aquifer in the United States. Depletion of aquifer storage from pumping has had substantial effects on irrigated agriculture in the High Plains, particularly in the southern half, where more than 50 percent of the saturated thickness has been dewatered in some areas. In Kansas, scientists have estimated the lifespan of the aquifer by projecting past trends into the future until the saturated thickness of the aquifer reaches a level

at which groundwater pumping for irrigation becomes impractical.[6] The results suggest that many areas in western Kansas have less than 50 years of usable groundwater remaining. Thirty feet of saturated thickness was the critical level in this study, although the researchers noted additional studies that suggest that 30 feet is not enough saturated thickness to provide sufficient well yields for irrigation.

Changes in groundwater levels throughout the High Plains aquifer are tracked annually through the cooperative effort of the U.S. Geological Survey and state and local agencies in the High Plains region (see Figure 2a). Despite the considerable effects of storage depletion in much of the High Plains, only 6 percent of the volume of water in the High Plains aquifer has been depleted since pumping began (see Figure 2b), illustrating how aggregated information about storage depletion over large areas can mask significant local effects.

Confined aquifers, which underlie low permeability confining systems, are filled by water under pressure and respond to pumping differently. The water for pumping is derived not from pore drainage but rather from aquifer compression and water expansion as the hydraulic pressure is reduced. Pumping from confined aquifers results in more rapid water-level declines covering much larger areas when compared to pumping the same quantity of water from unconfined aquifers. If water levels in an area are reduced to the point where an aquifer changes from a confined to an unconfined condition (becomes dewatered), the source of water becomes gravity drainage as in an unconfined aquifer. A major complication arises, however, because the drawdowns in the confined aquifer will induce leakage from adjacent confining units. Slow leakage over large areas can result in the confining unit supplying much, if not most, of the water derived from pumping.[7] Therefore, it is particularly difficult to relate

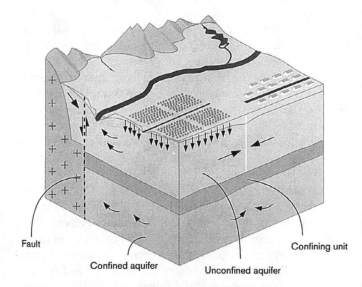

Fault   Confined aquifer   Unconfined aquifer   Confining unit

Figure 1 Hypothetical basin-fill aquifer system.

Note. The arrows show the direction of groundwater flow. Among the features shown are an unconfined aquifer overlying a confining unit and confined aquifer, a gaining stream, recharge from irrigated agriculture, and mountain-front recharge.

Source: Modified from S. A. Leake, *Modeling Ground-Water Flow with MODFLOW and Related programs,* U.S. Geological Survey Fact Sheet, 121–97 (Washington, DC. 1997).

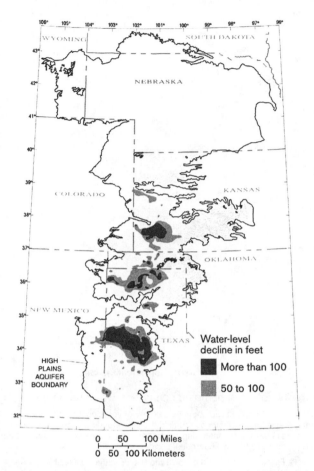

Figure 2a Changes in groundwater levels in the high plains aquifer from predevelopment to 2000

Source: V. L. McGuire et al., *Water in Storage and Approaches to Ground-Water Management, High Plains Aquifer, 2000*, U.S. Geological Survey Circular 1243 (Reston, VA, 2003).

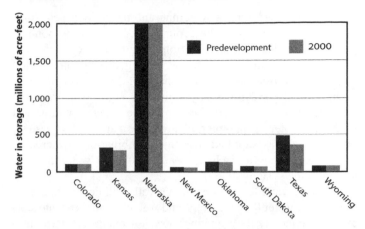

Figure 2b Comparison of predevelopment and 2000 groundwater in storage by state.

Source: V. L. McGuire et al., *Water in Storage and Approaches to Ground-Water Management, High Plains Aquifer, 2000*, U.S. Geological Survey Circular 1243 (Reston, VA, 2003).

Several key points arise from the examples and discussion thus far. First, measurement of storage depletion should be placed in the context of individual aquifer systems. For example, in evaluating water-level declines, one has to distinguish carefully between confined and unconfined aquifers, as the two respond very differently to pumping. Second, aquifer-wide estimates of recoverable water in storage have limited utility without considering the distribution of water-level changes and their effects. Finally, depletion of a small part of the total volume of water in storage can have substantial effects that become the limiting factors to development of the groundwater resource. These issues are further reinforced when one considers the response of surface-water bodies to groundwater pumping.

Interactions with Surface Water

Groundwater flows from areas of recharge to areas of discharge. Recharge includes water that naturally enters a groundwater system and water that enters the system at artificial recharge facilities or as a consequence of human activities such as irrigation and waste disposal. Discharge may occur to the atmosphere by transpiration; to streams, lakes, and other surface-water bodies; or through a pumping well. The balance between groundwater recharge and discharge controls groundwater levels and storage in a manner analogous to how deposits and withdrawals control savings in a bank account. If recharge exceeds discharge for some period, groundwater levels and storage will increase. Conversely, groundwater levels and storage will decline during periods when discharge exceeds recharge.

A common misperception is that the development of a groundwater system is "safe" if the average rate of groundwater withdrawal does not exceed the average annual rate of natural recharge. People sometimes make the erroneous assumption that natural recharge is equivalent to the basin sustainable yield.[10] Even further misinterpretations suggest that pumping at less than the recharge rate will not cause water levels and groundwater storage to decline.

estimates of the volume of groundwater in storage to the usable volume of groundwater in confined aquifers.

Further complications may arise for those aquifers with silt and clay layers that can permanently compact as a result of pumping. Consider, for example, the Central Valley aquifer in California, the nation's second-most-pumped aquifer. By 1977, about 28 percent of the decrease in aquifer storage of 60 million acre-feet was the result of permanent reduction of pore space by compaction, resulting in land subsidence throughout much of the area.[8] Farmers in the Central Valley have drawn on imported surface water as a major source of irrigation water to reduce groundwater depletion and associated subsidence. The decrease in aquifer storage of 60 million acre-feet, although very large, represented only a small part of the more than 800 million acre-feet of freshwater stored in the upper 1,000 feet of sediments in the Central Valley.[9]

Similarly, subsidence caused by groundwater pumping in the low-lying coastal environment of Houston, Texas, has increased its vulnerability to flooding and tidal surges. As a result, Houston has undertaken an expensive shift from sole reliance on its vast groundwater resource to partial reliance on surface water for its water supply.

To understand the fallacy inherent in these conclusions, one needs to consider how groundwater systems respond to pumping. Under natural conditions, a groundwater system is in long-term equilibrium. That is, averaged over some period (and in the absence of climate change), the amount of water recharging the system is approximately equal to the amount of water leaving (discharging from) the system. Withdrawal of groundwater by pumping changes the natural flow system, and the water is supplied by some combination of increased recharge, decreased discharge, and removal of water that was stored in the system.

Initially, water levels in pumping wells will decline to induce the flow of water to these wells, and water is removed from storage. Subsequently, the groundwater system readjusts to the pumping stress by "capturing" recharge or discharge. Also, the storage contribution to the water budget decreases with time for any given withdrawal. If the system can come to a new equilibrium, the changes in storage will cease (at a new reduced level of groundwater storage), and inflows will again balance outflows. Thus, the long-term source of water to discharging wells becomes a change in the inflow and outflow from the groundwater system.

The amount of groundwater available for use depends upon how the changes in inflow and outflow affect the surrounding environment and upon the extent to which society is willing to accept the resultant environmental changes. Consequences include reduced availability of water to riparian and aquatic ecosystems and reduced availability of surface water for use by humans. Further complicating matters, the effects of pumping on surface-water resources can be spread out over a long period of time, as illustrated by the alluvial aquifer example in Figure 3.

In many areas, the effects of groundwater pumping on surface-water resources, and importantly, the large uncertainties associated with these effects, become the limiting factors to groundwater development. For example, University of Arizona water law and policy expert Robert Glennon in his popular book *Water Follies* describes controversial situations from throughout the United States where groundwater pumping affects streams and lakes.[11] The effects on surface water can occur with relatively little depletion of the total amount of groundwater in storage. One of the areas Glennon discusses is the Upper San Pedro River Basin in southeastern Arizona, where concerns about streamflow depletion have caused conflicts between development and environment interests in this ecologically diverse riparian system. The health of the riparian system is dependent on the groundwater level and hydraulic gradient near the stream. A key question is how pumping in the basin affects these components of riparian system health. Congressionally mandated efforts are under way to reduce the annual storage depletion (overdraft) in the Sierra Vista area—a subwatershed of the Upper San Pedro Basin.[12] Current overdraft in the Sierra Vista subwatershed is about 10,000 acre-feet per year, which is small, relative to estimates ranging from 20 to 26 million acre-feet of total groundwater storage.[13] A monitoring plan is an important element for verifying the effectiveness of management measures in reducing overdraft in the Sierra Vista

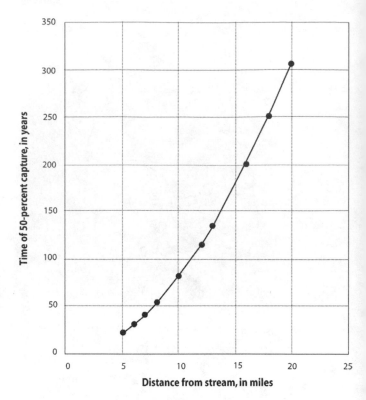

Figure 3 Effect of pumping on surface-water resources.

Note. Time of 50-percent capture is the number of years until 50 percent of the pumping rate is accounted for as reduced groundwater discharge to the stream. The relation is for a fully penetrating stream in an aquifer having a transmissivity to storage ratio of 110,000 square feet per day.

Source: C. Fillippone and S. A. Leake, "Time Scales in the Sustainable Management of Water Resources," *Southwest Hydrology* 4, no. 1 (2005): 17. © SAHRA–University of Arizona.

subwatershed, with the ultimate goal of mitigating impacts on the riparian system.

Water-Quality Limitations

Groundwater contamination from human activities clearly places constraints on groundwater availability. Likewise, water-quality constraints on groundwater availability can result from pumping. Perhaps best known are the many cases of saltwater intrusion from pumping groundwater along coastal areas. Groundwater pumping also can induce movement of saline water from underlying aquifers in inland areas. Likewise, shallow polluted groundwater may be induced or accelerated downward and throughout an aquifer by prolonged pumping, such that contaminated groundwater penetrates further and more quickly than otherwise anticipated. The removal of water from storage also changes the quality of the remaining groundwater because good quality water commonly is withdrawn first, and the residual often includes poorer quality groundwater from elsewhere in the aquifer or groundwater that has leaked into the aquifer from adjacent units in response to declining water levels. All these and other possible changes in water quality need to be considered in conjunction with information about changes in water levels and water in storage in evaluating the availability of groundwater. In some cases, the quality of groundwater will be suitable for some uses but not others. Water treatment may be necessary to meet some needs.

Groundwater Use

An average of 85 billion gallons of groundwater are withdrawn daily in the United States. More than 90 percent of these withdrawals are used for irrigation, public supply (deliveries to homes, businesses, industry), and self-supplied industrial uses. Irrigation is the largest use, accounting for about two-thirds of the amount. The percentage of total irrigation withdrawals provided by groundwater increased from 23 percent in 1950 to 42 percent in 2000. Groundwater provides about half the nation's drinking water with nearly all those in rural areas reliant upon groundwater.[14]

The importance of groundwater withdrawals in the United States is similar to that in the rest of the world, with some variations from country to country. Rapid expansion in groundwater use occurred between 1950 and 1975 in many industrial nations and subsequently in much of the developing world. The intensive use of groundwater for irrigation in arid and semi-arid countries has been called a "silent revolution" as millions of independent farmers worldwide have chosen to become increasingly dependent on the reliability of groundwater resources, reaping abundant social and economic benefits but with limited management controls by government water agencies.[15] Perhaps as many as two billion people worldwide depend directly upon groundwater for drinking water. The dependence on groundwater for drinking water is particularly high in Europe, where about 75 percent of the drinking-water supply is obtained from groundwater.[16]

Water-use data, when coupled with a scientific understanding of how aquifers respond to withdrawals, are crucial for water planning. Yet information on groundwater use is spotty and often inaccurate within the United States and worldwide. In the United States, practices for collecting water-use data vary significantly from state to state and from one water-use category to another, in response to laws regulating water use and interest in water-use data as an input for water management. Programs to collect water-use data in each state are summarized in a review by the National Research Council.[17]

Some water-use data, such as withdrawals for drinking water and other household uses and withdrawals by some industrial users are obtained by direct measurement, and some may be estimated as the amount reported or allowed by permit. Many uses, such as for self-supplied domestic use, agriculture, and some industries, are often estimated using coefficients relating water use to another characteristic, such as number of employees, number of units manufactured, irrigated acreage, or number of livestock. For example, self-supplied domestic water withdrawals are typically determined by multiplying an estimate of the self-supplied population by a per-capita use coefficient. Likewise, water use for a particular type of industry might be estimated using information on employment or production and estimates of gallons per day per employee or per unit of product. Ideally, coefficients used for water-use estimation are grounded

in representative data records. In practice, they are often derived empirically or developed using data that are sparsely sampled in time and space and perhaps extrapolated beyond the climatic, technological, and economic conditions for which they were originally developed. Other complications arise in these calculations because it may be difficult to separate surface-water and groundwater withdrawals without site-specific data and because small-scale use may be excluded from official statistics.

In determining the effects of pumping, it is important to recognize that not all the water pumped is necessarily consumed. For example, some of the water pumped for irrigation is lost to evapotranspiration, and some of the water returns to the groundwater system by infiltration, canal leakage, and other paths of irrigation return flow. Of course, water that is not used for consumption can undergo substantial changes in quality between withdrawal and recharge. Ideally, information on groundwater use includes estimates of consumptive use and return flow as well as withdrawals.

Groundwater Sustainability and Management

Achieving an acceptable tradeoff between groundwater use and the long-term effects of that use is a central theme in the evolving concept of groundwater sustainability.[18] Initially, people viewed groundwater as a convenient resource for general use, and they focused their attention on the economic aspects of groundwater development. Sustainability concerns, emerging in the early 1980s, have brought environmental viewpoints and an intergenerational perspective to the forefront in discussions about groundwater availability.

Groundwater sustainability is commonly defined in a broad context as the development and use of groundwater resources in a manner that can be maintained for an indefinite amount of time without causing unacceptable environmental, economic, or social consequences. The amount of time it takes for the effects of pumping to be manifested elsewhere in the environment reinforces the importance of sustainability as a concept for groundwater management but also makes sustainable solutions difficult to apply in practice. Application of sustainability concepts to water resources requires that the effects of many different human activities on water resources and the overall environment be understood and quantified to the greatest extent possible over the long term. Thus, sustainability likely requires an iterative process of continued monitoring, analysis, application of management practices, and revision. For some cases, particularly in arid areas, the groundwater resource is treated as nonsustainable.[19]

The tradeoff between the water used for consumption and the effects of groundwater withdrawals—on maintenance of instream-flow requirements for fish and other aquatic species, the health of riparian and wetland areas, and other environmental needs—is the driving force behind discussions about the sustainability of many groundwater systems. Considerable scientific uncertainty is associated with disputes over whether pumping will have a specific impact on a particular river or spring. Further complicating matters is the fact that although they are linked through the hydrologic cycle, groundwater and surface water are typically managed separately under different laws and administrative bodies.

Groundwater management strategies are composed of a small number of general approaches:[20]

- use of sources of water other than local groundwater, by shifting the local source of water (either completely or in part) from groundwater to surface water or importing water from outside the local water-system boundaries (the California Central Valley and Houston have implemented these approaches);
- changing rates or spatial patterns of groundwater pumping to minimize existing or potential unwanted effects (examples include moving well fields inland to avoid saltwater intrusion, shifting from deep to shallow groundwater or vice versa, and maintaining sufficient distances between wells to avoid excessive drawdown);
- control or regulation of groundwater pumping through implementation of guidelines, policies, taxes, or regulations by water management authorities (these imposed actions may include restrictions on some types of water use, limits on withdrawal volumes, or establishment of critical levels for aquifer hydraulic heads);
- artificial recharge through the deliberate introduction of local or imported surface water—whether potable, reclaimed, or waste-stream discharge—into the subsurface for purposes of augmenting or restoring the quantity of water stored in developed aquifers (options include infiltration from engineered impoundments, direct-well injection, and pumping designed to induce inflow of freshwater from surface waterways);
- use of groundwater and surface water through the coordinated and integrated use of the two sources to ensure optimum long-term economic and social benefits;
- conservation practices, techniques, and technologies that improve the efficiency of water use, often accompanied by public education programs on water conservation;
- reuse of wastewater (gray water) and treated wastewater (reclaimed water) for non-potable purposes such as irrigation of crops, lawns, and golf courses;
- desalination of brackish groundwater or treatment of otherwise impaired groundwater to reduce dependency on fresh groundwater sources.

These general approaches are not mutually exclusive; that is, the various approaches overlap, or the implementation of one approach will inevitably involve or cause the implementation of another. For example, many approaches involve combinations of surface water, groundwater, and artificial recharge. During periods of excess surface-water runoff, and when surface-water impoundments are at or near capacity, surplus surface water can be stored in aquifer systems through artificial recharge. Conversely, during droughts, increased groundwater pumping can be used to offset shortfalls in surface-water supplies. Depleted

aquifer systems can be seen as potential subsurface reservoirs for storing surplus imported or local surface water.

It is important to frame the hydrologic implications of various alternative development strategies in such a way that their long-term implications can be properly evaluated, including effects on the water budget. For example, changing the rates or patterns of groundwater pumping will lead to changes in the spatial patterns of recharge to or discharge from groundwater systems. As another example, in some areas of extensive use of artificial recharge, such as parts of southern California, water from artificial recharge may have replaced much of the native groundwater.

Monitoring Groundwater Reserves

Water-level measurements in observation wells provide the primary source of information about groundwater reserves. Water-level data collected over periods of days to months are useful for determining an aquifer's hydraulic properties; however, data collected over years to decades are required to monitor the long-term effects of aquifer development and management.

The amount of effort in collecting long-term water-level data varies greatly from state to state, and many long-term monitoring wells are clustered in certain areas.[21] Although they are difficult to track, the number of long-term observation wells appears to be declining because of limitations in funding and human resources. For example, the number of long-term observation wells monitored by the U.S. Geological Survey (USGS) declined by about half from the 1980s to 2000.

For many decades, hydrogeologists and others have been making periodic calls for a nationwide program to obtain more systematic and comprehensive records of water levels in observation wells. O. E. Meinzer, a longtime chief of the USGS Ground Water Division and considered by many to be the father of the science of hydrogeology, described the characteristics of such a program about 70 years ago:

The program should cover the water-bearing formations in all sections of the country; it should include beds with water-table conditions, deep artesian aquifers, and intermediate sources; moreover, it should include areas of heavy withdrawal by pumping or artesian flow, areas which are not affected by heavy withdrawal but in which the natural conditions of intake and discharge have been affected by deforestation or breaking up of prairie land, and, so far as possible, areas that still have primeval conditions. This nation-wide program should furnish a reliable basis for periodic inventories of the ground-water resources, in order that adequate provision may be made for our future water supplies. [22]

More recently, the Heinz Center report *The State of the Nation's Ecosystems* indicated that data on groundwater levels and rates of change are "not adequate for national reporting."[23] This report advocated supplementing existing networks to develop a national indicator of trends in groundwater levels.

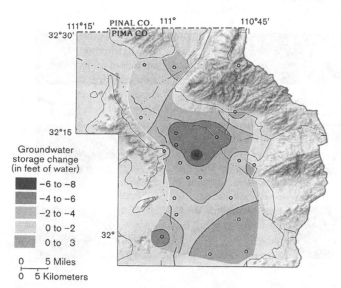

Figure 4 Change in groundwater storage in the Tucson Basin, 1989–1998.

Note. Change in storage was estimated using microgravity measurements.

Source: D. R. Pool, D. Winster, and K. C. Cole, *Land Subsidence and Ground-Water Storage Monitoring in the Tucson Active Management Area, Arizona,* U.S. Geological Survey Fact Sheet 084–00 (Reston, VA, 2000).

The U.S. Government Accountability Office noted that no federal agencies are collecting groundwater data on a national scale and only the USGS and National Park Service are collecting water-level data on a regional scale.[24]

Historically, water-level measurements were simply tabulated, recorded in a paper file, and possibly published in reports. Today, many agencies use the Internet to enhance users' access to current and historical monitoring data. Furthermore, continuous collection, processing, and transmission of water-level data on the Internet in "real time" (typically updated every few hours) is becoming more of a standard procedure. Real-time groundwater data are useful in formulating drought warnings, as they suggest potential effects on water levels in shallow domestic wells. Real-time capability can lead to improved data quality (from continual review of the data) as well as to increased interest in groundwater conditions on the part of the general public.

In addition to water-level monitoring, certain geophysical techniques can enhance the declination and interpretation of water-level changes over a region. For example, microgravity methods can be used to measure the small gravitational changes that result from changes in groundwater storage (including water stored in the unsaturated zone) over an area (see Figure 4). More recently, researchers have proposed satellite-based measurements of gravity to measure changes over areas the size of a large part of the High Plains aquifer.[25] Meeting the majority of needs for water-level information requires much finer detail than that which satellite measurements can provide, but future technologies may improve this technique. Land- and satellite-based gravity measurements provide area-wide information on changes in the volume of water in storage but do not provide information on vertical changes in heads (water levels) in aquifer systems.

A second geophysical technique, Interferometric Synthetic Aperture Radar (InSAR), uses repeated radar signals from space to measure land-surface uplifts or subsidence at high degrees of measurement resolution and spatial detail.[26] Like gravity methods, InSAR has the advantage of being able to make measurements over large areas and between monitoring wells. The InSAR information can provide additional insights into the areal extent of groundwater depletion where it is linked to subsidence, and can even detect uplift from artificial recharge. InSAR has been found to be particularly useful in identifying faults and geologic structures that may impede groundwater flow and affect the response of an aquifer system to pumping.

Integrated Monitoring and Assessment of Groundwater Reserves

As previously noted, the desire for a national network for monitoring groundwater levels has been discussed since the early 1900s but remains unfulfilled. Meanwhile, groundwater issues have evolved beyond early concerns focused on the hydraulics of individual wells and well field development to encompass many aspects of groundwater, including quantity, quality, and interactions with surface water. Technological advances have been made in sensors, communications, and electronic control systems to monitor groundwater, and computer modeling has become widely used to evaluate groundwater systems. From today's perspective, what might an ideal national program involve?

First and foremost, a national water-level monitoring program should be a collaborative process that involves discourse among local, state, and federal governmental agencies, nongovernmental organizations, and the public.[27] Ideally, data collected would serve double-duty by contributing to the larger regional and national picture while meeting local needs. There should be sufficient consistency in approach to describe the status of groundwater reserves across the country and to show how different constraints affect utilization of the nation's aquifers. A major early goal would be to identify critical gaps in existing coverage.[28]

Many of the primary issues affecting groundwater availability require analysis at the scale of aquifers to achieve a meaningful perspective. To that end, monitoring programs should be designed in the context of the specific characteristics of each aquifer system. A comprehensive national monitoring program should track major aquifers that are affected by groundwater pumping, areas of future groundwater development, and areas of groundwater recharge. Water levels should be measured in wells open to different depths and in the context of the three-dimensional groundwater-flow system.

A long-term record of water-level measurements should encompass the period between the natural and developed states of aquifer systems. Other approaches, such as gravity measurements to estimate subsurface-water storage changes, should be considered in conjunction with the water-level monitoring program. Establishing links between water-level and water-quality requires an understanding of groundwater-flow systems. Studies of natural mixing in aquifers suggest that existing damage to

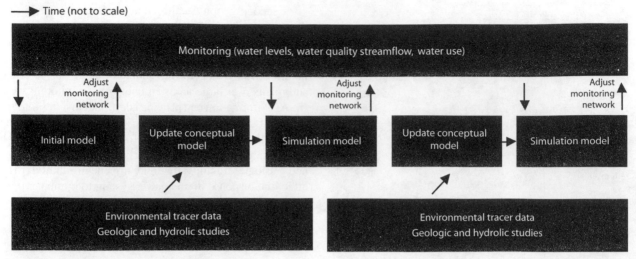

Figure 5   Tracking groundwater resources.

groundwater quality may be lasting and could gradually extend deeper into aquifer systems, thereby reducing further groundwater availability.[29]

Data on changes in groundwater levels provide essential information about changes in groundwater storage and provide the simplest way to convey the extent of groundwater depletion. However, changes in groundwater storage are only part of the story. As noted in previous examples, the status of groundwater reserves should be placed in the context of the complete water budget for that aquifer system. Thus, the monitoring of surface water and groundwater should be linked, particularly measurements of streamflow during low-flow periods when groundwater discharge is the primary component of streamflow. This might require an increase in the number of streamgaging stations in targeted basins to estimate the groundwater contribution to streamflow.

In addition to monitoring data on natural systems, estimation of water withdrawals and consumptive use is an essential part of computing a water budget for a developed aquifer system. Groundwater pumping is one component of the water budget that is physically possible to measure; yet it is commonly one of the most uncertain components of the water budget. Water-use information should be an integral part of evaluations of groundwater quantity and quality and other environmental conditions. Where multiple overlying aquifers are used, efforts should be made to estimate withdrawals from each.

Groundwater systems are dynamic and adjust over decades or more to pumping and other stresses. Many aquifer systems have undergone several decades of intensive development and may be far from equilibrium. Thus, it is challenging to place current conditions in the context of the dynamic but slow changes that may be taking place. A simple snapshot of current conditions may not indicate, for example, how future streamflow depletion will evolve from the pumping that has already occurred.

During the past several decades, computer models for simulating groundwater and surface-water systems have played an increasing role in the evaluation of groundwater development and management alternatives. Groundwater modeling serves as a quantitative means of evaluating the water balance of an aquifer, as it is affected by land use, climate, and groundwater withdrawals, and how these changes affect streamflow, lake levels, water quality, and other important variables. Generally, monitoring and computer modeling are treated as distinct activities, but to be most effective, the two should be linked. Such a framework is considered further below, and its essential elements are illustrated in Figure 5.

Monitoring groundwater reserves serves as primary information used in the development and calibration of computer models. Likewise, the process of model calibration and use provides insights into which components of the system are best known, which components are poorly known, and which components are more important than others. Thus, the experience gained from modeling should provide a basis for a periodic evaluation of the monitoring network.

As its basis, every simulation model has a conceptual model that represents the prevailing theory of how the groundwater system works. The appropriateness of this conceptual model is tested as a numerical model is built, and field observations are compared to the model simulations. Unfortunately, more often than not, data will fit more than one conceptual model, and good calibration of a model does not ensure a correct conceptual model. Thus, conceptual and numerical modeling should be viewed as an iterative process in which the conceptual model is continuously reformulated and updated as new information is acquired.[30] The importance of this approach and its link to monitoring data is recognized explicitly in Figure 5 as a key step prior to each new stage of groundwater modeling.

Additional scientific studies conducted at the time of modeling or during intervening periods can provide insights into the adequacy of the conceptual model that underlies the computer model as well as help in adjustment of model parameters. Such studies include use of environmental tracers, studies of the geologic framework, and geophysical studies. For example, an increasing number of chemical and isotopic substances are being measured in groundwater to identify water sources, trace directions of groundwater flow, and measure the age of the water (time since recharge). Comparison of the results from these environmental tracers with information from computer modeling can

Middle Rio Grande Basin

The Middle Rio Grande Basin encompasses about 38 percent of the population of New Mexico and is the primary source of water supply to the City of Albuquerque and surrounding area. Some of the most productive parts of the aquifer system are in eastern Albuquerque, where, coincidentally, most of the initial groundwater development occurred. This led to the popular belief that the entire Middle Rio Grande Basin was underlain by a highly productive aquifer that was equivalent to one of the Great Lakes. During the 1980s and early 1990s, a combination of large water-level declines measured in monitoring wells (greater than 150 feet in some areas) and new insights into the geologic framework of the basin led to serious questions about this paradigm. In 1995, the New Mexico State Engineer declared the Middle Rio Grande Basin a "critical basin" faced with rapid economic and population growth for which there is less than adequate technical information about the available groundwater supply.

To fill some of the gaps in information, an intensive 6-year effort was undertaken to improve understanding of the hydrogeology of the basin.[1] Geological, geophysical, and environmental tracer studies provided new insights into the source areas for recharge to different parts of the aquifer; indicated that mountain-front recharge is less than previously estimated; showed that the hydraulic connection between the Rio Grande and the aquifer is less than previously thought in some areas; identified new faults that may affect groundwater flow; and suggested that the aquifer is less productive in some areas than previously thought (see the figure at right). The new information was incorporated into a revised groundwater model for the region. In conjunction with the study, new monitoring wells were established in the Albuquerque area, generally as nests of several wells completed at different depths in the aquifer and located to minimize short-term fluctuations caused by nearby high-capacity production

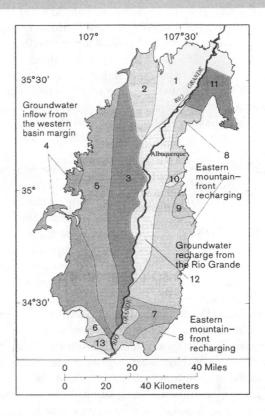

wells. The combined approach of monitoring, modeling, and scientific studies has been instrumental in helping the City of Albuquerque revise its water-use and future water-supply strategy.

1. J. R. Bartolino and J. C. Cole, *Ground-Water Resources of the Middle Rio Grande Basin, New Mexico,* U.S. Geological Survey Circular 1222 (Reston, VA, 2002), http://pubs.water.usgs.gov/circ1222.

lead to either increased confidence in the conceptual model of a groundwater system or recognition of the need for changes. The Middle Rio Grande Basin in central New Mexico (see box on page 170) provides an example of how long-term water-level monitoring combined with environmental and geologic studies has contributed to an evolving series of conceptual and simulation models used to help manage the groundwater resources of the basin.

Not all aquifer systems lend themselves to the exact same approach. For example, consolidated geological formations with fractures, joints, or solution cavities can be difficult to model, given the discontinuous nature of their permeability. These rocks commonly are highly vulnerable to contamination, and the wide range in water-level fluctuations can cause shallow domestic wells to go dry during extended droughts. Interpretation of water-level monitoring from individual wells is difficult in such terrain. It remains important, however, to have a conceptual model of the system as a driving force behind the monitoring network design

with a goal to quantify that model as knowledge of the system improves.

One should not infer that the simulation model in Figure 5 is always the same. Indeed, a stepwise approach may be used in which simpler analytic codes are used in the initial phases before constructing three-dimensional numerical models. Also, it is likely that further groundwater research will develop multiple models addressing different roles and objectives. Each model provides a means to reevaluate the monitoring network from a different perspective and to advance understanding of how the water balance of the aquifer system responds to human development.

Generalized long-term monitoring will provide critical information for many uses but will not offset the need for very specific monitoring to address more localized issues, such as the effects of pumping on the ecology of a stream reach. Ideally, the broader scale monitoring programs provide a hydrologic context for the design of such studies.

Conclusions

Groundwater monitoring data serve as a foundation that permits informed management decisions on many kinds of groundwater resource and sustainability issues. Unfortunately, data on groundwater conditions and trends are generally lacking worldwide: Groundwater is commonly undervalued, and there is a deceptive time lag between withdrawals and the resultant impacts of those withdrawals. Long-term groundwater data from individual wells are useful primarily as part of a broader analysis of aquifer systems; thus, the value of data from individual wells is often as invisible as the resource they represent.

Long-term water-level monitoring needs to be integrated with analysis of other monitoring data and an underlying model of the water budget of the aquifer system (typically a simulation model) as a means for interpreting monitoring results and guiding the design of monitoring networks. Regular reassessment of monitoring objectives is necessary to ensure that monitoring programs provide the information needed by groundwater users and those who manage water resources. To enhance the value of groundwater data, managers and policymakers must also ensure the continuity of data-collection programs over time.

Within the United States, one might summarize the current situation as one in which we have some ability to track groundwater levels and water use for many aquifers, generally have limited ability to place these data in the context of groundwater sustainability for most aquifers, and often lack an integrated approach with feedback among monitoring, simulation, scientific studies, and management approaches. Similar issues exist in managing groundwater resources throughout the world.[31] Fortunately, the ability to access groundwater data on the Internet and to portray them in a spatial context should continue to enhance their visibility and value in the coming years.

Notes

1. Single geologic units may define aquifers. Alternately, multiple aquifers and surrounding lower permeability units may be collectively referred to as "aquifer systems." The two terms are used somewhat interchangeably in this article, with "aquifer systems" used when an emphasis is placed on aquifers as hydrologic systems. The *Ground Water Atlas of the United States* describes many of the important aquifers of the nation and can be found at http://capp.water.usgs.gov/gwa/.

2. W. M. Alley, R. W. Healy, J. W. LaBaugh, and T. E. Reilly, "Flow and Storage in Groundwater Systems," *Science* 296, 5575 (14 June 2002): 1985–90.

3. National Research Council, *Confronting the Nation's Water Problems: The Role of Research* (Washington, DC: National Academies Press, 2004), 187; and L. F. Konikow and E. Kendy, "Groundwater Depletion: A Global Problem," *Hydrogeology Journal* 13, no. 1 (2005): 317–20.

4. G. M. Hornberger, "A Water Cycle Initiative," *Ground Water* 43, no. 6 (2005): 771.

5. Saturated thickness is the vertical thickness of the aquifer in which the pore spaces are filled (saturated) with water. Specific yield is the ratio of the volume of water that a saturated rock will yield by gravity drainage to the volume of the rock. Specific yield typically ranges from 0.05 to 0.3.

6. B. B. Wilson, D. P. Young, and R. W. Buddemeier, *Exploring Relationships Between Water Table Elevations, Reported Water Use, and Aquifer Subunit Delineations,* Kansas Geological Survey Open File Report 2002-25D (Lawrence, KS, 2002).

7. For example, in a well-known study, researchers found that most of the water pumped from the confined Dakota sandstone aquifer in South Dakota has come from confining beds. J. D. Bredehoeft, C. E. Neuzil, and P. C. D. Milly, *Regional Flow in the Dakota Aquifer: A Study of the Role of Confining Layers,* U.S. Geological Survey Water-Supply Paper 2237 (Washington, DC, 1983).

8. G. L. Bertoldi, R. H. Johnston, and K. D. Evenson, *Ground Water in the Central Valley, California—A Summary Report,* U.S. Geological Survey Professional Paper 1401-A (Washington, DC, 1991). Land subsidence is a gradual settling or sudden sinking of the Earth's surface. Several different processes can cause it. Most water-related subsidence occurs as a result of compaction of aquifer materials (as in the Central Valley), drainage and oxidation of organic soils, and the dissolution and collapse of limestone and other susceptible rocks forming sinkholes and similar features.

9. Ibid., page 27.

10. J. D. Bredehoeft, "Safe Yield and the Water Budget Myth," *Ground Water* 35, no. 6 (1997): 929.

11. R. J. Glennon, *Water Follies: Groundwater Pumping and the Fate of America's Fresh Waters* (Washington, DC: Island Press, 2004).

12. U.S. Department of the Interior, *Water Management of the Regional Aquifer in the Sierra Vista Subwatershed, Arizona—2004 Report to Congress,* prepared in consultation with the Secretaries of Agriculture and Defense and in cooperation with the Upper San Pedro Partnership in response to Public Law 108-136, Section 321, 30 March 2005.

13. Arizona Department of Water Resources, *Upper San Pedro Basin Active Management Area Review Report* (Phoenix, AZ, 2005) 3–25, http://www.azwater.gov/dwr/Content/Publications/files/UpperSanPedro/UpperSanPedroBasinAMAReviewReport.pdf.

14. M. A. Maupin and N. L. Barber, *Estimated Withdrawals from Principal Aquifers in the United States, 2000,* U.S. Geological Survey Circular 1279 (Reston, VA, 2005), http://pubs.water.usgs.gov/circ1279.

15. M. R. Llamas and P. Martinez-Santos, "Intensive Groundwater Use: A Silent Revolution that Cannot be Ignored," *Water Science and Technology* 51, no. 8 (2005): 167–74.

16. B. L. Morris et al., *Groundwater and its Susceptibility to Degradation: A Global Assessment of the Problem and Options for Management,* United Nations Environment Programme (UNEP) Early Warning and Assessment Report Series, RS 03-3 (Nairobi, Kenya, 2001).

17. National Research Council, *Estimating Water Use in the United States* (Washington, DC: National Academy Press, 2002).

18. W. M. Alley, T. E. Reilly, and O. L. Franke, *Sustainability of Ground-Water Resources,* U.S. Geological Survey Circular 1186 (Denver, CO, 1999), http://pubs.water.usgs.gov/circ1186; and W. M. Alley and S. A. Leake, "The Journey from Safe Yield to Sustainability," *Ground Water* 42, no. 1 (2004): 12–16.

19. W. A. Abderrahman, "Should Intensive Use of Non-renewable Groundwater Resources Always Be Rejected?" in R. Llamas and E. Custodio, eds., *Intensive Use of Groundwater: Challenges and Opportunities* (Lisse, Netherlands: A. A. Balkema, 2002), 191–203.

20. D. L. Galloway, W. M. Alley, P. M. Barlow, T. E. Reilly, and P. Tucci, *Evolving Issues and Practices in Managing Ground-Water Resources: Case Studies on the Role of Science,* U.S. Geological Survey Circular 1247 (Reston, VA, 2003), http://pubs.water.usgs.gov/circ1247.

21. C. J. Taylor and W. M. Alley, *Ground-Water-Level Monitoring and the Importance of Long-Term Water-Level Data,* U.S. Geological Survey Circular 1217 (Denver, CO, 2001), http://pubs.water.usgs.gov/circ1217.

22. O. E. Meinzer, "Introduction" in R. M. Leggette, et al., *Report of the Committee on Observation Wells, United States Geological Survey,* (unpublished manuscript on file in Reston, VA, 1935), 3.

23. H. John Heinz III Center for Science, Economics and the Environment, *The State of the Nation's Ecosystems: Measuring the Lands, Waters, and Living Resources of the United States* (Cambridge, UK: Cambridge University Press, 2002), http://www.heinzctr.org/ecosystems/report.html.

24. U.S. General Accountability Office, *Watershed Management: Better Coordination of Data Collection Efforts Needed to Support Key Decisions,* GAO-04-382 (Washington, DC, 2004).

25. M. Rodell and J. S. Famiglietti, "Detectability of Variations in Continental Water Storage from Satellite Observations of the Time Dependent Gravity Field," *Water Resources Research* 35 (1999): 2705–23.

26. For examples of the use of InSAR to understand groundwater systems, see G. W. Bawden, M. Sneed, S. V. Stork, and D. L. Galloway, *Measuring Human-Induced Land Subsidence from Space,* U.S. Geological Survey Fact Sheet 069-03 (Sacramento, CA, 2003), http://pubs.water.usgs.gov/fs-069-03/.

27. National Ground Water Association, *Ground Water Level and Quality Monitoring* (2005) http://www.ngwa.org/pdf/monitoring7.pdf (accessed 8 February 2006).

28. P. M. Barlow et al., *Concepts for National Assessment of Water Availability and Use,* U.S. Geological Survey Circular 1223 (Reston, VA, 2002), http://pubs.water.usgs.gov/circ1223.

29. G. E. Fogg, "Groundwater Quality Sustainability, Creeping Normalcy, and a Research Agenda," *Geological Society of America Abstracts with Programs* 37, no. 7 (2005): 247.

30. J. D. Bredehoeft, "The Conceptualization Model Problem—Surprise," *Hydrogeology Journal* 13, no. 1 (2005): 37–46.

31. Recently, the International Groundwater Resources Assessment Center (IGRAC) has been established to share data and information on groundwater resources worldwide, http://www.igrac.nl/.

WILLIAM M. ALLEY is chief of the Office of Ground Water at the U.S. Geological Survey. He is an active participant in groundwater conferences and has served on national and international committees for the American Geophysical Union, National Ground Water Association, UNESCO, and the National Research Council. He can be reached at walley@usgs.gov.

From *Environment,* April 2006, pp. 11–25. This article is in the public domain. Published by Heldref Publications, 1319 Eighteenth St., NW, Washington, DC 20036-1802.

How Much Is Clean Water Worth?

A lot, say researchers who are putting dollar values on wildlife and ecosystems—and proving that conservation pays

JIM MORRISON

The water that quenches thirsts in Queens and bubbles into bathtubs in Brooklyn begins about 125 miles north in a forest in the Catskill Mountains. It flows down distant hills through pastures and farmlands and eventually into giant aqueducts serving 9 million people with 1.3 billion gallons daily. Because it flows directly from the ground through reservoirs to the tap, this water—long regarded as the champagne of city drinking supplies—comes from what's often called the largest "unfiltered" system in the nation.

But that's not strictly true. Water percolating through the Catskills is filtered naturally—for free. Beneath the forest, fine roots and microorganisms break down contaminants. In streams, plants absorb nutrients from fertilizer and manure. And in meadows, wetlands filter nutrients while breaking down heavy metals.

New York City discovered how valuable these services were 15 years ago when a combination of unbridled development and failing septic systems in the Catskills began degrading the quality of the water that served Queens, Brooklyn and the other boroughs. By 1992, the U.S. Environmental Protection Agency (EPA) warned that unless water quality improved, it would require the city to build a filtration plant, estimated to cost between $6 and $8 billion and between $350 and $400 million a year to operate.

Instead, the city rolled the dice with nature in a historic experiment. Rather than building a filtration plant, officials decided to restore the health of the Catskills watershed, so it would do the job naturally.

What's this ecosystem worth to the city of New York? So far, $1.3 billion. That's what the city has committed to build sewage treatment plants upstate and to protect the watershed through a variety of incentive programs and land purchases. It's a lot of money. But it's a fraction of the cost of the filtration plant—a plant, city officials note, that wouldn't work as tirelessly or efficiently as nature.

"It was a stunning thing for the New York City council to think maybe we should invest in natural capital," says Stanford University researcher Gretchen Daily.

Daily is one of a growing number of academics—some from economics, some from ecology—who are putting dollar figures on the services that ecosystems provide. She and other "ecological economists" look not only at nature's products—food, shelter, raw materials—but at benefits such as clean water, clean air, flood control and storm mitigation, irreplaceable services that have been taken for granted throughout history. "Much of Mother Nature's labor has enormous and obvious value, which has failed to win respect in the marketplace until recently," Daily writes in the book *The New Economy of Nature: The Quest to Make Conservation Profitable.*

Ecological economist Geoffrey Heal, a professor of public policy and business responsibility at Columbia University, became interested in the field as an economist who was concerned about the environment. "The idea of ecosystem services is an interesting framework for thinking why the environment matters," says Heal, author of *Nature and the Marketplace: Capturing the Value of Ecosystem Services.* "The traditional argument for environmental conservation had been essentially aesthetic or ethical. It was beautiful or a moral responsibility. But there are powerful economic reasons for keeping things intact as well."

Daily notes that beyond providing clean water, the Catskills ecosystem has value for its beauty, as wildlife habitat and for recreation, particularly trout fishing. Such values are not inconsequential. While no one has assessed the total worth of the watershed, even a partial look reveals that habitat and wildlife are powerful economic engines.

Restored habitat for trout and other game fish, for example, attracts fishermen, and angling is big business in this state. According to a report by the U.S. Fish and Wildlife Service (FWS), more than 1.5 million people fished in New York during 2001, yielding an economic benefit to the state of more than $2 billion and generating the equivalent of 17,468 full-time jobs and more than $164 million in state, federal, sales and motor fuel taxes. Though not as easily measured, individual Catskills species also have value. Beavers, for instance, create wetlands that are vital to filtering water and to biodiversity.

Ecological economists maintain that ecosystems are capital assets that, if managed well, provide a stream of benefits just as any investment does. The FWS report, for example, notes that 66 million Americans spent more than $38 billion in 2001 observing, feeding or photographing wildlife. Those expenditures resulted in more than a million jobs with total wages and salaries of $27.8 billion. The analysis found that birders alone spent an estimated $32 billion on wildlife watching that year, generating $85 billion of economic benefits. In Yellowstone National Park, the reintroduction of gray wolves that began in 1995 has already increased revenues in surrounding communities by $10 million a year, with total benefits projected to reach $23 million annually as more visitors come to catch a glimpse of these charismatic predators.

When it comes to water quality, EPA projects that the United States will have to spend $140 billion over the next 20 years to maintain minimum required standards for drinking water quality. No wonder, then, that 140 U.S. cities have studied using an approach similar to New York's. Under that agreement, finalized in 1997, the city promised to pay farmers, landowners and businesses that abided by restrictions designed to protect the watershed. (The city owns less than 8 percent of the land in the 2,000-square-mile watershed; the vast majority is in private hands.) "In the case of the Catskills, it was a matter of coming up with a way to reward the stewards of the natural asset for something they had been providing for free," Daily says. "As soon as they got paid even a little bit, they were much happier and inclined to go about their stewardship." There's no guarantee this experiment will work, of course; it may be another decade before the city finds out.

Elsewhere, other governing bodies are also recognizing the value of ecosystem services. The U.S. Army Corps of Engineers, for example, bought 8,500 acres of wetlands along Massachusetts' Charles River for flood control. The land cost $10 million, a tenth of the $100 million the Corps estimated it would take to build the dam and levee originally proposed. To fight floods in Napa, California, county officials spent $250 million to reconnect the Napa River to its historical floodplains, allowing the river to meander as it once did. The cost was a fraction of the estimated $1.6 billion that would have been needed to repair flood damage over the next century without the project. Within a year, notes Daily, flood insurance rates in the county dropped 20 percent and real estate prices rose 20 percent, thanks to the flood protection now promised by nature.

Even insects supply vital ecosystem services. More than 218,000 of the world's 250,000 flowering plants, including 70 percent of all species of food plants, rely on pollinators for reproduction—and more than 100,000 of these pollinators are invertebrates, including bees, moths, butterflies, beetles and flies. Another 1,000 or more vertebrate species, including birds, mammals and reptiles, also pollinate plants. According to University of Arizona entomologist Stephen Buchmann, author of *The Forgotten Pollinators,* one of every three bites of food we eat comes courtesy of a pollinator.

A Cornell University study estimated the value of pollination by honeybees in the United States alone at $14.6 billion in 2000. Yet honeybee populations are dropping everywhere, as much as 25 percent since 1990, according to one study. Now many farms and orchards are paying to have the bees shipped in.

Today's interest in assigning dollar values to pollination and other ecosystem services was spawned by publication of a controversial 1997 report in *Nature* that estimated the total global contribution of ecosystems to be $33 trillion or more each year—roughly double the combined gross national product of all countries in the world. The study became a lightning rod. Detractors scoffed at the idea that one could put a dollar value on something people weren't willing to purchase. One report by researchers at the University of Maryland, Bowden College and Duke University called the estimate "absurd," noting that if taken literally, the figure suggests that a family earning $30,000 annually would pay $40,000 annually for ecosystem protection.

Other researchers, including Daily and Heal, charged that the $33 trillion figure greatly underestimates nature's value. "If you believe, as I do, that ecosystem services are necessary for human survival, they're invaluable really," Heal says. "We would pay anything we could pay."

Daily doesn't believe the absolute value of an ecosystem can ever be measured. Heal agrees, yet both scientists say that pricing ecosystem services is an important tool for making decisions about nature—and for making the case for conservation. "Valuation is just one step in the broader politics of decision making," she says. "We need to be creative and innovative in changing social institutions so we are aligning economic forces with conservation."

Indeed, as dollar values for nature's services become available, environmentalists increasingly use them to bolster arguments for conservation. One high-profile example is the contentious dispute over whether to tear down four dams on the lower Snake River in southeastern Washington to restore salmon habitat, and thus the region's lucrative salmon fishery. Ed Whitelaw, a professor of economics at the University of Oregon, notes that estimates of the economic impact of breaching

---

Nature's Services

How the Experts Categorize Them

- Ecosystem goods, the traditional measure of nature's products such as seafood, timber and agriculture
- Basic life support functions such as water purification, flood control, soil renewal and pollination
- Life-fulfilling functions, the beauty and inspiration we get from nature, including activities such as hiking and wildlife watching
- Basic insurance, the idea that nature's diversity contains something—like a new drug—the value of which isn't known today, but may be large in the future

Natural Capital

What's the Annual Dollar Value of . . .?

- Recreational saltwater fishing in the United States: $20 billion
- Wild bee pollinators to a single coffee farm in Costa Rica: $60,000
- Tourism to view bats in the city of Austin, Texas: $8 million
- Wildlife watching in the United States: $85 billion
- U.S. employment income generated by wildlife watching: $27.8 billion
- State and federal tax revenues from wildlife watching: $6.1 billion
- Natural pest control services by birds and other wildlife to U.S. farmers: $54 billion

the dams range from $300 million in net costs to $1.3 billion in net benefits, largely due to the wide range of projections about recreational spending.

A 2002 report by the respected, nonprofit think-tank RAND Corporation concluded the dams could be breached without hurting economic growth and employment. Energy lost as a result of the breaches could be replaced with more efficient sources, including natural gas, resulting in 15,000 new jobs. Further, the report noted that recreation, retail, restaurants and real estate would experience a marked growth. Recreational

activities alone would increase by an estimated $230 million over 20 years.

There's no question that returning the salmon runs would have a major impact on the region. When favorable ocean conditions increased the runs in 2001, Idaho's Department of Fish and Game estimated the salmon season that year alone generated more than $90 million of revenue in the state, most of it in rural communities that badly needed the funds.

"Some people think it sounds crass to put a price tag on something that's invaluable, careening down the slippery slope of the market economy," says Daily. "In fact, the idea is to do something elegant but tricky: to finesse the economic system, the system that drives so much of our individual and collective behavior, so that without even thinking it makes natural sense to invest in and protect our natural assets, our ecosystem capital."

What Daily and other ecological economists want is to insinuate consideration of ecosystem services into daily decision making, whether it takes the form of financial incentive or penalty. "At a practical level, decisions are made at the margin, not at the 'should we sterilize the Earth' level," she says. "It's in all the little decisions—whether to farm here or leave a few trees, whether to build the shopping mall there or leave the wetland, whether to buy an SUV or a Prius—that ecosystem service values need to be incorporated."

Heal agrees. "Although ecosystem services have been with us for millennia," he says, "the scale of human activity is now sufficiently great that we can no longer take their continuation for granted."

Virginia journalist **JIM MORRISON** wrote about polar bears and global warming in the February/March 2004 issue.

Searching for Sustainability

Forest Policies, Smallholders, and the Trans-Amazon Highway

Eirivelthon Lima et al.

It is a powerful and disturbing image: loggers driving roads deep into the forest to remove a few mahogany trees, with slash-and-burn settlers following closely on their heels. However, it no longer captures the whole picture of logging in the Brazilian Amazon. So, then, what is the role of logging in the impoverishment or potential conservation of the Amazon rainforest? The answer to this question is deceptively complex: To achieve a sustainable future in Amazon forestry, policymakers and stakeholders must understand the physical, economic and political dimensions of competing land use options and economic interests. They must provide effective governance for multiple agendas that require individual oversight.

For simplicity's sake, suppose that forest governance can be approached from two angles: a preservation approach in which the land is tucked away, never to be used again; and a "use-it-or-lose-it" approach in which a well-managed forest estate becomes part of a sustainable economic development scenario and competes successfully with other land use options. In fact, 28 percent of the Brazilian Amazon is already listed as some form of park, or as a protected or indigenous area.[1] But what of the forest without protection, found mainly on private lands or on as-yet undesignated government lands? For many, selective logging of these forests is a form of forest impoverishment that is only slightly less devastating than forest clear-cutting.[2] For others, the selective harvest of timber is the best way to make the long-term protection of standing forests economically and politically viable.[3]

Opponents base their argument on two points: The long-term selective logging of primary tropical forest is financially impracticable, and selective logging is the first step in a vicious cycle of degradation that includes settlement and land clearing.[4] Advocates say that selective logging, when done well (called "reduced impact logging"), is renewable, economically viable, and may provide an important stream of revenue for government and private landowners that would encourage the maintenance of forest cover.[5] These proponents contend that if tropical forestry is to compete successfully with other land use options and essentially push back against the encroaching line of deforestation, some conditions must be met: the removal of subsidies to other land use options; the breakdown of barriers to entry, such as complex forest management plans; the dissemination of information on forestry to all potential market participants;

and the elimination of perverse incentives for deforestation—in particular the establishment of land titles through clearing to demonstrate active use. Furthermore, if forestry is deemed the least cost-effective approach to maintaining forest cover outside parks and protected areas, then subsidies to forest management activities might also be appropriate.

There are vast forested areas in the Brazilian Amazon located outside parks and protected areas, and a multitude of landowners, including state and federal governments, are controlling that forest. As a result, policies to manage forest resources must necessarily be comprehensive, flexible, and appropriate to varying conditions and agents. The Brazilian government has recently identified, and is now beginning to implement, a strategy of timber concessions that should help to corral some part of the industry into a controlled region. This should make it easier to monitor and will hopefully reduce illegal logging. The policy, however, mostly ignores the sticky issue of forestry on private land, which, although complex, could provide the engine for sustainable economic development among the disenfranchised settlers of the Amazon frontiers. The settlers may straggle onto the frontier individually, but they eventually form communities, control large areas of land, and become an increasingly important component of the timber industry.

A major economic corridor in the Brazilian Amazon—the Trans-Amazon Highway—illustrates how logging can be transformed from a force driving forest impoverishment to one driving forest conservation, and how this transition, in turn, carries important potential benefits for the semi-subsistence farmers who live along this corridor. To fully describe this transformation, it is necessary to place it in light of the history and current context of the timber industry of the Amazon, and with the understanding that the complexity inherent in the largest and most diverse tropical forest in the world makes forest governance a mighty task.

A Brief History of the Amazon Timber Industry

Understanding logging along the Trans-Amazon Highway depends upon the historical context of the timber industry in the Amazon, which can be roughly divided into three periods.[6]

The early production period lasted from the 1950s to the early 1970s and was followed by a transition or boom period, which lasted from the mid 1970s to the late 1980s. A third period, industry consolidation and migration to new frontiers, started in the early 1990s but is now coming to an end. The current timber industry is in such disarray from political mismanagement that in October 2005 the federal police temporarily suspended the transport of all logs from the Amazon.

Early Days (1950s to Mid-1970s)

In the 1950s, the island region of the Amazon delta in the state of Pará was the center of the wood industry in the Amazon. Through the 1960s, there were three large plywood mills and six large sawmills that controlled production. With no connection to the large domestic markets of southeastern Brazil and the dependence on fluvial transport to access raw materials and deliver products, these mills produced only for the export market. Limited shipping capacity and irregular delivery schedules hindered sales to ports in northeastern Brazil, which could be reached by ship along the Atlantic coast. The primary source of raw material was smallscale landowners who sold logs along the banks of rivers. The environmental impact of logging was minimal, as timber extraction was an integrated part of diverse smallscale family farming systems on the Amazon River floodplain. The two popular tree species harvested were Virola (*Virola surinamensis*) for plywood and Andiroba (*Carapa guianensis*) for sawnwood (the first stage of the log processing sequence in which logs are cut into boards, but not planed).

In the early and mid-1970s, a number of smaller sawmills began to appear in the island region and farther up along the upper Amazon River. Into the mid-1970s, the Amazon remained disconnected from domestic markets but the export market flourished. Estimated log consumption was in the region of 2.5 million cubic meters per year—all harvested by axe. Early reports on timber production in the Brazilian Amazon suggest this was a period of poor market access, poor quality of laborers, obsolete equipment, insufficient knowledge of local tree species, and scarce information on prices and markets for products.[7]

Transition Period (Late 1970s to Early 1990s)

A period of dramatic transition in the timber sector began in the late 1970s to early 1980s. Several highways were completed to link the Amazon to domestic southeastern and northeastern Brazilian markets. The states of Rondônia, Mato Grosso, and Pará became connected through the BR364, BR163, and BR010 highways. Large public investment programs for the construction of dams, hydropower plants, a railroad for the Carajás mining program, and the settlement of migrants from southern and northeastern Brazil changed the interfluvial forests of the Amazon, passively protected until that time by their inaccessibility.

Deforestation during this time was largely a response to government actions that either directly promoted or enabled land conversion from forests to other uses. The number and size of sawmills increased in response to the inexpensive primary resource and newly accessible markets, growing local demand, and the availability of cheap labor. Mechanization of harvesting, transport, and processing also contributed to the growth of sawnwood output.

By the early 1980s, Paragominas (a city in Pará) became the most important mill center in the Amazon, producing mostly for the domestic market. The state of Mato Grosso also produced lumber for the domestic market, with important logging centers appearing in the towns of Sinop and Alta Floresta, Meanwhile, the island region continued to produce for the export market. In all, the transition period during the 1970s and mid-to-late 1980s was a turning point in the timber industry of the Brazilian Amazon.

Consolidation and Migration (Mid-1990s to 2000s)

After the transition period, another (less dramatic) period of consolidation and expansion ensued along old and new logging frontiers.[8] Old frontiers can now be found in eastern Pará (Paragominas and Tailandia) and in northern Mato Grosso (Sinop). In these areas, virgin forests have become increasingly scarce, and the logging industry became more diverse and efficient. The more inefficient logging firms exited the market, and those that remained became vertically integrated in an effort to capture value added in downstream processing.

Access to the old frontiers is generally good given the high density of paved roads. In contrast, new frontiers are characterized by a rapid inflow of mills and producers from the old frontier, poor government regulation, and high transport costs. The notable new logging frontier is in western Pará along the northern section of the Santarém-Cuiaba Highway, the BR163.

The Industry Today

The current volume of wood produced in the Legal Amazon is between 20 and 30 million cubic meters, of which more than 50 percent is sold in the domestic Brazilian market.[9] Prior to 2003, legal timber harvest was possible through the preparation of a forest management plan submitted to the government agency and approved with a temporary land title. All that was required by IBAMA, the Brazilian government environmental agency, was proof that a firm or individual had initiated a land legalization process with Brazilian land titling institutions such as the Institute of Colonization and Agrarian Reform (INCRA).[10] Generally, land titling procedures took years, and they did not always result in legalization. By the time the land titling institution had made its decision, the harvest was already complete and the loggers had moved on to the next native forest stocks.

In 2003, the Brazilian government abruptly decided that management plans could no longer be approved on lands where property rights were not well established. That year, nearly all forest management plans were rejected.[11] The government, however, did not have an alternative readily available for the nearly 2,500 logging companies based in the Amazon, and an unintended side effect of the policy has been that more companies now simply operate illegally in such areas. Conflicts, protests, and widespread unregistered logging are now the norm.

To solve the problem of legalizing timber harvest and controlling the timber industry, the Brazilian government has proposed implementing forest concessions on public lands.[12] While this approach has some merit, and indeed has been debated extensively in the Brazilian public arena, large concessions controlled by a few companies may not be the best economic option in the regions where smallholders and other private landowners, including a large number of migrant settlers, are the predominant land users.[13]

The Case of the Trans-Amazon Highway

For two weeks in August 2003, the Trans-Amazon Highway was impassable: Angry loggers had blocked the road, stopping traffic to protest a government-imposed timber shortage. A similar display occurred outside the town of Santarém, Pará, in January 2005 and recently on the BR163 Highway in western Pará. Tragically, access to timber was also one of the underlying reasons for the murder of Sister Dorothy Stang in the municipality of Anapú.[14] Timber scarcity is a startling concept for the Amazon. How is it that a resource so apparently abundant can be the root cause of violent conflicts and protests?[15] The answer lies partially in the sudden requirement by the Brazilian government that loggers provide proper legal documentation for land rights in areas where logs are extracted. But who owns the forests and logs along this frontier highway?

Built by General Emílio Garrastazu Médici (president of Brazil from 1969–1974), the main part of the Trans-Amazon Highway stretches approximately 1,000 kilometers from the town of Marabá to Itaituba on the banks of the Tapajós River.[16] The highway is largely unpaved and virtually impassable for four months of the year during the rainy season. Homesteaders are usually allocated demarcated lots of 100 hectares apiece (approximately 250 acres) and then often battle the elements and wealthy land speculators to continue occupying the land.[17] Still, migration to the region is relentless, as a constant stream of formal and informal land control followed early colonization projects in the late 1970s.[18] INCRA, the federal land settlement agency, has formally settled approximately 30,000 families and an unknown number of informal squatters.

While it is commonly accepted that smallholders control vast areas of land along the Trans-Amazon Highway, the exact quantity of land is debatable. This question is taken up under the auspices of the Green Highways Project, an international multi-institutional project, led by the Brazilian nongovernmental organization Instituto de Pesquisa Ambiental da Amazônia (IPAM, Amazon Institute of Environmental Research) with the support of the Massachusetts-based Woods Hole Research Center (WHRC).[19] An area 100 kilometers (km) on either side of the Trans-Amazon Highway from the municipality of Itupiranga to Placas was mapped using satellite imagery and secondary data from Brazilian government sources. Land distribution

was mapped and deforestation measured using 30-meter spatial resolution satellite images and secondary data from INCRA and the Brazilian Institute of Geography and Statistics (IBGE).[20] Images were classified into forest and non-forest classes by supervised classification and visual interpretation.[21] The objective was to identify where smallholders are located and where they will be located in the future.

Of the total 15.7 million hectares located within this buffer, 7.9 million are under the control of or are promised to smallholders. Of the total area within the 100-km study area, the land distribution is: 1.1 percent in demarcated settlements, 5.4 percent in current settlements, 11.4 percent as squatters (*posseiros*), 13.2 percent in old colonization projects, and 19.5 percent destined for future settlements by INCRA. Four percent of the land is in conservation areas, 7.6 percent in informal medium and large-scale land holdings, 15.4 percent in indigenous reserves, and a final 21.2 percent is unclaimed government land.[22] The number of smallholders currently residing in the 100-km zone was estimated by summing the area with active settlements, which includes current settlements, colonization, and squatters, and then dividing by an 82.6-hectare average lot size from survey results (see below), giving a total area of approximately 4.7 million hectares held by 57,000 smallholder families.

Given the observed distribution of smallholders from the spatial analysis, the next logical question for the Green Highways Project was whether these agents could potentially supply the timber industry with wood. Demand for timber in the area is strong; the demand for logs on the Trans-Amazon Highway more than doubled over 12 years, increasing from roughly 340,000 cubic meters in 1990 to approximately 840,000 cubic meters in 2002. To determine whether smallholders can provide this quantity it is important to first estimate the growing stock potential of the forest held by smallholders, assuming that smallholders will in fact sell wood (this assumption will be revisited below). Using conservative (high) deforestation assumptions (for example, a range of 60 percent deforested for old colonization areas to 15 percent deforested for INCRA land allocated to future settlement) and a conservative stand volume of ten cubic meters per hectare, forest stock in active settlement areas is estimated to be 25.8 million cubic meters.[23] Using a harvest cycle of 30 years, this would give a sustained harvest volume of approximately 860,000 cubic meters, which matches current demand. At an estimated stumpage price of 10 Reais (R$10) per cubic meter of standing trees (approximately US$3.33 per cubic meter), this volume would generate R$8.6 million per year.[24] To put this in perspective, if the smallholder forests within current settlements were used to their full potential right now, and the benefits distributed evenly to every family (recall there are an estimated 57,000), each smallholder household could receive R$150 per year—a large sum given the discussion below.

Assuming that smallholders will eventually settle in areas set aside by INCRA, there will be an estimated forest stock of 52.6 million cubic meters, which could render a sustainable harvest of approximately 1.7 million cubic meters per year, more than double the current regional demand. Thus, there appears to be sufficient potential forest stock to meet the demand, and a tremendous opportunity for a redistribution of wealth to the poor, should smallholders have an unhampered market to sell wood [see Table 1 below].[25]

However, one needs to ask if these estimates based upon government census data are consistent with data on the ground. To answer this question we make use of data generated from a recent comprehensive socioeconomic survey of smallholders along the Trans-Amazon Highway. Between June and December 2003, a total of nearly 3,000 families were interviewed, of which 2,441 lived within the 100-km zone.[26]

In the survey, smallholders were asked about their forest production, and socioeconomic data were collected. The results add to the discussion above, showing that 26 percent had sold wood, and those sales had occurred largely within the last 5 years. There had been only one sale per lot. Ninety-six percent of the smallholders sold standing trees, and the average number of trees sold was 20 per smallholder, which corresponds to a harvest rate of approximately 1 tree per 5 hectares and, assuming an average volume of 5 cubic meters of log per tree, an average sale volume of 100 cubic meters. The average total sale value was R$173, which corresponds to R$8.65 per tree or R$1.73 per cubic meter.[27]

Comparing these observations with the results from the geospatial analysis above, based on timber produced through legal deforestation and harvest of legal forest reserves (the area of smallholder land prohibited from clearing for crops), smallholders are selling approximately 1 cubic meter per hectare, and only 26 percent of them actually sell wood. At this harvest volume, it would take the harvest from 10,000 families per year—about 18 percent of the estimated total smallholder families—to sustain current demand from the area industry at current prices. This amounts to a harvest volume that is only 4 percent of current estimates for Amazon timber production from other studies. At a harvest intensity of 10 cubic meters per hectare, this participation requirement would be drastically reduced to only 1,000 families per year (which represents 1.8 percent of all families). This level of participation could be easily achieved without undue

change in the smallholder system by subcontracting the timber industry to do much of the technical work associated with logging. The production of logs is dramatically low on smallholder lots because smallholders have limited knowledge of the forest potential and limited access to the financial resources required to manage the forests. This barrier can be overcome with a partnership between smallholders and the timber industry. For the successful implementation of such a partnership, however, it will be important for smallholders to understand the logging process and have adequate access to production information so that they can maintain a check on their industrial partners.

Holding Back the Tide of Smallholder Forestry

From the perspective of community foresters, the current ideal is that individuals within the communities must work collectively and must control the entire chain of production through to sales of the final product. Formal interaction with the timber industry is anathema. Also, there is still the idea that forest management must happen in large, undisturbed, contiguous tracts of forests. This closely held and restrictive view has undermined the potential of community forestry in the Amazon. The reality is that there are more than 500,000 settlement families in the Brazilian Amazon who work individually or in community associations and who specialize (though perhaps not yet efficiently) in the supply of standing timber by working closely with logging companies.

However, most of the community-based forestry operations have two key problems. First, when dealing with smallholders on an individual basis, the loggers hold all the cards. They have more information about the species and value of timber, and they exploit the immediate financial needs of cash-poor smallholders. Second, logging on smallholder lots is legal only under two premises: smallholders have deforestation licenses that allow the clearing of 3 hectares per year and the sale of

Table 1 Timber Potential from Smallholder Lots on the Trans-Amazon Highway

Smallholders	Total Area (hectares (ha))	Percent Land Area	Forest Cover (percent)	Total Forest Area (ha)	Timber Stock (m³)	Potential Timber Flow (m³/year)
Future settlement projects	3,055,000	19.5	85	2,596,000	25,965,000	865,000
Colonization projects	2,063,000	13.2	40	825,000	8,252,000	275,000
Informal settlement	1,792,000	11.4	60	1,075,000	10,750,000	358,000
INCRA settlements	852,000	5.4	80	682,000	6,815,000	227,000
Demarcated settlements	169,000	1.1	50	85,000	847,000	28,000
Total smallholders	7,931,000	50.6	–	5,263,000	52,629,000	1,753,000

Note. INCRA is Brazil's National Institute of Colonization and Agrarian Reform. The entire buffer area is 15,643,000 ha. The area not occupied by smallholders is comprised of unclaimed government land (21.2 percent), indigenous land (15.4 percent), medium and large informal settlement (7.6 percent), and conservation units (4.2 percent).

Source: Instituto de Pesquisa Ambiental da Amazônia (the Amazon Institute of Environmental Research).

60 cubic meters per year (up to 20 percent of the land area owned), or they may have the option to develop a forest management plan that must be approved by IBAMA. Of the sales registered in the surveys, 26 percent came from deforestation permits, and a startling 79 percent came from the "legal reserve" on each plot.[28] Because no formal forest management plans have been developed for these smallholder systems, this would imply that nearly 80 percent of log sales from smallholders are currently illegal by government rules; in addition, few smallholders get legal deforestation permits. Why are there no formal plans? A forest management plan requires that the landowner hold legal title, and although 95 percent of smallholders surveyed claimed to be the landowner, we found only 26 percent held formal title; a statistic supported by previous research in the region.[29] This lack of coordination between agencies and resource users is a major barrier to overcoming illegal logging within smallholder systems and to the integration of smallholders into the formal timber market.

Small-Farm Family Forestry in the Amazon

Coordination between ministries is not an impossible task, however. For example, IBAMA, INCRA, and the Ministry of Public Works of the town of Santarém (in Pará) operating with limited resources but in partnership with loggers and smallholders, found a creative solution to this problem in the form of an equitable partnership between industry and smallholders. In this case, the community associations subcontract the loggers to plan and implement harvesting, while the government ministries have the responsibility of expediting title and management approval. The land is owned individually, and management plans are done for each private 100-hectare lot, but the negotiations are between the logger and the community association. The community can demand higher prices by selling as a group, and the logger is assured of a long-term supply of timber. As a result, legal forest operations are taking place and smallholders are capturing a fair share of the benefits from the timber harvest on their land [see box on next page].[30]

However, changes in government personnel and extreme inefficiency (the project industry coordinator has had management plans under review at IBAMA for more than a year) has made even this promising partnership tenuous. These types of projects are in danger of failing because government oversight is inefficient, inadequate, corrupt, and contradictory.[31] There may be a partial solution to be found in timber concessions, but even with successful concessions, the large-scale problems of illegal logging will not disappear. Indeed, the problems of illegal logging will never be solved if IBAMA cannot control the industry or support it effectively, but there is no indication so far that IBAMA can do it alone. It is reasonable to assume, however, that the economic benefits of timber production on their private landholdings will stimulate smallholders to manage their forests and help control illegal activities.

What Does the Future Hold?

What do the results of the Green Highways Project have to say about the issue of loggers and forest policies? As mentioned above, the main thrust of the new forest policy centers around timber concessions on public lands with some allowances given to communities. This is an effective program for a portion of the industry, but there are two problems with the idea. First, the evidence presented above indicates that this approach is inadequate for some major economic corridors where there are many smallholders, such as in the case of the region surrounding the Trans-Amazon Highway; of the 80 percent of land available for harvest (for example, excluding conservation units and indigenous areas), the Green Highways Project shows that 64 percent is under the control of, or is promised to, smallholders. Second, it also shows that forestry is highly underutilized in these smallholder systems. This and the fact that there are more than 500,000 families settled in the Amazon region mean these results imply a very large economic loss to Brazilian society from not capturing a potential timber supply that would almost do away with the need for timber concessions on public lands.

Further, by excluding smallholders from access to the timber industry through current management plan requirements, smallholders are denied what could amount to a substantial and vital source of economic development. In some settlements, research has shown that the value of a single harvest can equal more than 15 years of agricultural production.[32] And finally, even if only some portion of the demand for logs is met by concessions harvesting on public government lands, it may have a negative socioeconomic impact on the potential for small farm forestry by depressing overall prices.

To promote sustainable forestry, the evidence indicates that the government has to realistically deal with land titling, facilitate institutional coordination, and commit to stopping illegal logging through better enforcement. Invariably, the causes of policy failure and poor governance are related to corruption and political auction of important positions in government institutions. An intricate net of political obligation, to the detriment of technical decisions, is commonplace, and even those individuals fiercely committed to their tasks (and there are many) struggle to make quality strategies a reality. A lack of efficiency in government agencies, whether through poor coordination or delays, increases transaction costs and makes formal forest management difficult. Also, by neglecting secure property rights, or making these difficult and costly to obtain, the government inadvertently creates incentives for smallholders and loggers to engage in illegal logging.

Forest management projects on smallholder settlement lots in the Brazilian Amazon will, if widely adopted, help move the region toward equitable forest-based economic development and a peaceful resolution to the problems now facing migrant families. This is not the only solution for the Amazon, but it is a step forward and one well within the reach of the current administration. Without change, however, we can expect further illegal degradation of the forest and a continuing struggle for economic development and social justice on the Amazon frontier.

André Da Silva Dias Reflects on the Forest Families Program

Forest management models that can contribute to the social, environmental, and economic development of smallholders and traditional populations have been the subject of many recent initiatives in the Amazon. The "Forest Families" program [in Santarém] works with a specific relationship that appears to be very common but little studied: smallholders and the timber industry. It is interesting to note some of the fundamental characteristics around which the program is built: the relationship between the smallholder and the industry already exists; its foundation is market-based; its actors are well-defined; and [it] is based on uncommonly strong legal and ethical rigor. The last characteristic alone makes one pay attention.

One can question whether this is community forest management or not, A pertinent doubt, but, in the end, there exists a forest and its resources and a people organized, or organizing, in communities. In fact, the smallholders are not directly managing their forests: they delegate this activity to a subcontractor and his team. And when they delegate they relinquish some personal control of the forest. However, they exercise their rights to the forest in a free manner, in a negotiation process that strengthens the local organization, generates collective responsibility, creates a commonly used infrastructure, provides income and, most importantly, gives value to the standing forest. All of which are the principles that underlie community forest management.

It is possible to imagine a scenario in which they should manage their own forests in accordance with their capacity, limitation, abilities, and interests. Perhaps this will happen one day. But for right now, the reality is different. No better and no worse, this is just different than many other community forest management initiatives where the local residents play the role of managers. The fact is that they, the owners, are who should say whether this is how it should be. And they seem to be making this [decision] in an informed way, understanding their limitations, and identifying opportunities. It is interesting to observe a community and its people (in this case Santo Antonio) started barely two years ago by families of different origin, who until this point never knew each other, but who already have solid development plans and a growing autonomy in the formulation of local projects, rather than just hope of better days.

I believe that one of the principal contributions that this program can lend to the discussion of local forest management is to define criteria and indicators of a healthy and egalitarian relationship between smallholders and the timber industry. To get there, some challenges that deserve more attention are:

- Improving local knowledge of good forest management practices.
- Identifying the impact of timber harvest on the supply of hunting and non-timber forest products.
- Analyzing the socioeconomic impact of the timber income on the smallholder systems.

Source: André da Silva Dias, Executive Manager, Fundação Floresta Tropical, December 2003. This box was translated from the Portuguese by Frank Merry and first published in a report by Instituto de Pesquisa Ambiental da Amazônia (IPAM) for the International Institute for Environment and Development (IIED) as part of the IIED Power Tools Initiative: Sharpening Policy Tools for Marginalized Managers of Natural Resources. F. Merry, E. Lima, G. Amacher, O. Almeida, A. Alves, and M. Guimares, *Overcoming Marginalization in the Brazilian Amazon Through Community Association: Case Studies of Forests and Fisheries,* (Edinburgh, UK, 2004). It is reprinted with permission.

Notes

1. This approach has been plied very successfully by large conservation organizations, in particular Conservation International, which solicits funds to buy up biodiversity "hotspots." For an analysis of the effects of parks and protected areas on fires in the Amazon, see D. Nepstad et al., "Inhibition of Amazon Deforestation and Fire by Parks and Indigenous Reserves," *Conservation Biology.* In press, expected publication February 2006.

2. I. Bowles, R. E. Rice, R. A. Mittermeier, and G. A. B. da Fonseca, "Logging On in the Rain Forests," *Science,* 4 September 1998, 1453–58; R. Rice, C. Sugal, and I. Bowles, *Sustainable Forest Management: A Review of the Current Conventional Wisdom.* (Washington, DC: Conservation International, 1998); R. Rice, R. Gullison, and J. Reid, "Can Sustainable Management Save Tropical Forests?" *Scientific American,* April 1997, 34–39.

3. D. Pearce, F. E. Putz, and J. Vanclay, "Sustainable Forestry in the Tropics: Panacea or Folly?" *Forest Ecology and Management* 172, no. 2 (2003): 229–247; M. Verissimo, A. Cochrane, and C. Sousa Jr., "National Forests in the Amazon," *Science,* 30 August 2002, 1478; F. E. Putz, K. H. Redford, J. G. Robinson, R. Fimbel,

and G. Blate, *Biodiversity Conservation in the Context of Tropical Forest Management* (Washington, DC: Biodiversity Studies, The World Bank, 2000), http://world-bank.org/biodiversity.

4. G. Asner, et al., "Selective Logging in the Amazon," *Science,* 21 October 2005: 480–481. The Asner study claimed that the selective logging of the Amazon is far more widespread than previously thought. The authors suggest that the source of logs in the Amazon is not slash-and-burn deforestation—those logs are simply burned—but conventional poor-quality selective logging and that this is the first step in the economic and ecological degradation of the forest. According to the data, this is more widely practiced and perhaps more damaging than previously thought.

5. Forest management and reduced impact logging (FM-RIL) guidelines are available from many sources: the Suriname Agricultural Training Center (CELOS); the International Tropical Timber Organization (ITTO); the Food and Agricultural Organization of the United Nations (FAO); the Institute of Humans and the Environment of the Amazon (IMAZON); and the Fundação Floresta Tropical (FFT, Tropical Forest Foundation). In addition, field models in Brazil demonstrate the improvements of FM-RIL practices over conventional selective logging. See the FFT website at

http://www.fft.org.br and click "Research." There have been several studies on the economic benefits of reduced impact logging and comparisons with "conventional" selective logging. For a few examples see: S. Armstrong and C. J. Inglis, "RIL For Real: Introducing Reduced Impact Logging Techniques into a Commercial Forestry Operation in Guyana," *International Forestry Review* 2, (2000): 264–72; F. Boltz, D. R. Carter, T. P. Holmes, and R. Perreira Jr., "Financial Returns Under Uncertainty for Conventional and Reduced-Impact Logging in Permanent Production Forests of the Brazilian Amazon," *Ecological Economics* 39 (2001): 387–98; P. Barreto, P. Amaral, E. Vidal, and C. Uhl, "Costs and Benefits of Forest Management for Timber Production in Eastern Amazonia," *Forest Ecology and Management* 108, no. 1 (1998): 9–26; and T. P. Holmes et al., *Financial Costs and Benefits of Reduced Impact Logging Relative to Conventional Logging in the Eastern Amazon* (Washington, DC: Tropical Forest Foundation, 1999).

6. Thanks to Johan Zweede of the Instituto Florestal Tropical in Belém, Brazil, and Benno Pokorny of the University of Freiburg, Germany, for valuable comments on the history and context of the timber industry.

7. For an excellent review, see I. Sholtz, *Overexploitation or Sustainable Management: Action Patterns of the Tropical Timber Industry: The Case of Pará, Brazil,* 1960–1997 (London: Frank Cass Publishers, 2001).

8. By definition the term "frontier," when applied to forests, implies the point at which new logging occurs. It is, however, common in the literature of logging in the Amazon to differentiate frontiers by age. This is done partially out of custom, but also because logging on all "frontiers" is relatively new; even old frontiers are less than 30 years old.

9. The "Legal Amazon" is a geo-political definition of the Amazon region in Brazil and comprises the states of Amapá, Amazonas, Acre, Maranhão, Mato Grosso, Pará, Rondônia, and Tocantins. The volume of sawnwood destined for export is different across frontiers. More than 60 percent of logs from new frontiers are destined for the export market, whereas on the intermediate and old frontiers, that level dips to 50 and 15 percent, respectively, according to F. Merry et al., "Industrial Development on Logging Frontiers in the Brazilian Amazon," *International Journal of Sustainable Development,* in review. For a recent discussion of production volumes, see G. Asner et al., note 4 above.

10. The Instituto Brasileiro do Meio Ambiente e dos Recursos Naturais Renováveis (http://www.ibama.gov.br) is the Brazilian government's environmental agency responsible for the forest sector and all issues of environmental control in the country. The federal land-titling agency is the Institute of Colonization and Agrarian Reform (INCRA). For more information, see http://www.incra.gov.br. Each state also has a local agency.

11. The forest management process includes a formal management plan that essentially states that the company intends to harvest in a given area (with accompanying maps and documentation) and subsequently an annual operating plan that delivers the details of each year's harvest operation. The term "forest management plan" includes both of these components of logging.

12. For more information on forest concessions, see A. Veríssimo, M. A. Cochrane, and C. Sousa Jr., National Forests in the Amazon," Science, 30 August 2002, 1478.

13. The forest concessions issue has long been debated in the scientific literature. Both sides of the argument for Brazil can be explored in F. D. Merry et al., "A Risky Forest Policy in the Amazon?" *Science,* 21 March 2003, 1843 and in F. D. Merry et al., "Some Doubts About Concessions in Brazil," *Tropical Forestry Update* 13, no. 3 (2003): 7–9 (see http://www.itto.or.jp/live/contents/download/tfu/TFU.2003.03.English.pdf). See also F. D. Merry and G. S. Amacher, "Forest Taxes, Timber Concessions, and Policy Choices in the Amazon," *Journal of Sustainable Forestry* 20, no. 2 (2005): 15–44; and Veríssimo, Cochrane, and Sousa, note 12 above. For earlier discussion on concessions see J.A. Gray, *Forestry Revenue Systems in Developing Countries,* FAO Forestry Paper 43 (Rome, 1983); R. Repetto and M. Gillis, eds., *Public Policies and the Misuse of Forest Resources,* (Cambridge, UK: Cambridge University Press, 1988); J. R. Vincent, "The Tropical Timber Trade and Sustainable Development," *Science,* 19 June, 1992, 1651–1655; and J. A. Gray, "Underpricing and Overexploitation of Tropical Forests: Forest Pricing in the Management, Conservation and Preservation of Tropical Forests," *Journal of Sustainable Forestry* 4, no. 1/2 (1997): 75–97. The Ministry of Environment has created a new law on public forest management (Law 4776/05), which was approved by Brazil's Chamber of Representatives in July, is still awaiting the vote of the Senate. This law would create the national forest service, the forest development fund, and would regulate timber harvest on public lands. Three kinds of harvest are sought for production forests: direct government management of conservation units (such as national forests); local community use (such as extractive reserves); and forest concessions.

14. Dorothy Stang, a 73-year-old nun from Dayton, Ohio, a practitioner of liberation theology, and an ardent supporter of local settlers, was assassinated in broad daylight in February 2005 in a remote farm community near her home of 25 years in Anapú on the Trans-Amazon Highway. Her battle for equal rights for the poor, including legal land and resource ownership, brought her in direct conflict with loggers and ranchers. Her death triggered an avalanche of government response. Two thousand soldiers were sent to the region to crack down on illegal loggers and land speculators, and five million hectares of forest (an area the size of Costa Rica) were designated as parks and reserves in what may be the world's single greatest act of tropical rainforest conservation.

15. The estimate of forest stock for the Amazon is approximately 60 billion cubic meters. There are varying estimates of the flow from the forest: The IBGE, which is the government institute of geography and statistics (http://ibge.gov.br), estimates log demand in the north of Brazil to be about 17 million cubic meters; IBAMA, the environmental regulation agency of Brazil, estimates it to be around 25 million; and IMAZON, a local nongovernmental research organization, estimates it at about 24 million—down from 28 million in 1999.

16. The entire Trans-Amazon Highway runs approximately 3,300 kilometers, connecting the state of Tocantins to the state of Acre near the Peruvian border. Continuing westward from Itaituba to the town of Humaitá (a stretch which lies to the west of the Tapajós River) is virtually uninhabited, but may be the future frontier on which this story is replayed some years hence.

17. For an excellent discussion on property rights, violence and settlement on the Trans-Amazon Highway see L. J. Alston, G. D. Libecap, and B. Mueller, *Titles, Conflict, and Land Use:*

The Development of Property Rights and Land Reform on the Brazilian Amazon Frontier (Ann Arbor: The University of Michigan Press, 1999); and L. G. Alston, G. D. Libecap, and B. Mueller, "Land Reform Policies, the Sources of Violent Conflict in the Brazilian Amazon," *Journal of Environmental Economics and Management* 39, no. 2 (2000): 162–188.

18. For more discussion of smallholder settlement in new and old settlements and the roles of community associations in economic development in migrant settlements see F. Merry and D. J. Macqueen, *Collective Market Engagement* (Edinburgh, International Institute for Environment and Development, 2004), http://www.iied.org/docs/flu/PT7_collective_market_engagement.pdf.

19. Other institutions working on the Trans-Amazon Highway within the Green Highways Project are the Fundação Viver, Produzir e Preservar (FVPP) and the Instituto Floresta Tropical (IFT).

20. The principal source of government statistics for Brazil is the Brazilian Institute of Geography and Statistics (Instituto Brasiliero de Geografia e Estatistica, IBGE). Their website can be accessed at http://www.ibge.gov.br.

21. A supervised classification is a procedure for identifying spectrally similar areas on an image by pinpointing training sites of known targets and then extrapolating those spectral signatures to other areas of unknown targets. The signatures are quantitative measures of the spectral properties at one or several wavelength intervals. These measures include class maximum, minimum, mean and covariance matrix values. Training areas, usually small and discrete compared to the full image, are identified through visual interpretation and used to "train" the classification algorithm to recognize land cover classes based on their spectral signatures, as found in the image. The training areas for any one land cover class need to fully represent the variability of that class within the image.

22. The total area for squatters was 19 percent of the buffer zone, of which local extension agents estimated 60 percent to be smallholders. The remaining 40 percent were said to be medium- and large-size holdings.

23. The evidence also indicated that only one percent of the buffer area is currently deforested, so these estimates could be considered very conservative for deforestation.

24. The price of R$10 is based on a conservative estimate of a formal logging contract between smallholders and the industry near the town of Santarém and the example of a forest concession (3-year cutting contract) in the Tapajós national forest—an ITTO project run by IBAMA—where the average stumpage fee for three price categories in 2003 was R$11.73. The exchange rate for the period of the survey was approximately R$3 per US$1, but is now at R$2.2 per US$1. For further commentary on the timber markets of Brazil, see A. Veríssimo and R. Smeraldi, *Acertando O Alvo: Consumo da Madeira no Mercado Interno Brasileiro a Promocao da Certificacao Florestal* (Finding the Target: Consumption of Wood in the Brazilian Domestic Market and the Promotion of Forest Certification) and M. Lentini, A. Verissimo, and L. Sobral, *Fatos Florestais da Amazônia* (Forest Facts of the Amazon) (Belém, Brazil: Imazon, 2003); E. Lima, and F. Merry, "Views of Brazilian Producers—Increasing and Sustaining Exports," in D. Macqueen, ed., *Growing Timber Exports: The Brazilian Tropical Timber Industry and International Markets* (London: IIED, 2003), 82–102.

25. For an economic model of smallholder decision-making, production, and labor allocation, see F. D. Merry and G. S. Amacher, "Emerging Smallholder Forest Management Contracts in the Brazilian Amazon: Labor Supply and Productivity Effects," Environment and Development Economics. Invited to revise and resubmit, expected publication 2006.

26. The preliminary results of the survey were presented in seminars to the smallholders in June 2004. Further details of this survey are available from the authors.

27. In comparison, the estimated price for logs at the mill gate in 2002 on the Trans-Amazon was R$58 per cubic meter, and an unadjusted five-year average price for logs from 1998 to 2002 was R$39 per cubic meter, but this is before accounting for harvest costs—which for intermediate frontiers such as the Trans-Amazon can run between 30 and 40 Reais per cubic meter and transportation costs; transport distances can run as far as 80 or 90 kilometers from log deck to mill.

28. The legal reserve (Reserva Legal) of a smallholder lot, or for that matter any private land holding in the Brazilian Amazon, is 80 percent of the total land area. This "reserve" area can only be used for forestry with approved forest management plans or the collection of non-timber forest products.

29. Alston, Libecap, and Mueller, note 17 above. Only 11 percent of land owners hold formal title. In our survey, individuals were asked whether they held "definitive title," not formal records.

30. This example is well documented. See D. Nepstad et al., "Managing the Amazon Timber Industry," *Conservation Biology* 18, no. 2 (2004): 575–577; D. Nepstad et al., "Governing the Amazon Timber Industry," in D. Zarin, J. R. R. Alavalapati, F. E. Putz, and M. Schmink, eds., *Working Forests in the American Tropics: Conservation through Sustainable Management?* (New York: Columbia University Press, 2004), 388–414.

31. Another example is the project Safra Legal (Legal Harvest) on the Trans-Amazon Highway. The objective of this project was to make use of the legal deforestation options available to smallholders. The idea of this project came from the forest management projects near Santarém and presented a wonderful alternative to smallholders who would have simply burned the trees where they planned to conduit agricultural activities. The project, however, has recently become embroiled in scandal as a conduit of illegal logging, see L. Coutinho, "More Petista Mud in the Ibama," *VEJA*, 15 June 2005, 70. The problems behind the Safra Legal program were also described in L. Rohter, "Loggers, Scorning the Law, Ravage the Amazon Jungle," *The New York Times*, 16 October 2005. These articles illustrate the far-reaching negative effects of corrupt government on the sustainable management of natural resources.

32. F. Merry et al., "Collective Action Without Collective Ownership: the Role of Formal Logging Contracts in Community Associations on the Brazilian Amazon Frontier," *International Forestry Review*, in review. Drafts available from the authors.

Eirivelthon Lima is an associate researcher at the Instituto de Pesquisa Ambiental da Amazônia (IPAM, the Amazon Institute of Environmental Research), headquartered in Belém, Pará, Brazil, and doctoral student in Forest Economics at the Virginia Polytechnic Institute and State University (Virginia Tech). **Frank Merry** is an associate researcher at IPAM, research fellow in environmental studies at Dartmouth College, and visiting assistant scientist at the Woods Hole Research Center (WHRC). **Daniel Nepstad** is a senior researcher at IPAM and senior scientist at WHRC. **Gregory Amacher** is an associate professor of forest economics at Virginia Tech and associate researcher at IPAM. **Cláudia Azevedo-Ramos** is a senior researcher at IPAM. **Paul Lefebvre** is a senior research associate at WHRC. **Felipe Resque Jr.** is a GIS technician at IPAM. We gratefully acknowledge funding from (in alphabetical order) the European Union; the Gordon and Betty Moore Foundation; the William and Flora Hewlett Foundation; NASA Large Scale Biosphere and Atmosphere Project; the National Science Foundation; and the United States Agency for International Development–Brazil program.

UNIT 6

The Hazards of Growth: Pollution and Climate Change

Unit Selections

Key Points to Consider

- How has one of the atmospheric scientific community's most respected voices been systematically discredited? What is the current nature of the scientific consensus over global warming and to what extent does it represent political opinion or scientific examination of atmospheric phenomena?

- Given the strength and nearly universal agreement between climate scientists on the future of our planet's climate, why has there been so little action on the part of the United States? At the present time, the United States is the only developed country that has refused to sign the Kyoto accord. Why, in light of the universal agreement between countries, are we still so uncertain about future climate change?

- Can we ever reach consensus on global warming? Republicans consistently block environmental measures, citing a lack of scientific consensus. When will we ever achieve consensus, or is it impossible?

Student Web Site

www.mhcls.com/online

Internet References

Further information regarding these Web sites may be found in this book's preface or online.

Persistent Organic Pollutants (POP)
http://www.chem.unep.ch/pops/
School of Labor and Industrial Relations (SLIR): Hot Links
http://www.lir.msu.edu/hotlinks/
Space Research Institute
http://arc.iki.rssi.ru/eng/index.htm
Worldwatch Institute
http://www.worldwatch.org

Anthony Ise/Getty Images

A chemical company once had a slogan: "Without chemicals, life itself would be impossible." Depending on how the word 'chemical' is defined, one could say that everything in our own bodies is composed of chemicals, so this phrase is certainly correct on one level. What is more relevant in the context of this book is that degree to which society has changed because of the production of synthetic chemicals. The accomplishments of chemistry have truly been remarkable. Many life-saving drugs have been created in the laboratory. The vast majority of crops on America's farms are fertilized with synthetic chemicals, raising yields to unheard of levels, and insect infestations are eradicated with synthetic pesticides. But a small fraction of these chemicals are ultimately found to have harmful side-effects that were never imagined by their developers, or worse, were understood but specifically hidden from the public. The classic example of a harmful chemical is DDT. Starting around 1940, DDT (dichloro-diphenyl-trichloroethane) was used as an incredibly effective pesticide. The complete elimination of malaria from North Africa can be tied directly to DDT's effectiveness as a pesticide. But unanticipated side-effects to DDT began to emerge. Highlighted by Rachel Carson's famous book "Silent Spring," our nation was alerted to the apparent link between DDT usage and the decline in the osprey and bald eagle population. Carson's book led to the banning of DDT and the eventual resurgence of the bald eagle. It also was the first time America's conscience was directed toward environmental stewardship.

The list of harmful chemicals is miniscule compared to those that have societal benefits, but it is the harmful ones that often receive (and are worthy of) the most attention. Chlorofluorocarbons or CFCs were used widely as refrigerants and propellants. CFCs are completely inert and harmless to human beings and animals. Unfortunately, CFCs are directly responsible for the generation of the ozone hole over Antarctica. The loss of ozone allows for ultraviolet radiation to reach the Earth's surface, causing cancer and cataracts. Fortunately, once recognized, the production of CFCs has been mostly banned, so that we can hope for the ozone depletion to decrease in the coming decades.

The battle between industry and environmentalists is nowhere more evident than with regards to CO_2 emissions and their linkage to global warming. There is now an overwhelming majority of scientists who recognize that climate change is tied to increasing CO_2 levels, and that the increase in CO_2 is due to the burning of fossil fuels. Yet even after decades of study, there is still a public uncertainty with regards to the certainties of global warming. In a shocking revelation, "The Truth about Denial," Sharon Begley of Newsweek exposes the efforts of powerful oil lobbies to stoke the propaganda machine against global warming.

Their goal appears to be to sow doubt, to make the science appear inexact. Even with a nearly unanimous voice from the scientific community regarding the seriousness of global warming, a lone voice—often funded by the oil industry—carries far more weight than any one of a hundred voices from the unified scientific community.

This somber point is driven home in "Swift Boating, Stealth Budgeting, and Unitary Executives" by James Hansen, director of NASA's Goddard Institute for Space Studies, and one of the first to recognize the dangers of global warming. Hansen talks about not only private industry's efforts to discredit those warning of the consequences of rising CO_2 levels, but also of the Government's efforts to cripple the efforts of NASA, the very agency entrusted to "understand and protect our home planet."

These two articles make the reader aware that our realization of the seriousness of this issue has been delayed by at least a decade. Every year we wait, the problems grow more serious. Every year we wait, we come closer to the 'tipping point,' the point from which we are unable to reverse the effects of global warming. Tragic as efforts to derail climate change legislation have been, we can be thankful that the American public is now overwhelmingly in favor of drastic change to reduce the deleterious effects of global warming. While it is nearly certain that no changes will occur in the final year of the Bush presidency, we can hope and look forward to a new leader—either Democrat or Republican—who will lead the United States and the world to redirect our efforts at preventing further destruction to our planet from the burning of fossil fuels.

Swift Boating, Stealth Budgeting, and Unitary Executives

JAMES HANSEN

The American Revolution launched the radical proposition that the commonest of men should have a vote equal in weight to that of the richest, most powerful citizen. Our forefathers devised a remarkable Constitution, with checks and balances, to guard against the return of despotic governance and subversion of the democratic principle for the sake of the powerful few with special interests. They were well aware of the difficulties that would be faced, however, placing their hopes in the presumption of an educated and honestly informed citizenry.

I have sometimes wondered how our forefathers would view our situation today. On the positive side, as a scientist, I like to imagine how Benjamin Franklin would view the capabilities we have built for scientific investigation. Franklin speculated that an atmospheric "dry fog" produced by a large volcano had reduced the Sun's heating of the Earth so as to cause unusually cold weather in the early 1780s; he noted that the enfeebled solar rays, when collected in the focus of a "burning glass," could "scarce kindle brown paper." As brilliant as Franklin's insights may have been, they were only speculation as he lacked the tools for quantitative investigation. No doubt Franklin would marvel at the capabilities provided by Earth-encircling satellites and super-computers that he could scarcely have imagined.

Yet Franklin, Jefferson, and the other revolutionaries would surely be distraught by recent tendencies in America, specifically the increasing power of special interests in our government, concerted efforts to deceive the public, and arbitrary actions of government executives that arise from increasing concentration of authority in a unitary executive, in defiance of the aims of our Constitution's framers. These tendencies are illustrated well by a couple of incidents that I have been involved in recently.

In the first incident, my own work was distorted for the purposes of misinforming the public and protecting special interests. In the second incident, the mission of the National Aeronautics and Space Administration (NASA) was altered surreptitiously by executive action, thus subverting constitutional division of power. These incidents help to paint a picture that reveals consequences for society far greater than simple enrichment of special interests. The effect is to keep the public in the dark about increasing risks to our society and our home planet.

The first incident prompted *New York Times* columnist Paul Krugman to argue not long ago that I must respond to "swift boaters"—those who distort the record to impugn someone's credibility. I have had reservations about doing so, stemming from the perceptive advice of Professor Henk van de Hulst, who said, when I was a post-doc at Leiden University, "Your success will depend upon choosing what not to work on." Unfortunately, given the shrinking fuse on the global warming time bomb, Krugman is probably right: we cannot afford the luxury of ignoring swift boaters and focusing only on science.

Pat Michaels, a swift boater to whom Krugman refers, is sometimes described as a "contrarian." Contrarians address global warming as if they were lawyers, not scientists. A lawyer's job often is to defend a client, not seek the truth. Instead of following Richard Feynman's dictum on scientific objectivity ("The only way to have real success in science . . . is to describe the evidence very carefully without regard to the way you feel it should be"), contrarians present only evidence that supports their desired conclusion.

Skepticism, an inherent aspect of scientific inquiry, should be carefully distinguished from contrarianism. Skepticism, and the objective weighing of evidence, are essential for scientific success. Skepticism about the existence of global warming and the principal role of human-made greenhouse gases has diminished as empirical evidence and our understanding have advanced. However, many aspects of global warming need to be understood better, including the best ways to minimize climate change and its consequences. Legitimate skepticism will always have an important role to play.

However, hard-core global warming contrarians have an agenda other than scientific truth. Their target is the public. Their goal is to create an impression that global warming or its causes are uncertain. Debating a contrarian leaves an impression with today's public of an argument among theorists. Sophistical contrarians do not need to win the scientific debate to advance their cause.

Science Fiction

Consider, for example, Pat Michaels' deceit (in a 2000 article in *Social Epistemology*) in portraying climate "predictions" that I made in 1988 as being in error by "450 percent." This distortion is old news, but by sheer repetition has become received wisdom among climate-change deniers. In fact, science fiction writer Michael Crichton was duped by Michaels, although Crichton reduced my "error" to "wrong by 300 percent" in his 2004 novel *State of Fear.*

People acquainted with this topic are aware that Michaels, in comparing global warming predictions made with the GISS (Goddard Institute for Space Studies) climate model with observations, played a dirty trick by showing model calculations for only one of the three scenarios (not predictions!) that I presented in 1988. Here's why this trick has a big impact.

The three scenarios (see figure, next page) were intended to bracket the range of likely future climate forcings (changes imposed on the Earth's energy balance that tend to alter global temperature either way). Scenario C had the smallest greenhouse gas forcing: it extended recent greenhouse gas growth rates to the year 2000 and thereafter kept greenhouse gas amounts constant, i.e., it assumed that after 2000 human sources of these gases would be just large enough to balance removal of these gases by the "sinks." Scenario B continued approximately linear growth of greenhouse gases beyond 2000. Scenario A showed exponential growth of greenhouse gases and included a substantial allowance for trace gases that were suspected of increasing but were unmeasured.

Scenarios A, B, and C also differed in their assumptions about future volcanic eruptions. Scenarios B and C included occasional eruptions of large volcanoes, at a frequency similar to that of the real world in the previous few decades. Scenario A, intended to yield the largest plausible warming, included no volcanic eruptions, as it is not uncommon to have no large eruptions for extended periods, such as the half century between the Katmai eruption in 1912 and the Agung eruption in 1963.

Multiple scenarios are used to provide a range of plausible climate outcomes, but also so that we can learn something by comparing real-world outcomes with model predictions. How well the model succeeds in simulating the real world depends upon the realism of both the assumed forcing and the climate sensitivity (the global temperature response to a standard climate forcing) of the model.

As it turned out, in the real world the largest climate forcing in the decade after 1988, by far, was caused by the Mount Pinatubo volcanic eruption, the greatest volcanic eruption of the past century. Forcings are measured in watt-years per square meter ($W\text{-}yr/m^2$) averaged over the surface of the Earth (1 $W\text{-}yr/m^2$ is a heating of 1 W/m^2 over the entire planet maintained for one year). The small particles injected into the Earth's stratosphere by Pinatubo reflected sunlight back to space, causing a negative (cooling) climate forcing of about -5 $W\text{-}yr/m^2$. In contrast, the added greenhouse gas climate forcings ranged from about $+1.6 W\text{-}yr/m^2$ in scenario C to about $+2.3 W\text{-}yr/m^2$ in scenario A.

So of the four scenarios (A, B, C, and the real world) only scenario A had no large volcanic eruption. The volcanic activity modeled in scenarios B and C was somewhat weaker than in the real world and was misplaced by a few years, but by good fortune it was such as to have a cooling effect pretty similar to that of Pinatubo. Despite the fact that scenario A omitted the largest climate forcing, Michaels chose to compare scenario A—and *only* scenario A—with the real world. Is this a case of scientific idiocy or is there something else at work? Perhaps Michaels is just not very interested in learning about the real world.

Although less important for the temperature change between 1997 and 1988 that Michaels examined, measured real-world greenhouse gas changes in carbon dioxide (CO_2), methane (CH_4), nitrous oxide (N_2O), and chlorofluorocarbons (CFCs) yielded a forcing similar to those in scenarios B and C. The reason for the slow real-world growth rate was that both CH_4 and CO_2 growth rates decreased in the early 1990s (the slowdowns may have been associated with Pinatubo; in any case the CO_2 growth rate has subsequently accelerated rapidly).

An astute reader may wonder why the world showed any warming during the period 1988–97, given that the negative (cooling) forcing by Pinatubo exceeded the positive (warming) forcing by greenhouse gases added in that period. The reason is that the climate system was also being pushed by the planetary "energy imbalance" that existed in 1988. The climate system had not yet fully responded to greenhouse gases added to the atmosphere before then. The observed continued decadal warming, despite the very large negative volcanic forcing, provides some confirmation of that planetary energy imbalance.

Noise and Distortion

Michaels' trick of comparing the real world only with the inappropriate scenario A accounts for his specious, incorrect conclusions. However, a second unscientific aspect of his method is also worth pointing out.

Scientists seek to learn something by comparing the real world with climate model calculations. Climate sensitivity is of special interest, as future climate change depends strongly upon it. In principal, we can extract climate sensitivity if we have accurate knowledge of the net forcing that drove climate change, and the global temperature change that occurred in response to that change. However, even if these demanding conditions are met, it is necessary to compare the magnitude of the calculated changes with the magnitude of "noise," including errors in the measurements and chaotic (unforced) variability in the model and real-world climate changes.

If Michaels had examined the noise question he would have realized that a nine-year change is insufficient to determine the real-world temperature trend or distinguish among the model runs. Even the period 1988–2005 is too brief for most purposes. Within several years the differences among scenarios A, B, and C, and comparisons with the real world, will become more meaningful.

Michaels' latest tomfoolery, repeated on several occasions, is the charge that I approve of exaggeration of potential consequences of future global warming. This is more unadulterated hogwash. Michaels quotes me as saying, "Emphasis on extreme scenarios may have been appropriate at one time, when the public and decision-makers were relatively unaware of the global warming issue."

What trick did Michaels use to create the impression that I advocate exaggeration? He took the above sentence out of context from a paragraph in which I was being gently critical of a tendency of Intergovernmental Panel on Climate Change climate simulations to emphasize only cases with very large increases of climate forcings. My entire paragraph (from a June 2003 presentation to the Council on Environmental Quality) read as follows:

> *Summary opinion on scenarios. Emphasis on extreme scenarios may have been appropriate at one time, when the public and decision-makers were relatively unaware of the global warming issue, and energy sources such as "synfuels," shale oil, and tar sands were receiving strong consideration. Now, however, the need is for demonstrably objective climate forcing scenarios consistent with what is realistic under current conditions. Scenarios that accurately fit recent and near future observations have the best chance of bringing all of the important players into the discussion, and they also are what is needed for the purpose of providing policy-makers the most effective and efficient options to stop global warming.*

Would an intelligent reader who read the entire paragraph (or even the entire sentence; by chopping off half of the sentence Michaels brings quoting-out-of-context to a new low) infer that I was advocating exaggeration? On the contrary. Perhaps I should take it as a compliment that anyone would search my writing so hard to find something that can be quoted out of context.

Having taken this trouble to refute Michaels' claims, I still wonder about the wisdom of arguing with contrarians as a strategy. Many of them, including Michaels, receive support from special interests such as fossil fuel and automotive companies. It is understandable that special interests gravitated, early on, to scientists who had a message they preferred to hear. But now that global warming and its impacts are clearer, it is time for business people to reconsider their position—and scientists, rather than debating contrarians, may do better to communicate with business leaders. The latter did not attain their positions without being astute and capable of changing. We need to make clear to them the legal and moral liabilities that accrue with continued denial of

global warming. It is time for business leaders to chuck contrarians and focus on the business challenges and opportunities.

Stealth Budgets & Unitary Executives

The second incident involved NASA's budget. Many people are aware that something bad happened to the NASA Earth Science budget this year, yet the severity of the cuts and their long-term implications are not universally recognized. In part this is because of a stealth budgeting maneuver.

When annual budgets for the coming fiscal year are announced, the differences in growth from the previous year, for agencies and their divisions, are typically a few percent. An agency with +3 percent growth may crow happily, in comparison to agencies receiving +1 percent. Small differences are important because every agency has fixed costs (civil service salaries, buildings, other infrastructure), so new programs or initiatives are strongly dependent upon any budget growth and how that growth compares with inflation.

When the administration announced its planned fiscal 2007 budget, NASA science was listed as having typical changes of 1 percent or so. However, Earth Science research actually had a staggering reduction of about 20 percent from the 2006 budget. How could that be accomplished? Simple enough: reduce the 2006 research budget retroactively by 20 percent! One-third of the way into fiscal year 2006, NASA Earth Science was told to go figure out how to live with a 20-percent loss of the current year's funds.

The Earth Science budget is almost a going-out-of-business budget. From the taxpayers' point of view it makes no sense. An 80-percent budget must be used mainly to support infrastructure (practically speaking, you cannot fire civil servants; buildings at large facilities such as Goddard Space Flight Center will not be bulldozed to the ground; and the grass at the centers

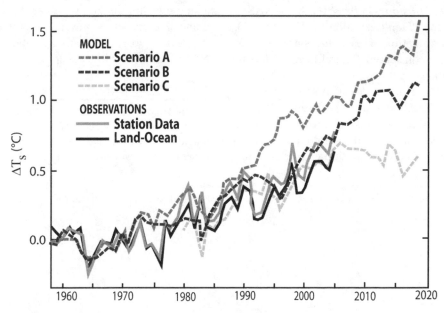

Figure 1 Annual mean global temperature change.

must continue to be cut). But the budget cuts wipe off the books most planned new satellite missions (some may be kept on the books, but only with a date so far in the future that no money needs to be spent now), and support for contractors, young scientists, and students disappears, with dire implications for future capabilities.

Bizarrely, this is happening just when NASA data are yielding spectacular and startling results. Two small satellites that measure the Earth's gravitational field with remarkable precision found that the mass of Greenland decreased by the equivalent of 200 cubic kilometers of ice in 2005. The area on Greenland with summer melting has increased 50 percent, the major ice streams on Greenland (portions of the ice sheet moving most rapidly toward the ocean and discharging icebergs) have doubled in flow speed, and the area in the Arctic Ocean with summer sea ice has decreased 20 percent in the last 25 years.

One way to avoid bad news: stop the measurements! Only hitch: the first line of the NASA mission is "to understand and protect our home planet." Maybe that can be changed to ". . . protect special interests' backside."

I should say that the mission statement *used* to read "to understand and protect our home planet." That part has been deleted—a shocking loss to me, as I had been using the phrase since December 2005 to justify speaking out about the dangers of global warming. The quoted mission statement had been constructed in 2001 and 2002 via an inclusive procedure involving representatives from the NASA Centers and e-mail interactions with NASA employees. In contrast, elimination of the "home planet" phrase occurred in a spending report delivered to Congress in February 2006, the same report that retroactively slashed the Earth Science research budget. In July 2006 I asked dozens of NASA employees and management people (including my boss) if they were aware of the change. Not one of them was. Several expressed concern that such management changes by fiat would have a bad effect on organization morale.

The budgetary goings-on in Washington have been noted, e.g., in editorials of *The Boston Globe:* "Earth to NASA: Help!" (June 15, 2006) and "Don't ask; don't ask" (June 22), both decrying the near-termination of Earth measurements. Of course, the *Globe* might be considered "liberal media," so their editorials may not raise many eyebrows.

But it is conservatives and moderates who should be most upset, and I consider myself a moderate conservative. When I was in school we learned that Congress controlled the purse strings; it is in the Constitution. But it does not really seem to work that way, not if the Bush administration can jerk the sci-

ence budget the way they have, in the middle of a fiscal year no less. It seems more like David Baltimore's "Theory of the Unitary Executive" (the legal theory that the president can do pretty much whatever he wants) is being practiced successfully. My impression is that conservatives and moderates would prefer that the government work as described in the Constitution, and that they prefer to obtain their information on how the Earth is doing from real observations, not from convenient science fiction.

Congress is putting up some resistance to the budget manipulation. The House restored a fraction of the fiscal year 2007 cuts to science and is attempting to restore planning for some planetary missions. But the corrective changes are moderate. You may want to check your children's textbooks for the way the U.S. government works. If their books still say that Congress controls the purse strings, some updating is needed.

The NASA Mission
To understand and protect our home planet, To explore the universe and search for life, To inspire the next generation of explorers . . . as only NASA can.

But may it be that this is all a bad dream? I will stand accused of being as wistful as the boy who cried out, "Joe, say it ain't so!" to the fallen Shoeless Joe Jackson of the 1919 Chicago Black Sox, yet I maintain the hope that NASA's dismissal of "home planet" is not a case of either shooting the messenger or a too-small growth of the total NASA budget, but simply an error of transcription. Those who have labored in the humid, murky environs of Washington are aware of the unappetizing forms of life that abound there. Perhaps the NASA playbook was left open late one day, and by chance the line "to understand and protect our home planet" was erased by the slimy belly of a slug crawling in the night. For the sake of our children and grandchildren, let us pray that this is the true explanation for the devious loss, and that our home planet's rightful place in NASA's mission will be restored.

JAMES HANSEN is an adjunct professor at the Columbia University Earth Institute and director of NASA's Goddard Institute for Space Studies in New York. He expresses his opinions here as a private citizen under the protection of the First Amendment.

The Truth About Denial

Sharon Begley

Sen. Barbara Boxer had been chair of the Senate's Environment Committee for less than a month when the verdict landed last February. "Warming of the climate system is unequivocal," concluded a report by 600 scientists from governments, academia, green groups and businesses in 40 countries. Worse, there was now at least a 90 percent likelihood that the release of greenhouse gases from the burning of fossil fuels is causing longer droughts, more flood-causing downpours and worse heat waves, way up from earlier studies. Those who doubt the reality of human-caused climate change have spent decades disputing that. But Boxer figured that with "the overwhelming science out there, the deniers' days were numbered." As she left a meeting with the head of the international climate panel, however, a staffer had some news for her. A conservative think tank long funded by ExxonMobil, she told Boxer, had offered scientists $10,000 to write articles undercutting the new report and the computer-based climate models it is based on. "I realized," says Boxer, "there was a movement behind this that just wasn't giving up."

If you think those who have long challenged the mainstream scientific findings about global warming recognize that the game is over, think again. Yes, 19 million people watched the "Live Earth" concerts last month, titans of corporate America are calling for laws mandating greenhouse cuts, "green" magazines fill newsstands, and the film based on Al Gore's best-selling book, "An Inconvenient Truth," won an Oscar. But outside Hollywood, Manhattan and other habitats of the chattering classes, the denial machine is running at full throttle—and continuing to shape both government policy and public opinion.

Since the late 1980s, this well-coordinated, well-funded campaign by contrarian scientists, free-market think tanks and industry has created a paralyzing fog of doubt around climate change. Through advertisements, op-eds, lobbying and media attention, greenhouse doubters (they hate being called deniers) argued first that the world is not warming; measurements indicating otherwise are flawed, they said. Then they claimed that any warming is natural, not caused by human activities. Now they contend that the looming warming will be minuscule and harmless. "They patterned what they did after the tobacco industry," says former senator Tim Wirth, who spearheaded environmental issues as an under secretary of State in the Clinton administration. "Both figured, sow enough doubt, call the science uncertain and in dispute. That's had a huge impact on both the public and Congress."

Just last year, polls found that 64 percent of Americans thought there was "a lot" of scientific disagreement on climate change; only one third thought planetary warming was "mainly caused by things people do." In contrast, majorities in Europe and Japan recognize a broad consensus among climate experts that greenhouse gases—mostly from the burning of coal, oil and natural gas to power the world's economies—are altering climate. A new NEWSWEEK Poll finds that the influence of the denial machine remains strong. Although the figure is less than in earlier polls, 39 percent of those asked say there is "a lot of disagreement among climate scientists" on the basic question of whether the planet is warming; 42 percent say there is a lot of disagreement that human activities are a major cause of global warming. Only 46 percent say the greenhouse effect is being felt today.

As a result of the undermining of the science, all the recent talk about addressing climate change has produced little in the way of actual action. Yes, last September Gov. Arnold Schwarzenegger signed a landmark law committing California to reduce statewide emissions of carbon dioxide to 1990 levels by 2020 and 80 percent more by 2050. And this year both Minnesota and New Jersey passed laws requiring their states to reduce greenhouse emissions 80 percent below recent levels by 2050. In January, nine leading corporations—including Alcoa, Caterpillar, Duke Energy, Du Pont and General Electric—called on Congress to "enact strong national legislation" to reduce greenhouse gases. But although at least eight bills to require reductions in greenhouse gases have been introduced in Congress, their fate is decidedly murky. The Democratic leadership in the House of Representatives decided last week not even to bring to a vote a requirement that automakers improve vehicle mileage, an obvious step toward reducing greenhouse emissions. Nor has there been much public pressure to do so. Instead, every time the scientific case got stronger, "the American public yawned and bought bigger cars," Rep. Rush Holt, a New Jersey congressman and physicist, recently wrote in the journal Science; politicians "shrugged, said there is too much doubt among scientists, and did nothing."

It was 98 degrees in Washington on Thursday, June 23, 1988, and climate change was bursting into public consciousness. The Amazon was burning, wildfires raged in the United States, crops in the Midwest were scorched and it was shaping up to be the hottest year on record worldwide. A Senate committee, including Gore, had invited NASA climatologist James

Hansen to testify about the greenhouse effect, and the members were not above a little stagecraft. The night before, staffers had opened windows in the hearing room. When Hansen began his testimony, the air conditioning was struggling, and sweat dotted his brow. It was the perfect image for the revelation to come. He was 99 percent sure, Hansen told the panel, that "the greenhouse effect has been detected, and it is changing our climate now."

The reaction from industries most responsible for greenhouse emissions was immediate. "As soon as the scientific community began to come together on the science of climate change, the pushback began," says historian Naomi Oreskes of the University of California, San Diego. Individual companies and industry associations—representing petroleum, steel, autos and utilities, for instance—formed lobbying groups with names like the Global Climate Coalition and the Information Council on the Environment. ICE's game plan called for enlisting greenhouse doubters to "reposition global warming as theory rather than fact," and to sow doubt about climate research just as cigarette makers had about smoking research. ICE ads asked, "If the earth is getting warmer, why is Minneapolis [or Kentucky, or some other site] getting colder?" This sounded what would become a recurring theme for naysayers: that global temperature data are flat-out wrong. For one thing, they argued, the data reflect urbanization (many temperature stations are in or near cities), not true global warming.

Shaping public opinion was only one goal of the industry groups, for soon after Hansen's sweat-drenched testimony they faced a more tangible threat: international proposals to address global warming. The United Nations had scheduled an "Earth Summit" for 1992 in Rio de Janeiro, and climate change was high on an agenda that included saving endangered species and rain forests. ICE and the Global Climate Coalition lobbied hard against a global treaty to curb greenhouse gases, and were joined by a central cog in the denial machine: the George C. Marshall Institute, a conservative think tank. Barely two months before Rio, it released a study concluding that models of the greenhouse effect had "substantially exaggerated its importance." The small amount of global warming that might be occurring, it argued, actually reflected a simple fact: the Sun is putting out more energy. The idea of a "variable Sun" has remained a constant in the naysayers' arsenal to this day, even though the tiny increase in solar output over recent decades falls far short of explaining the extent or details of the observed warming.

In what would become a key tactic of the denial machine—think tanks linking up with like-minded, contrarian researchers—the report was endorsed in a letter to President George H.W. Bush by MIT meteorologist Richard Lindzen. Lindzen, whose parents had fled Hitler's Germany, is described by old friends as the kind of man who, if you're in the minority, opts to be with you. "I thought it was important to make it clear that the science was at an early and primitive stage and that there was little basis for consensus and much reason for skepticism," he told Scientific American magazine. "I did feel a moral obligation."

Bush was torn. The head of his Environmental Protection Agency, William Reilly, supported binding cuts in greenhouse emissions. Political advisers insisted on nothing more than voluntary cuts. Bush's chief of staff, John Sununu, had a PhD in engineering from MIT and "knew computers," recalls Reilly. Sununu frequently logged on to a computer model of climate, Reilly says, and "vigorously critiqued" its assumptions and projections.

Sununu's side won. The Rio treaty called for countries to voluntarily stabilize their greenhouse emissions by returning them to 1990 levels by 2000. (As it turned out, U.S. emissions in 2000 were 14 percent higher than in 1990.) Avoiding mandatory cuts was a huge victory for industry. But Rio was also a setback for climate contrarians, says UCSD's Oreskes: "It was one thing when Al Gore said there's global warming, but quite another when George Bush signed a convention saying so." And the doubters faced a newly powerful nemesis. Just months after he signed the Rio pact, Bush lost to Bill Clinton—whose vice president, Gore, had made climate change his signature issue.

Groups that opposed greenhouse curbs ramped up. They "settled on the 'science isn't there' argument because they didn't believe they'd be able to convince the public to do nothing if climate change were real," says David Goldston, who served as Republican chief of staff for the House of Representatives science committee until 2006. Industry found a friend in Patrick Michaels, a climatologist at the University of Virginia who keeps a small farm where he raises prize-winning pumpkins and whose favorite weather, he once told a reporter, is "anything severe." Michaels had written several popular articles on climate change, including an op-ed in The Washington Post in 1989 warning of "apocalyptic environmentalism," which he called "the most popular new religion to come along since Marxism." The coal industry's Western Fuels Association paid Michaels to produce a newsletter called World Climate Report, which has regularly trashed mainstream climate science. (At a 1995 hearing in Minnesota on coal-fired power plants, Michaels admitted that he received more than $165,000 from industry; he now declines to comment on his industry funding, asking, "What is this, a hatchet job?")

The road from Rio led to an international meeting in Kyoto, Japan, where more than 100 nations would negotiate a treaty on making Rio's voluntary—and largely ignored—greenhouse curbs mandatory. The coal and oil industries, worried that Kyoto could lead to binding greenhouse cuts that would imperil their profits, ramped up their message that there was too much scientific uncertainty to justify any such cuts. There was just one little problem. The Intergovernmental Panel on Climate Change, or IPCC—the international body that periodically assesses climate research—had just issued its second report, and the conclusion of its 2,500 scientists looked devastating for greenhouse doubters. Although both natural swings and changes in the Sun's output might be contributing to climate change, it concluded, "the balance of evidence suggests a discernible human influence on climate."

Faced with this emerging consensus, the denial machine hardly blinked. There is too much "scientific uncertainty" to justify curbs on greenhouse emissions, William O'Keefe, then a vice president of the American Petroleum Institute and leader of the Global Climate Coalition, suggested in 1996. Virginia's Michaels echoed that idea in a 1997 op-ed in The

Washington Post, describing "a growing contingent of scientists who are increasingly unhappy with the glib forecasts of gloom and doom." To reinforce the appearance of uncertainty and disagreement, the denial machine churned out white papers and "studies" (not empirical research, but critiques of others' work). The Marshall Institute, for instance, issued reports by a Harvard University astrophysicist it supported pointing to satellite data showing "no significant warming" of the atmosphere, contrary to the surface warming. The predicted warming, she wrote, "simply isn't happening according to the satellite[s]." At the time, there was a legitimate case that satellites were more accurate than ground stations, which might be skewed by the unusual warmth of cities where many are sited.

"There was an extraordinary campaign by the denial machine to find and hire scientists to sow dissent and make it appear that the research community was deeply divided," says Dan Becker of the Sierra Club. Those recruits blitzed the media. Driven by notions of fairness and objectivity, the press "qualified every mention of human influence on climate change with 'some scientists believe,' where the reality is that the vast preponderance of scientific opinion accepts that human-caused [greenhouse] emissions are contributing to warming," says Reilly, the former EPA chief. "The pursuit of balance has not done justice" to the science. Talk radio goes further, with Rush Limbaugh telling listeners this year that "more carbon dioxide in the atmosphere is not likely to significantly contribute to the greenhouse effect. It's just all part of the hoax." In the new *NEWSWEEK* Poll, 42 percent said the press "exaggerates the threat of climate change."

Now naysayers tried a new tactic: lists and petitions meant to portray science as hopelessly divided. Just before Kyoto, S. Fred Singer released the "Leipzig Declaration on Global Climate Change." Singer, who fled Nazi-occupied Austria as a boy, had run the U.S. weather-satellite program in the early 1960s. In the Leipzig petition, just over 100 scientists and others, including TV weathermen, said they "cannot subscribe to the politically inspired world view that envisages climate catastrophes." Unfortunately, few of the Leipzig signers actually did climate research; they just kibitzed about other people's. Scientific truth is not decided by majority vote, of course (ask Galileo), but the number of researchers whose empirical studies find that the world is warming and that human activity is partly responsible numbered in the thousands even then. The IPCC report issued this year, for instance, was written by more than 800 climate researchers and vetted by 2,500 scientists from 130 nations.

Although Clinton did not even try to get the Senate to ratify the Kyoto treaty (he knew a hopeless cause when he saw one), industry was taking no chances. In April 1998 a dozen people from the denial machine—including the Marshall Institute, Fred Singer's group and Exxon—met at the American Petroleum Institute's Washington headquarters. They proposed a $5 million campaign, according to a leaked eight-page memo, to convince the public that the science of global warming is riddled with controversy and uncertainty. The plan was to train up to 20 "respected climate scientists" on media—and public—outreach with the aim of "raising questions about and undercutting the 'prevailing scientific wisdom' " and, in particular, "the Kyoto treaty's scientific underpinnings" so that elected officials "will seek to prevent progress toward implementation." The plan, once exposed in the press, "was never implemented as policy," says Marshall's William O'Keefe, who was then at API.

The GOP control of Congress for six of Clinton's eight years in office meant the denial machine had a receptive audience. Although Republicans such as Sens. John McCain, Jim Jeffords and Lincoln Chafee spurned the denial camp, and Democrats such as Congressman John Dingell adamantly oppose greenhouse curbs that might hurt the auto and other industries, for the most part climate change has been a bitterly partisan issue. Republicans have also received significantly more campaign cash from the energy and other industries that dispute climate science. Every proposed climate bill "ran into a buzz saw of denialism," says Manik Roy of the Pew Center on Climate Change, a research and advocacy group, who was a Senate staffer at the time. "There was no rational debate in Congress on climate change."

The reason for the inaction was clear. "The questioning of the science made it to the Hill through senators who parroted reports funded by the American Petroleum Institute and other advocacy groups whose entire purpose was to confuse people on the science of global warming," says Sen. John Kerry. "There would be ads challenging the science right around the time we were trying to pass legislation. It was pure, raw pressure combined with false facts." Nor were states stepping where Washington feared to tread. "I did a lot of testifying before state legislatures—in Pennsylvania, Rhode Island, Alaska—that thought about taking action," says Singer. "I said that the observed warming was and would be much, much less than climate models calculated, and therefore nothing to worry about."

But the science was shifting under the denial machine. In January 2000, the National Academy of Sciences skewered its strongest argument. Contrary to the claim that satellites finding no warming are right and ground stations showing warming are wrong, it turns out that the satellites are off. (Basically, engineers failed to properly correct for changes in their orbit.) The planet is indeed warming, and at a rate since 1980 much greater than in the past.

Just months after the Academy report, Singer told a Senate panel that "the Earth's atmosphere is not warming and fears about human-induced storms, sea-level rise and other disasters are misplaced." And as studies fingering humans as a cause of climate change piled up, he had a new argument: a cabal was silencing good scientists who disagreed with the "alarmist" reports. "Global warming has become an article of faith for many, with its own theology and orthodoxy," Singer wrote in The Washington Times. "Its believers are quite fearful of any scientific dissent."

With the Inauguration of George W. Bush in 2001, the denial machine expected to have friends in the White House. But despite Bush's oil-patch roots, naysayers weren't sure they could count on him: as a candidate, he had pledged to cap carbon dioxide emissions. Just weeks into his term, the Competitive Enterprise Institute heard rumors that the draft of a speech Bush was preparing included a passage reiterating that pledge. CEI's Myron Ebell called conservative pundit Robert Novak, who had booked Bush's EPA chief, Christie Todd Whitman, on CNN's "Crossfire." He asked her about the line, and within hours the possibility of a carbon cap was the talk of the Beltway. "We alerted anyone we thought could have influence and get the line, if it was in the speech, out," says CEI president Fred Smith, who counts this as another notch in CEI's belt. The White House declines to comment.

Bush not only disavowed his campaign pledge. In March, he withdrew from the Kyoto treaty. After the about-face, MIT's Lindzen told NEWSWEEK in 2001, he was summoned to the White House. He told Bush he'd done the right thing. Even if you accept the doomsday forecasts, Lindzen said, Kyoto would hardly touch the rise in temperatures. The treaty, he said, would "do nothing, at great expense."

Bush's reversal came just weeks after the IPCC released its third assessment of the burgeoning studies of climate change. Its conclusion: the 1990s were very likely the warmest decade on record, and recent climate change is partly "attributable to human activities." The weather itself seemed to be conspiring against the skeptics. The early years of the new millennium were setting heat records. The summer of 2003 was especially brutal, with a heat wave in Europe killing tens of thousands of people. Consultant Frank Luntz, who had been instrumental in the GOP takeover of Congress in 1994, suggested a solution to the PR mess. In a memo to his GOP clients, he advised them that to deal with global warming, "you need to continue to make the lack of scientific certainty a primary issue." They should "challenge the science," he wrote, by "recruiting experts who are sympathetic to your view." Although few of the experts did empirical research of their own (MIT's Lindzen was an exception), the public didn't notice. To most civilians, a scientist is a scientist.

Challenging the science wasn't a hard sell on Capitol Hill. "In the House, the leadership generally viewed it as impermissible to go along with anything that would even imply that climate change was genuine," says Goldston, the former Republican staffer. "There was a belief on the part of many members that the science was fraudulent, even a Democratic fantasy. A lot of the information they got was from conservative think tanks and industry." When in 2003 the Senate called for a national strategy to cut greenhouse gases, for instance, climate naysayers were "giving briefings and talking to staff," says Goldston. "There was a constant flow of information—largely misinformation." Since the House version of that bill included no climate provisions, the two had to be reconciled. "The House leadership staff basically said, 'You know we're not going to accept this,' and [Senate staffers] said, 'Yeah, we know,' and the whole thing disappeared relatively jovially without much notice," says Goldston. "It was such a foregone conclusion."

Especially when the denial machine had a new friend in a powerful place. In 2003 James Inhofe of Oklahoma took over as chairman of the environment committee. That summer he took to the Senate floor and, in a two-hour speech, disputed the claim of scientific consensus on climate change. Despite the discovery that satellite data showing no warming were wrong, he argued that "satellites, widely considered the most accurate measure of global temperatures, have confirmed" the absence of atmospheric warming. Might global warming, he asked, be "the greatest hoax ever perpetrated on the American people?" Inhofe made his mark holding hearing after hearing to suggest that the answer is yes. For one, on a study finding a dramatic increase in global temperatures unprecedented in the last 1,000 years, he invited a scientist who challenged that conclusion (in a study partly underwritten with $53,000 from the American Petroleum Institute), one other doubter and the scientist who concluded that recent global temperatures were spiking. Just as Luntz had suggested, the witness table presented a tableau of scientific disagreement.

Every effort to pass climate legislation during the George W. Bush years was stopped in its tracks. When Senators McCain and Joe Lieberman were fishing for votes for their bipartisan effort in 2003, a staff member for Sen. Ted Stevens of Alaska explained to her counterpart in Lieberman's office that Stevens "is aware there is warming in Alaska, but he's not sure how much it's caused by human activity or natural cycles," recalls Tim Profeta, now director of an environmental-policy institute at Duke University. "I was hearing the basic argument of the skeptics—a brilliant strategy to go after the science. And it was working." Stevens voted against the bill, which failed 43–55. When the bill came up again the next year, "we were contacted by a lot of lobbyists from API and Exxon-Mobil," says Mark Helmke, the climate aide to GOP Sen. Richard Lugar. "They'd bring up how the science wasn't certain, how there were a lot of skeptics out there." It went down to defeat again.

Killing bills in Congress was only one prong of the denial machine's campaign. It also had to keep public opinion from demanding action on greenhouse emissions, and that meant careful management of what federal scientists and officials wrote and said. "If they presented the science honestly, it would have brought public pressure for action," says Rick Piltz, who joined the federal Climate Science Program in 1995. By appointing former coal and oil lobbyists to key jobs overseeing climate policy, he found, the administration made sure that didn't happen. Following the playbook laid out at the 1998 meeting at the American Petroleum Institute, officials made sure that every report and speech cast climate science as dodgy, uncertain, controversial—and therefore no basis for making policy. Ex-oil lobbyist Philip Cooney, working for the White House Council on Environmental Quality, edited a 2002 report on climate science by sprinkling it with phrases such as "lack of understanding" and "considerable uncertainty." A short section on climate in another report was cut entirely. The White House "directed us to remove all mentions of it," says Piltz, who resigned in protest. An oil lobbyist faxed Cooney, "You are doing a great job."

The response to the international climate panel's latest report, in February, showed that greenhouse doubters have a lot of fight left in them. In addition to offering $10,000 to scientists willing to attack the report, which so angered Boxer, they are emphasizing a new theme. Even if the world is warming now, and even if that warming is due in part to the greenhouse gases emitted by burning fossil fuels, there's nothing to worry about. As Lindzen wrote in a guest editorial in NEWSWEEK International in April, "There is no compelling evidence that the warming trend we've seen will amount to anything close to catastrophe."

To some extent, greenhouse denial is now running on automatic pilot. "Some members of Congress have completely internalized this," says Pew's Roy, and therefore need no coaching from the think tanks and contrarian scientists who for 20 years kept them stoked with arguments. At a hearing last month on the Kyoto treaty, GOP Congressman Dana Rohrabacher asked whether "changes in the Earth's temperature in the past—all of these glaciers moving back and forth—and the changes that we see now" might be "a natural occurrence." (Hundreds of studies have ruled that out.) "I think it's a bit grandiose for us to believe . . . that [human activities are] going to change some major climate cycle that's going on." Inhofe has told allies he will filibuster any climate bill that mandates greenhouse cuts.

Still, like a great beast that has been wounded, the denial machine is not what it once was. In the *NEWSWEEK* Poll, 38 percent of those surveyed identified climate change as the nation's gravest environmental threat, three times the number in 2000. After ExxonMobil was chastised by senators for giving $19 million over the years to the Competitive Enterprise Institute and others who are "producing very questionable data" on climate change, as Sen. Jay Rockefeller said, the company has cut back its support for such groups. In June, a spokesman said ExxonMobil did not doubt the risks posed by climate change, telling reporters, "We're very much not a denier." In yet another shock, Bush announced at the weekend that he would convene a global-warming summit next month, with a 2008 goal of cutting greenhouse emissions. That astonished the remaining naysayers. "I just can't imagine the administration would look to mandatory [emissions caps] after what we had with Kyoto," said a GOP Senate staffer, who did not want to be named criticizing the president. "I mean, what a disaster!"

With its change of heart, ExxonMobil is more likely to win a place at the negotiating table as Congress debates climate legislation. That will be crucially important to industry especially in 2009, when naysayers may no longer be able to count on a friend in the White House nixing mandatory greenhouse curbs. All the Democratic presidential contenders have called global warming a real threat, and promise to push for cuts similar to those being passed by California and other states. In the GOP field, only McCain—long a leader on the issue—supports that policy. Fred Thompson belittles findings that human activities are changing the climate, and Rudy Giuliani backs the all-volunteer greenhouse curbs of (both) Presidents Bush.

Look for the next round of debate to center on what Americans are willing to pay and do to stave off the worst of global warming. So far the answer seems to be, not much. The *NEWSWEEK* Poll finds less than half in favor of requiring high-mileage cars or energy-efficient appliances and buildings. No amount of white papers, reports and studies is likely to change that. If anything can, it will be the climate itself. This summer, Texas was hit by exactly the kind of downpours and flooding expected in a greenhouse world, and Las Vegas and other cities broiled in record triple-digit temperatures. Just last week the most accurate study to date concluded that the length of heat waves in Europe has doubled, and their frequency nearly tripled, in the past century. The frequency of Atlantic hurricanes has already doubled in the last century. Snowpack whose water is crucial to both cities and farms is diminishing. It's enough to make you wish that climate change were a hoax, rather than the reality it is.

In *"The Truth About Denial"* (Aug. 13), we said that Congressman John Dingell "adamantly oppose[s] greenhouse curbs that might hurt the auto and other industries." While Dingell has long opposed greenhouse curbs, he is now a co-sponsor of a bill aimed at improving fuel efficiency.

With Eve Conant, Sam Stein and Eleanor Clift in Washington and Matthew Philips in New York.

How Science Makes Environmental Controversies Worse

I use the example of the 2000 US Presidential election to show that political controversies with technical underpinnings are not resolved by technical means. Then, drawing from examples such as climate change, genetically modified foods, and nuclear waste disposal, I explore the idea that scientific inquiry is inherently and unavoidably subject to becoming politicized in environmental controversies. I discuss three reasons for this. First, science supplies contesting parties with their own bodies of relevant, legitimated facts about nature, chosen in part because they help make sense of, and are made sensible by, particular interests and normative frameworks. Second, competing disciplinary approaches to understanding the scientific bases of an environmental controversy may be causally tied to competing value-based political or ethical positions. The necessity of looking at nature through a variety of disciplinary lenses brings with it a variety of normative lenses, as well. Third, it follows from the foregoing that scientific uncertainty, which so often occupies a central place in environmental controversies, can be understood not as a lack of scientific understanding but as the lack of coherence among competing scientific understandings, amplified by the various political, cultural, and institutional contexts within which science is carried out.

In light of these observations, I briefly explore the problem of why some types of political controversies become "scientized" and others do not, and conclude that the value bases of disputes underlying environmental controversies must be fully articulated and adjudicated through political means before science can play an effective role in resolving environmental problems.

DANIEL SAREWITZ

Introduction

One may or may not find believable the claim by Bjorn Lomborg, author of *The Skeptical Environmentalist,* that, starting out as "an old left-wing Greenpeace member" gloom-and-doom environmentalist (Lomborg, 2001, p. xix) he gradually convinced himself, through the power of statistical analysis, that the environmental conditions upon which humanity depends for its well-being were not getting worse, but were actually getting better. Whether or not Lomborg did undergo a data-induced perceptual transformation, his underlying claim is a familiar and comfortable one. Our commitments to acting in the world must be based on a foundation of fact, and when a conflict arises between the two, then our commitments must accordingly change. Thomas Lovejoy, in a sharply critical review of Lomborg's book, nevertheless supports a similar view, where appropriate action is determined by scientific inquiry: "researchers identify a potential problem, scientific examination tests the various hypotheses, understanding of the problem often becomes more complex, researchers suggest remedial policies—and *then* the situation improves" (Lovejoy, 2002, p. 12; emphasis in original). David Pimentel, another Lomborg critic, argues in the same vein: "As an agricultural scientist and

ecologist, I wish I could share Lomborg's optimism, but my investigations and those of countless other scientists lead me to a more wary outlook" (Pimentel, 2002, p. 297).

So Lomborg and his critics share the old-fashioned idea that scientific facts build the appropriate foundation for knowing how to act in the world. How, then, are we to understand the radical divergence of the supposedly science-based views held by opposing sides in the controversy? If we accept the arguments of the critics, the divergence is simply a reflection of Lomborg's (perhaps willful) misunderstanding of the data. Yet, as Harrison (this issue) amply documents, Lomborg also has his supporters within the community of scientists. Are we instead witnessing a debate that exists because the science is incomplete, and thus allows for different interpretations? Stephen Schneider (2002, p. 1), another of Lomborg's critics, notes: "I readily confess a lingering frustration: uncertainties so infuse the issue of climate change that it is still impossible to rule out either mild or catastrophic outcomes, let alone provide confident probabilities for all the claims and counterclaims made about environmental problems."

There is an obvious problem of causation here. If the science is insufficiently certain to dictate a shared commitment to a particular line of action, from where do these commitments spring? For Lovejoy, the process starts when "researchers identify a potential problem," but the recognition that something is a "problem" demands a pre-existing framework of values and interests within which problems can be recognized. And Pimentel's "wary outlook" (not to mention Lomborg's rosy one) presupposes some expectations of what the world ought to look like in the first place.

This paper thus confronts a well-known empirical problem. In areas as diverse as climate change, nuclear waste disposal, endangered species and biodiversity, forest management, air and water pollution, and agricultural biotechnology, the growth of considerable bodies of scientific knowledge, created especially to resolve political dispute and enable effective decision making, has often been accompanied instead by growing political controversy and gridlock. Science typically lies at the center of the debate, where those who advocate some line of action are likely to claim a scientific justification for their position, while those opposing the action will either invoke scientific uncertainty or competing scientific results to support their opposition.[1]

A significant body of literature both documents and seeks to understand this dynamic (see, e.g., the admirable synthesis by Jasanoff and Wynne, 1998). This literature is characterized, for example, by the understanding that scientific facts cannot overcome, and may reinforce, value disputes and competing interests (e.g., Nelkin, 1975; Nelkin, 1979; Collingridge and Reeve, 1986), that scientific knowledge is not independent of political context but is co-produced by scientists and the society within which they are embedded (e.g., Jasanoff, 1996a), that different stakeholders in environmental problems possess different bodies of contextually validated knowledge(e.g., Wynne, 1989), and that the boundaries between science and policy or politics are constantly being renegotiated as part of the political process (e.g., Jasanoff, 1987; Jasanoff, 1990).

This work adds up to a deeply textured portrayal of the troubled relationship between science and decision making in the realm of the environment. Yet, as the Lomborg controversy highlights to a degree that is almost painful, high-profile public discourse surrounding environmental disputes remains stubbornly innocent of this past 20 or more years of constructivist scholarship. The notion that science is a source of facts and theories about reality that can and should settle disputes and guide political action remains a core operating principle of partisans on *both* sides of the Lomborg case and other environmental controversies.[2]

Much, perhaps most, of the recent literature grounds its critique in the difficulties associated with the first component of this pervasive and strongly held notion—that science is a source of verifiable facts and theories about reality. In this paper, I treat this realist notion not as a contestable idea but as an initial condition of the policy context—a starting point for further analysis. My goal is to offer an interpretation of the current, lamentable state of affairs whose acceptance by political actors does not require an abandonment of fundamental cosmologies. Thus, I look for explanation not in the social construction of science, but precisely in "the fact that scientists do exactly what they claim to do," (Hull, 1988, p. 31) and argue that the

fulfillment of this promise is what gets us deeper into the hot water—science does its job all too well. The argument, in brief, is this: nature itself—the reality out there—is sufficiently rich and complex to support a science enterprise of enormous methodological, disciplinary, and institutional diversity. I will argue that science, in doing its job well, presents this richness, through a proliferation of facts assembled via a variety of disciplinary lenses, in ways that can legitimately support, and are causally indistinguishable from, a range of competing, value-based political positions. I then show that, from this perspective, scientific uncertainty, which so often occupies a central place in environmental controversies, can be understood not as a lack of scientific understanding but as the lack of coherence among competing scientific understandings. These considerations lead me finally to consider the question of why environmental controversies tend to become highly "scientized," and to speculate about what might happen if we could "de-scientize" them.

But first, as a sort of control case, it might be helpful to visit a major political controversy that was not resolved through resort to scientific research: the 2000 US Presidential election.

Determining an Integer

The front page of the 6 May 2000 *Washington Post* reported that political scientists were using mathematical models to predict the winner of the forthcoming US Presidential election between Democratic candidate Al Gore and Republican George W. Bush (Kaiser, 2000). The models, which integrated such factors as the state of the economy and public opinion data, indicated that Gore would handily win the election. But by the time the polls in most states had closed, it was apparent that victory in this remarkably close election would depend on the outcome of the vote in the populous and tightly contested state of Florida, with its 25 electoral votes. At about 8:00 P.M. on election night, the major television networks famously declared Gore the winner on the basis of data from Florida exit polls. But as the actual Florida returns came in, it soon became clear that the race was too close to call, and the networks rescinded their initial prediction of a Gore victory. Early the next morning, the networks named Bush the victor, but they soon learned that the closeness of the vote would trigger an automatic recount, so again they had to reverse themselves. A day later, with all precincts reporting, the initial vote count for Florida indicated a difference between Gore and Bush of about 1800 votes out of almost six million cast—a margin of less than three hundredths of a percentage point.

With so much at stake, the vote count was of course furiously contested, with claims of irregularities, miscounts, misvotes, machine failure, and even voter intimidation, and demands for recounts. Yet the basic contention focused on the determination of a single, apparently simply fact: how many votes did each candidate receive?

This is a remarkably straightforward-seeming problem, with an apparently clear path to resolution: count all votes and determine the winner. Yet, in the end, the winner of the Florida vote was not determined by an assertion of fact, it was determined through the resolution of a legal battle between lawyers representing the interests of each candidate, and decided by the Supreme Court of the United States (2000). The decision overturned an earlier Florida Supreme Court ruling that would have allowed additional recounts. In so doing, it accepted a vote count that had been previously certified by the state of Florida, which showed Bush to be the winner (now by 537 votes) despite ongoing uncertainties about the actual tally. In other words, the Court asserted that the final answer to the question "Who got more votes in Florida?" was appropriately determined by legal and political processes.

A Thought Experiment

Would not it have made more sense to simply get a definitive count of the votes to objectively determine the winner? Imagine that the contesting parties had agreed that the problem was a technical one, not a legal or political one, and that they had turned the vote counting over to a team of disinterested experts whose job would be to determine the correct result. On its face, it is hard to imagine a problem more suited to a strictly technical approach. The system under consideration—an election—is in principle a closed one, with a finite number of system components (voters; voting machines; vote counting procedures) obeying simple decision rules (each voter votes once for one candidate based on personal preference; all votes for each candidate are added up) within clearly defined spatial (the state of Florida) and temporal (the time period during which the polls were open) boundaries. The correct answer is known to be an integer, and it is derived through the simplest possible arithmetic process.

Of course the political debate over the vote count revealed system complexities. Overvotes (ballots that were not counted because they appeared to contain votes for more than one candidate) and undervotes (ballots that were not counted because they appeared not to indicate a vote for any candidate) totaled more than 175,000—more than three hundred times the final, certified 537 vote differential between Bush and Gore (Merzer, 2001). Many of these ballots seemed to indicate the intent of the voter (for example, in cases where voters had not pushed the vote-punch machine hard enough to fully separate the chad from the card; or where voters had both punched and written in a vote for the same candidate). Confusing ballot graphics may have led some people to vote for the wrong candidate. More generally, the reliability of various voting technologies was called into question.

So the team of experts tasked with coming up with the final vote tally might need to be drawn from several disciplines. For example, there would certainly need to be a statistics group to develop models of reliability for different types of voting machines. Statisticians would also need to develop rigorous error analyses that would characterize the probabilities that a given tally actually determined the real victor. A technology evaluation group would need to assess the sources of failure for different types of machines, while a cognitive neuroscience group might look at visual perception problems to try to understand the extent to which ballot design could contribute to wrong votes. Fundamental research could address such questions as chad behavior under different states of compressive

stress (material science), the relation between the physical strength of the voter and chad behavior (physiology), the variable behavior of vote-punch machines (mechanical engineering), and the causes of overvotes (psychology).

The first result of these analyses would likely be reports with technical-sounding titles such as "Florida's Residual Votes, Voting Technology, and the 2000 Election," or "Elections: Statistical Analysis of Factors that Affected Uncounted Votes in the 2000 Presidential Election," or "The Butterfly Did It: The Aberrant Vote for Buchanan in Palm Beach County, Florida."[3] Once the experts began to make their results known, other experts would need to review them, and disagreements—over methodology, data, and conclusions—would undoubtedly emerge. Studies from different disciplines would need to be integrated, and even then the final calculations would have to be governed by rules about what constitutes a valid vote. Normative questions (such as whether the vote count should capture the intent of all voters, or simply record "actual" votes, which would in turn raise questions about what "actual" meant) would thus govern both the design of studies and the interpretation of results, and we could expect that the political backgrounds and affiliations of the experts would be exhaustively scrutinized to try to sniff out potential conflicts of interest. To make matters yet more complex, ideological affinity would probably influence what type of science one was willing to accept, because different approaches to counting would have different implications for the election results. In this light, because so many of the uncounted overvotes and undervotes were from precincts with Democratic majorities,[4] Bowker and Star (2001, p. 422) recognized "a party political divide [that] aligned the purity of numbers with the Republican right and a faith in statistics with the Liberal left."

Can we really imagine that such a technical process would have led to a swift determination of the "real winner" in a timely fashion, and in a manner that preserved the legitimacy of the electoral system and the eventual winner? Would the experts have been able to arrive at a number—a simple integer—that everyone could agree on as the "right" answer, and that all contesting interests would have accepted? Indeed, the Miami Herald sponsored an unofficial recount of over- and undervotes which showed that either Gore or Bush could have been the winner depending on criteria used to judge the validity of ballots.

In contrast, it should be remembered that the political/judicial process that actually was followed did accomplish these goals, conferring a final decision in 36 days, and yielding a new president who was broadly accepted despite the fact that more than half of the nation's voters had opposed him, and in the absence of any agreement of what the final vote tally "actually" was.

This story and thought experiment are meant to highlight four points.

1. Even apparently simple systems can display unprecedented, surprising behavior. Fifty-two previous presidential elections were not enough to reveal all possible system behaviors and permit accurate prediction of future outcomes.

2. The same uncertainties that were revealed in the Florida election no doubt exist in all elections. They became significant because the election was so close (both sides could reasonably visualize themselves as potential winners—or losers), and because the stakes were enormously high—the presidency would be determined by its outcome.

3. The dispute was not resolved by addressing the technical aspects of the vote count, but by subjecting the vote count process to political and judicial mediation procedures that were legitimated by their capacity not to arrive at "truth" but to transparently negotiate among competing players. But because this system was broadly accepted as legitimate—that is, people understood and agreed on the rules—its results were also broadly accepted. Moreover, the result is not permanent—it will be revisited in the next election.

4. A purely technical approach, aimed at overcoming uncertainty about the vote count, and subject both to the strictures of scientific method and the close attention of the public, could not have achieved resolution as quickly, as decisively, or as legitimately, as the political/legal approach.

Excess of Objectivity

I want to explore the possibility that environmental controversies typically bear a much greater likeness to the 2000 Florida election controversy than might at first seem apparent. To do so I begin by considering why facts[5] often fail to behave in the manner that both Lomborg and his critics claim they should.

In July 2003, two conservative think-tanks, the Hoover Institution on War, Revolution and Peace at Stanford University, and the George C. Marshall Institute in Washington, DC, published a book entitled *Politicizing Science: The Alchemy of Policymaking* (Gough, 2003). The book visited a number of examples, from a right-wing perspective, of how science had been manipulated, distorted, or suppressed, mostly in support of liberal causes, mostly related to the environment. The underlying theme of the book was that science could guide politics only when it was free from ideology. "The more that political considerations dominate scientific considerations, the greater the potential for policy driven by ideology and less based on strong scientific underpinnings" (2003, p. 3). The point, of course, is that "policy driven by ideology" is supposed to be undesirable.

The next month, Congressman Henry Waxman, a liberal Democrat, released a report entitled "Politics and Science in the Bush Administration" (United States House of Representatives, 2003), which pointed to numerous examples of how the Administration "manipulated the scientific process and distorted or suppressed scientific findings" to yield results that favored the interests of its supporters.

While neither of these publications were works of scholarly research, and both made some points that seem reasonable and others that are less so, the more interesting observation is that, in coming from strongly contrasting ideological positions, they (along with the combatants in the Lomborg controversy) shared

a view of science as a disinterested force that could guide political decision making by providing appropriate facts—so long as it was kept separate from politics. Yet the simultaneous appearance of these two products amusingly highlights what neither side was willing or able to contemplate: if everyone politicizes the science, maybe there is something about science that lends itself to being politicized?

Consider climate change, which may variously be understood as a "problem" of climate impacts, weather impacts, biodiversity, land use, energy production and consumption, agricultural productivity, public health, economic development patterns, material wealth, demographic patterns, etc. Each of these ways of looking at the problem of climate change involves a variety of interests and values, and each may call on a body of relevant knowledge to help understand and respond to the problem. Not only may the interests, values, and knowledge relevant to one way of understanding the problem be, in small part or large, different from those associated with another way, but they may also be contradictory. Conversely, those holding different value perspectives may see in the huge and diverse body of scientific information relevant to climate change different facts, theories, and hypothesis relevant to and consistent with their own normative frameworks. This condition may be termed an "excess of objectivity," because the obstacle to achieving any type of shared scientific understanding of what climate change (or any other complex environmental problem) "means," and thus what it may imply for human action, is not a lack of scientific knowledge so much as the contrary—a huge body of knowledge whose components can be legitimately assembled and interpreted in different ways to yield competing views of the "problem" and of how society should respond. Put simply, for a given value-based position in an environmental controversy, it is often possible to compile a supporting set of scientifically legitimated facts.

A familiar illustration is the documentation of global warming. While the observation of a global warming signal over the past 20 years is well accepted, discrepancies between surface temperature measurements and lower-to-middle troposphere temperature data from satellites and radiosondes continue to offer scientifically credible facts for global warming "contrarians." A National Academy of Sciences report (NRC, 2000) failed to resolve this conflict, concluding that, on the one hand, "the warming trend in global-mean surface temperature observations during the past 20 years is undoubtedly real" and also that "the troposphere actually may have warmed much less rapidly than the surface from 1979 into the late 1990s" (NRC, 2000, p. 2).[6]

Yet resolving the discrepancy to everyone's satisfaction would not really solve anything. After all, the fact of global warming is not itself inherently problematic; what worries us is the possibility that warming will cause a variety of undesirable outcomes. When global warming is considered in terms of its specific potential social consequences, however, the availability of competing facts and scientific perspectives quickly spirals out of control. Consider the following chain of logic: human greenhouse gas emissions are causing global warming; global warming will lead to increased frequency and severity of extreme weather events; reducing greenhouse emissions can thus help reduce the impacts of extreme weather events. Each link in this chain is saturated with the potential for competing, fact-based perspectives. For example, climate models and knowledge of atmospheric dynamics suggest that increased warming may contribute to a rising incidence and magnitude of extreme weather events (Houghton et al., 2001, p. 575); but observations of weather patterns over the past century do not show clear evidence of such increases, while model results are still ambiguous, and "data continue to be lacking to make conclusive cases" (Houghton et al., 2001, p. 774). While economists can show how tradable permit schemes combined with mandated emissions targets can reduce greenhouse gas emissions (Chichilnisky and Heal, 1995), they cannot agree on plausible future rates of emissions increase (The Economist Print Edition, 2003). Furthermore, perspectives on the history and economics of innovation suggest that decarbonization is likely to depend primarily on technology evolution and diffusion, not policies governing consumption (Ausubel, 1991; Nakicenovic, 1996). Social science research on natural hazards suggests that socioeconomic factors (such as land use patterns, population density, and economic growth), rather than changing magnitude of frequency of hazards, are responsible for increasing societal losses from extreme events (Pielke et al., 2003; Changnon et al., 2001). And in any case, climate scientists disagree about the extent to which greenhouse gases are responsible for warming trends, given that other phenomena, such as land use patterns, may also strongly influence global climate (e.g., Marland et al., 2003). Finally, climate models that as yet have no capacity to accurately predict regional variability in extreme events are thus even further from providing useful information about how greenhouse gas emissions reductions might influence future incidence and magnitude of extreme events. Each level of analysis is not only associated with its own competing bodies of contestable knowledge and facts, but is also dependent on how one views the other levels of analysis. Facts can be assembled to support entirely different interpretations of what is going on, and entirely different courses of action for how to address what is going on.

As Michael (1995, p. 473) explained, in the context not of global warming but ecosystem management: "More information provides an ever-larger pool out of which interested parties can fish differing positions on the history of what has led to current circumstances, on what is now happening, on what needs to be done, and on what the consequences will be. And more information often stimulates the creation of more options, resulting in the creation of still more information" (also see Herrick and Jamieson, 1995; Herrick and Sarewitz, 2000; Sarewitz, 2000).

As an explanation for the complexity of science in the political decision making process, the "excess of objectivity" argument views science as extracting from nature innumerable facts from which different pictures of reality can be assembled, depending in part on the social, institutional, or political context within which those doing the assembling are operating. This is more than a matter of selective use of facts to support a pre-existing position.[7] The point is that, when cause-and-effect relations are not simple or well-established, *all* uses of facts are selective. Since there is no way to "add up" all the facts relevant to a complex problem like global change to yield a "complete" picture of "the problem," choices must be made. Particular sets of facts may stand out as particularly compelling, coherent, and useful in the context of one set of values and interests, yet in another appear irrelevant to the point of triviality.[8]

Value in Discipline

While this argument may help make clear why "more science" often stokes, rather than quenches, environmental controversies, I believe it does not go far enough. It seems to me that there is likely to be a causal connection between the ways that we have organized scientific inquiry into nature, and the ways we organize human action (and thus political decision making) related to the environment.

For example, consider the controversy over the Acoustic Thermometry of Ocean Climate (ATOC) experiment. The history of this controversy is discussed in detail in Oreskes (2004). In brief, oceanographers at the Scripps Institution of Oceanography designed a clever experiment to measure changes in the global average temperature by monitoring how the velocity of sound waves traveling over long distances in the ocean were changing over time. ATOC was promoted as the experiment that could finally settle the question of whether, and how fast, global warming was actually occurring. However, an alliance of environmentalists and biologists opposed the experiment because of concerns about its effects on whales and other marine mammals. While designers of the experiment sought to assure the opponents that the experiment would not harm marine mammals, they lacked both the scientific legitimacy, and the institutional disinterest, to make a convincing case. A National Research Council (NRC) study was commissioned to resolve the dispute, and while it was unable to confirm the potential for ATOC to harm marine mammals, neither could it entirely discount the possibility. Oceanographers working on the experiment interpreted the report as a green light for ATOC, while biologists saw it as a vindication of their opposition.

Oceanographers were primarily concerned about conducting an oceanographic experiment that would document global warming. Biologists were primarily concerned about the effects of acoustic transmissions on the well-being of marine mammals already under assault from human activities. These positions are not reconcilable because there is nothing to reconcile—they recognize and respond to different problems. They also point to a direct connection between scientific perspectives and values. Oceanographers chose to interpret the uncertainty associated with the National Research Council study as an endorsement of the safety of ATOC. Biologists interpreted the same study as an affirmation of potential harm. Scientific orientation helped to determine one's assessment of the level risk posed by ATOC,[9] one's willingness to face that risk, and one's view about the potential benefits of ATOC (Oreskes, 2004). There is a notable incoherence in this debate—an incommensurability of contesting fact-value positions ("contradictory certainties," to use Schwarz and Thompson's (1990, p. 60) memorable term). The benefits of performing ATOC, as understood and articulated by physical oceanographers, had no bearing on the well-being of marine mammals, as understood by biologists. To put it bluntly, but perhaps not too simplistically, oceanographers' values were represented by the conduct and outputs of oceanography; biologists' values were not.

Could scientific orientation be related to the values that one holds? Science divides up the environment partly by disciplinary orientations that are characterized by particular methods, hypotheses, standards of proof, subjects of interest, etc. My point is certainly not that disciplines are associated with monolithic worldviews and value systems. But, while some see a grand unification of all knowledge as an inevitable product of scientific advance (Wilson, 1998), thus far the growth of disciplinary scientific methods and bodies of knowledge results in an increasing disunity that translates into a multitude of different yet equally legitimate scientific lenses for understanding and interpreting nature (e.g., Dupré, 1993; Rosenberg, 1994; Cartwright, 1999; Kitcher, 2001, chapter 6). Similarly, individual humans can have only the most partial of understandings of the world in which they reside, and it is in the context of these always-incomplete understandings that they make their decisions and judgments (e.g., Simon, 1997 and Simon, 1983). Without saying anything about direction of causation, it seems entirely plausible to suggest that the formal intellectual framework used by a scientist to understand some slice of the world may be causally related to that scientist's normative framework for interpreting and acting in the world.

The ongoing debate over genetically modified organisms (GMOs) in agriculture helps to illustrate the idea. There are many ways to "understand" GMOs: in terms of their connection to food production, human health, economic development, ecosystem dynamics, biotechnology innovation processes, plant genetic diversity, even culinary arts. Each of these perspectives is associated in part with separate disciplinary perspectives. What might such diverse perspectives mean for how one views the "value" or "risks" of GMOs? Environmental benefits typically attributed to GMOs include: better resistance to environmental stresses; more agricultural productivity; fewer agrochemical inputs such as pesticides; and design of crops that can actually remediate polluted soils and aquifers. Environmental risks typically attributed to GMOs include: uncontrolled introgression of genes into other varieties or species; harmful mutations of inserted genes; competition and breeding with wild species; negative effects of insecticidal GMOs on beneficial non-target organisms such as birds and pollinating insects; and growing resistance of insects to insecticidal GMOs (Wolfenbarger and Phifer, 2000; Food and Agriculture Organization, 2003a,b). One obvious attribute of these two lists is that the putative benefits derive from straightforward cause-and-effect relations that reflect the intent of scientists working on GMOs, whereas the putative risks arise from more complex interactions that are largely unintended. It thus seems reasonable to expect that scientists from disciplines involved in design and application of GMOs, such as plant geneticists and molecular biologists, would be potentially more inclined to view GMOs in terms of their planned benefits, and ecologists or population biologists would be more sensitized to the possibility of unplanned risks at a systemic level.

These sorts of relations become palpable in trying to unravel the vicious debate sparked by the publication of a paper in *Nature* by Ignacio Chapela, a microbial ecologist, and his graduate student, David Quist, documenting the occurrence of transgenic corn in Mexico, and the introgression of the insecticidal (Bt) transgenes into native maize varieties (Quist and Chapela, 2001). The original article, which went through *Nature's* standard peer review process, was attacked vociferously by numerous scientists, including former Berkeley colleagues of Quist and Chapela. Both the methods and the conclusions of the original paper were strongly criticized. One scientist called the paper "a testimony to technical incompetence," another termed it "so outlandish as to be pathetic," and a third dismissed it as "trash and indefensible" (Lepkowski, 2002a). These and other critics initially insisted that the issue was simply one of the quality of the science, but in reality the dispute was inextricably intertwined in the larger controversy over biotechnology. If the original results were correct it meant that GM corn had made its way into Mexico despite the fact that it was banned by the Mexican government, and, more damningly, that genes from the GM corn had moved into native Mexican varieties that are the original source of the world's genetic diversity for corn. If these conclusions turned out to be true, it could damage the prospects of the agricultural biotechnology industry, because it would indicate that the ecological and genetic impacts of GM corn were not predictable and could not be controlled.

An underlying theme of the debate was that Quist and Chapela's attackers were in the pocket of the biotechnology industry, or, from the opposite perspective, that Quist, Chapela, and their allies were shills for the anti-biotechnology lobby. Environmental and industry groups mobilized their constituencies on behalf of the scientists who best represented their interests. Scientists themselves traded accusations about the political motives and economic interests of those whose science they were attacking (Lepkowski, 2002a; Nature, 2002). If nothing else, the high stakes of the debate ensured that it would attract much more attention than a disagreement that was merely "scientific." As *Nature* editor Maxim Clarke observed: "scientists with strong interests scrutinize published papers more intently than they would otherwise do . . . because they are very motivated to find any flaws which can be used to undermine or support the conclusions of the paper" (Lepkowski, 2002b).

Yet on another level that was never discussed, the disciplinary structure and disunity of science itself was at the roots of the controversy. The two sides of the debate represented two contrasting scientific views of nature—one concerned about complexity, interconnectedness, and lack of predictability, the other concerned with controlling the attributes of specific organisms for human benefit. In disciplinary terms, these competing views map onto two distinctive intellectual schools in life science—ecology and molecular genetics (e.g., see Holling, 1998).

Thus, it is not surprising that Quist and Chapela's strongest scientific critics were those whose own research focused on the genetic engineering of individual plant varieties, while their scientific supporters focused on ecosystem behavior (as did Quist and Chapela). Those representing the molecular genetics perspective aimed their critique at flaws in Quist and Chapela's techniques and the ambiguity of their results (Metz and Fütterer, 2002; Kaplinsky et al., 2002). Wayne Parrott of the University of Georgia, one of their most aggressive attackers, said:

"[W]hen we do our work, we run a PCR [polymerase chain reaction] first. Then we take our positive samples and do a more reliable test on them. Chapela used it in its entirety. He could have taken his positive samples and followed them up with something more definitive, such as spraying the things with a herbicide. Or he could have looked for a protein. There are many things he could have done that would have taken maybe a couple more weeks. No one would have questioned it. The thing is that he tried to get into a top journal by using a preliminary test. Then he makes all sorts of claims based on this. He used the wrong enzyme, he used the wrong extraction procedure, everything he did was wrong. And it's not worth the paper it's written on" (Lepkowski, 2002a).

But Allison Snow, a researcher at Ohio State University who studies gene flow in the environment, had a more generous view, despite acknowledging the methodological flaws:

"I don't think the science in the second half of their paper was very good. They said there were multiple insertions of transgenes, where they were going in the genome wasn't predictable, and that therefore that there was something scary about transgenes. But the first half of the paper, while you could always have asked them to do a better job, I thought was well supported. And anyway, a lot of people already believe that transgression has already happened and the Mexican government has confirmed it and talked about it in several news releases. What was interesting was Chapela's positive control with the grain from the local store. That had been shipped in from the United States as animal feed and was definitely transgenic. It was not for human consumption but people are planting it. So there are all those different parts to this puzzle" (Lepkowski, 2002a).

Parrott, whose analytical frame of reference is the gene, assessed Quist and Chapela's work strictly in terms of its adherence to the standards necessary for genetic engineering. Failing to pass muster from that perspective, he deemed the work worthless. Snow, whose focus is on the ecosystem scale, could acknowledge these flaws but still recognize that parts of the research had important implications for ecosystem behavior, and as well that the research reflected such scientific virtues as replicability of results and the clever identification of a control case.

The implications of these competing perspectives are readily apparent in the ways that Parrott and Snow describe their own work. Parrott's website says: "Our laboratory conducts research on crop genetic engineering, although its members also dabble in molecular markers. The bulk of the work deals with the development of protocols for . . . genetic transformation of soybean, peanut, alfalfa, and maize" (Parrot Lab, 2003). In contrast, Snow says of her research: "I study microevolutionary processes in plant populations, with an emphasis on breeding systems, pollination ecology, and conservation biology . . . Most recently, my research focuses on the applied question of how gene flow from cultivated species affects the evolutionary ecology of weedy relatives" (Snow, 2003). Parrott's concerns end with the "genetic transformation" of specific crops, while this transformation is the starting point for Snow's work.

In this context, there is nothing inherently implausible in the claims of scientists on both sides that their positions were scientific, not political or economic. Two of Quist and Chapela's critics, accused in a letter to *Nature* of having conflicts of interest because their research was partly funded by the biotechnology industry (Nature, 2002, p. 898) defended themselves in the following manner: "We are not unlike many scientists in that we have shared research and funding with industry at some point. In stating that we have 'compromised positions,' [our critics] wrongly imply that private-sector funding strips us of integrity and legitimacy in the arena of scientific discourse." One may accept this argument and still see a connection between the type of science being conducted, a worldview compatible with that science, and the interests of those who might find the science compelling and valuable.[10]

This alignment of disciplinary perspective and worldly interests is critically important in understanding environmental controversies, because it shows that stripping out conflicts of interest and ideological commitments to look at "what the science is really telling us" can be a meaningless exercise. Even the most apparently apolitical, disinterested scientist may, by virtue of disciplinary orientation, view the world in a way that is more amenable to some value systems than others. That is, disciplinary perspective itself can be viewed as a sort of conflict of interest that can never be evaded. In cases such as the Mexican corn controversy, it might be most accurate to look at the scientific debate not as tainted by values and interests, but as an explicit—if arcane—negotiation of the conflict between competing values and interests embodied by competing disciplines.

From a similar perspective, the economist Richard Norgaard (2002) assessed Lomborg's *The Skeptical Environmentalist*. Norgaard notes that economists have been generally sympathetic with Lomborg's optimistic evaluation of the state of the world's environment, while ecologists and other environmental scientists have been largely outraged.[11] The reason, he suggests, is that "the thinking of economists requires the existence of scarcity," (2002, p. 288) and the history of industrial economies is one of overcoming scarcity through innovation. Progress through innovation and economic growth is a first principle underlying conventional economic dogma, and this principle dictates that current scarcity of environmental assets will be overcome in a similar manner. One presumes, although Norgaard is not explicit about this, that he sees environmental scientists as inherently less inclined toward an optimistic view of the future, perhaps because their disciplines do not include the faith in the inevitability of progress that he attributes to economics.

To summarize thus far: central to the idea that science can help resolve environmental controversies is the expectation that science can help us understand current conditions under which our decisions are being made, and the potential future consequences of those decisions. This expectation must confront the proliferation of available facts that can be used to build competing pictures of current and future conditions, and the embeddedness of such facts in disciplinary perspectives that carry with them normative implications. These problems are in part a reflection of the diversity of human values and interests, but they also reflect the richness of nature, and the consequent incapacity of science (at least in this stage of its evolution) to develop a coherent, unified picture of "the environment" that all can agree on. This lack of coherence goes by the name of "uncertainty."

Origins of Uncertainty

Reduction of uncertainty is a central, perhaps the central, goal of scientific research carried out in the context of environmental controversies ranging from climate change to ecosystem restoration, as variously articulated in innumerable policy documents, research reports, and scientific articles. The standard model, of course, is that if uncertainty surrounding the relevant scientific facts can be reduced, then the correct course of action will become more apparent. Uncertainty is thus portrayed as the cause of inaction. But the notion of a clearly demarcated body of relevant fact is highly problematic. And, as the 2000 election story shows, uncertainty about facts need not be an impediment to political resolution of heated controversy. The standard model will hold up. To begin to develop a more satisfactory alternative, I examine how estimates of uncertainty have changed in the arenas of earthquake prediction, nuclear waste disposal, and climate change. Based on these examples, I present the idea that uncertainty in environmental controversies is a manifestation of scientific disunity (excess of objectivity; disciplinary diversity) and political conflict.

The Parkfield Prediction

In 1985, seismologists from the US Geological Survey estimated with 95% probability that a mid-size earthquake along the Parkfield segment of the San Andreas fault would occur by the year 1993. The 95% certainty level assigned to the event was derived from a statistical analysis of the recurrence interval of past earthquakes along the Parkfield segment (Bakun and Lindh, 1985) and it was endorsed by the scientific judgment of the seismological community as a whole, as expressed by expert oversight bodies at the state and national level (Nigg, 2000).

The earthquake, however, did not take place, and by the end of 2003 had still not occurred. One possible explanation is that reality is occupying the tail of the probability curve—that is, the 95% probability was "correct," and the non-occurrence of the earthquake was indeed a highly unlikely event reflecting aberrant behavior of the fault system (analogous, perhaps, to the uniquely rare confluence of events in the 2000 Florida election). If this were true, we would expect, for example, that if similar predictions were made for nineteen other, similar fault segments, earthquakes would occur in all cases. But the state of seismological knowledge has only rarely allowed scientists to issue earthquake predictions with confidence, and even more rarely have those predictions been borne out (Nigg, 2000). Indeed, subsequent analysis has shown that the Parkfield prediction was based on insufficient analysis of available data

and incomplete understanding of the fault's behavior (Kagan, 1997; Roeloffs and Langbein, 1994). From this perspective, the uncertainty estimate needs to be recognized as a statement not about the actual behavior of a natural phenomenon, but about the state of scientific understanding of that phenomenon at a particular time, and the state of confidence that scientists had in that state of understanding at that time.

Uncertainty estimates, that is, are in part a measure of the psychological state of those making the estimates, which is in turn influenced by political context within which the science is carried out. In the case of the Parkfield prediction, an important aspect of the story was that the fault segment ran through a sparsely populated agricultural region of California. Thus, the political and economic stakes of a false prediction (or, for that matter, an accurate one) were low. If seismologists had arrived at similar probabilities for an earthquake in San Francisco, the consequences of both the prediction itself, and the predicted event, would have been considerably greater. Under such circumstances, scientific and political scrutiny of the prediction would have greatly intensified, the pressure on the scientists to be "right" would have been intense, and the population of scientists and perhaps of disciplines involved in the prediction process would have expanded. It is difficult to imagine that such conditions would not have influenced the certainty levels expressed by scientists, or undercut the unanimity of opinion surrounding particular statements of certainty. If the stakes had been higher, certainty would have been lower.

Water Flowing Underground

One key attribute that determines the performance of any nuclear waste site is its hydrological system. If water flows through a site, it may accelerate degradation of waste containment vessels that could in turn lead to mobilization of radionuclides and contamination of water supplies and the environment adjacent to the site. Because radioactive waste decays over periods of tens of thousands of years, assessing the behavior of a potential site involves efforts to understand how the hydrological system might evolve over long time frames.[12]

Since the early 1980s, hydrologists have been estimating percolation flux, or the rate at which a volume of water flows through a unit area of rock, at the proposed US high level nuclear waste site at Yucca Mountain, Nevada. Initial estimates, made in the early 1980s based on field studies, indicated a flux of between 4 and 10 mm per year, but further research reduced these estimates to between 0.1 and 1 mm per year, an uncertainty range that was reinforced by additional studies over the next 12 years. These estimates, based on combinations of expert judgment, numerical models, and laboratory experiments, were a crucial input for integrated performance assessment models of overall repository site behavior. Indeed, the 0.1–1 mm per year range allowed such performance assessments to conclude that the site was sufficiently dry to meet safety standards set by the US Environmental Protection Agency. As a result of this combined scientific stability and political desirability, by the mid-1990s, "thinking about percolation flux had almost achieved the status of conventional wisdom" (Metlay, 2000, p. 210).

However, the Yucca Mountain site was the focus of intense political controversy, and the scientific results that issued from the Department of Energy (DoE), which had responsibility for the site, were under constant fire. DoE was also subject to the oversight of two external bodies that reviewed the science and made recommendations for further research.[13] In this politically contentious environment, DoE was pushed to drill a tunnel that would enable direct sampling of rocks at the actual level of the proposed repository. Subsequent analysis of water in those rocks indicated the presence of radioactive isotopes generated from atmospheric nuclear weapons tests in the early years of the Cold War. That water containing these isotopes had made it from the surface to the repository site—300 meters beneath the surface—in less than 50 years was evidence that percolation flux was perhaps ten times faster than indicated in modeling studies over the previous decade. Following this discovery, an aggregation of estimates from seven outside experts concluded that the 95% probability range for percolation flux lay between 1 and 30 mm per year—far higher rates than were encompassed by the "conventional wisdom" born during the prior decade or more of research. To complicate the story even further, efforts to reproduce the isotopic analysis of the repository water have yielded highly inconsistent results (Nuclear Waste Technical Review Board, 2000). After 20 years of research, uncertainties surrounding percolation flux seem only to have increased.

A key aspect of this story is that the decade or more of research reinforcing the belief that percolation flux lay between 0.1 and 1 mm per year was sponsored by the agency which had general responsibility for developing the repository site and strong institutional and political motivations to keep the project moving forward. As Metlay (2000, p. 211) observed, "when faced with the need to resolve uncertainty about percolation flux, the scientists [at DoE laboratories] had little organizational incentive to settle on a higher value or, more important, to question whether a lower value was correct. This approach to addressing uncertainty need not have been adopted consciously; in fact, it probably was not. More likely, it arose simply because organizational norms and culture have a well-documented and pervasive effect on individuals' actions and judgments."[14] Moreover, initial estimates of percolation flux emerged from an organizational and scientific context that was relatively homogeneous in terms of both scientific and political goals. As the research process opened up to more diverse scientific and political players, a greater diversity of values of interests were implicated, leading to the introduction of new sources of uncertainty.

Climate Change

In climate change science, one closely scrutinized area of uncertainty is climate sensitivity, or the average global temperature increase associated with a doubling of atmospheric carbon dioxide. More than a century ago the Swedish chemist Arrhenius estimated this value at 5.5 °C (Rayner, 2000), a number remarkably close to the likely temperature range of 1.5–4.5 °C estimated by modern climate scientists using highly sophisticated numerical models, and adopted by the Intergovernmental Panel on Climate Change (IPCC) (Houghton et al., 2001, p. 67). While this latter temperature spread is very commonly used as

an indication of the uncertainty range associated with climate sensitivity, the spread itself is not a probability range—that is, the probability of any particular temperature increase within this range is unspecified (as is the probability of the doubling temperature falling outside this range). Rather, the uncertainty range purportedly reflects the difference between the smallest and largest predicted temperature increases generated by a suite of 15 climate models (Houghton et al., 2001, p. 561). Yet, as van der Sluijs et al. (1998) have pointed out, a notable attribute of the canonical, IPCC-endorsed uncertainty range is that, for more than two decades, it has not changed, despite huge increases in the sophistication of climate models over that time—a fact that they explain in terms of an ongoing process of evolving judgment and negotiation among climate modelers working in a politically heated area of science, where significant changes in scientific conclusions could have considerable political repercussions.

Outside the IPCC process, however, the uncertainty associated with climate sensitivity has been expressed in several different ways. One effort (Morgan and Keith, 1995) elicited probabilities from 16 climate experts, and arrived at a mean sensitivity of 2.6 °C with a mean standard deviation of 1.4 °C. This type of "subjective probability" presumes that all experts are providing equally "probable" estimates. Another method estimates climate sensitivity based on a simple climate/ocean model that simulates the observed hemispheric-mean near-surface temperature changes for the past century or so (Andronova and Schlesinger, 2001). This approach concludes that the 90% confidence interval for climate sensitivity is 1.0–9.3 °C—much wider than the more familiar model uncertainty spread used by the IPCC. Other recent studies (Knutti et al., 2002; Forest et al., 2002) also indicate considerably wider spreads than the IPCC estimate. Thus, as in the Yucca Mountain case, an expansion of the institutional and scientific players destabilizes estimates of uncertainty.

This phenomenon threatens the claim that scientific research will help resolve scientific controversy through reduction of uncertainty. My own experience on the Climate Research Committee of the National Research Council illustrates how institutions central to the mainstream of scientific research may nevertheless seek to buttress this claim. During 2001–2003 I was part of a panel writing a report that was originally to be entitled *Climate Change Feedbacks: Characterizing and Reducing Uncertainties*. The panel's formal tasks in the report were to:

1. characterize the uncertainty associated with climate change feedbacks[15] that are important for projecting the evolution of the Earth's climate over the next 100 years; and

2. define a research strategy to reduce the uncertainty associated with these feedbacks. . . .

Because the report was to deal directly with the problem of uncertainty in climate science, the panel decided to include a section discussing some of the complexities surrounding the concept of uncertainty in science and policy. This decision was particularly notable because such an approach had not been taken before. Numerous previous NRC reports on climate science, while frequently using the word "uncertainty," and asserting the importance of reducing it, did not make a serious effort to distinguish among the various meanings of the word, even while repeatedly making the claim that more research would reduce uncertainty, and reduced uncertainty would aid policy makers.[16]

A draft of the report was circulated to outside reviewers. The draft included a brief discussion of the concept of uncertainty, illustrated by a discussion of differing approaches to estimating the uncertainty associated with climate sensitivity. The draft also included the statements that "characterizing the uncertainty is not the same as reducing it," and that "there is no guarantee that further research will soon reduce the uncertainty in climate projections."[17]

Reviewers were extremely critical of this discussion, noting especially that the word "uncertainty" was used in many different ways without clearly defining them. "This is not a trivial issue," one reviewer acutely noted, "because it is a matter of objectives. Each definition of uncertainty represents a different definition of objectives, which leads to a different definition of metrics." Reviewers also strongly criticized the report's focus on "model uncertainty," which was deemed "unacceptable for institutional reasons. It states the objective of a multibillion dollar program in strictly insider's terms, i.e., understandable and of interest only to scientists, and makes a point of saying that societal links cannot be established."[18]

Yet the reviewers did not suggest that a revised draft include a more careful definition and delineation of the various uses of the word "uncertainty," but rather recommended the opposite—that the specific discussion of uncertainty be omitted. This seems to indicate that the problem was not the various meanings and uses of the word throughout the report, but the calling attention to them in the introductory discussion. Indeed, prior NRC climate reports also used "uncertainty" in a similar variety of ways without defining or distinguishing them (e.g., NRC, 1999a; NRC, 2000; NRC, 2001).

The panel's subsequent set of revisions included a very reduced section on uncertainty, but did not eliminate the discussion of climate sensitivity, which was felt to be an important illustration of how uncertainty could be characterized in different ways. Statements about the difficulties of reducing uncertainty were also left in. Reviewers again objected, and it was made clear to the panel that unless the offending language was removed, the report would not be published.[19] Thus, in the final report all discussion of uncertainty was removed. Even the word "uncertainty" was stripped from the title, which was changed to *Understanding Climate Change Feedbacks* (NRC, 2003). The multiplicities of meaning and use of the word "uncertainty" remain (unacknowledged) in the report,[20] as does the promise, both explicit and implicit, that more research, and better models, will reduce uncertainties.[21] Absent, however, is any discussion of these issues.

The conspicuous contradiction between the reviewers' comments and their suggested changes makes it very difficult to understand the review process as anything other than an effort to reinforce what Shackley and Wynne (1996, p. 285) termed the "condensation" of uncertainty's many meanings and complexities into "one undifferentiated category" that allows broad claims

to be made about how the key to a given problem is more research and more time. Moreover, by presenting uncertainty as a vague but putatively coherent concept that is "reduced" through more research, the scientific community assures that the phenomenon of uncertainty remains located in our imperfect (but always-improving) understanding of nature, and is not an attribute of nature itself, of the structure of disciplinary science, or of the social and political context within which research is conducted. In this way, scientists can maintain control over the management of uncertainty while also, in the words of Shackley and Wynne (1996, p. 287), "strengthening the authority of science [that] in turn reinforces a particular policy order."

As it pertains to environmental controversy, the word "uncertainty" refers most generally to the disparity between what is known and what actually *is* or *will be*. Uncertainty, that is, reflects our incomplete and imperfect characterization of current conditions relevant to an environmental problem, and our incomplete and imperfect knowledge of the future consequences of these conditions. For a well bounded problem, these insufficiencies can to some extent be addressed (although never eliminated) through additional research, but there are many reasons why such an approach might not succeed, for example, when additional research reveals heretofore unknown complexities in natural systems, or highlights the differences between competing disciplinary perspectives, and thus expands the realm of what is known to be unknown.

But as the previous examples show, the characterization of uncertainty also reflects the political and institutional contexts within which science is conducted and debated, the diversity of scientific practice, and the psychological states of those making the characterizations.[22] Uncertainty is in part a manifestation of the disunity of science and the plurality of institutional and political players (and their competing value commitments) involved in the conduct and interpretation of scientific research. It is the location where conflicts between competing sets of facts and disciplinary perspectives reside.

One simple way to think about these relations is shown in Fig. 1. When political stakes associated with a controversy are relatively low, high certainty is more permissible than when the

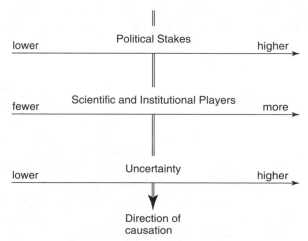

Figure 1 Schematic illustration of influence of political context on level of uncertainty.

stakes are high (e.g., Collingridge and Reeve, 1986). Fewer disciplines, institutions, and stakeholders are likely to have strong and competing interests in any particular assertion of uncertainty levels. This relation is illustrated by the Parkfield prediction. But when the costs and benefits associated with action on a controversy begin to emerge and implicate a variety of interests, both political and scientific scrutiny of the problem will increase, as will sources of uncertainty, as shown by the climate sensitivity and nuclear waste cases.[23] Moreover, when political controversy exists, the whole idea of "reducing uncertainty" through more research is incoherent because there will never be a single problem for which a single, optimizable research strategy or solution path can be identified, let alone characterized through a single approach to determining uncertainty. Instead, there will be many different problems defined in terms of many competing value frameworks and studied via many disciplinary approaches.

Recent developments in the Yucca Mountain story starkly illustrate the implications of these observations. In July 2002 President Bush signed a Congressional resolution that allows the US Department of Energy to apply for a license to actually construct the nuclear waste repository at Yuccá Mountain (Holt, 2003). This crucial political step was taken even though uncertainty about site behavior has increased significantly in recent years, due both to controversy over the hydrogeology, discussed above, and new insights into the effects of corrosion on waste containment vessels (Nuclear Waste Technical Review Board, 2003). What allowed political action to take place was the consolidation of national political power in the hands of the Republican party, which is sympathetic to the interests of the nuclear power industry, and thus supports moving ahead with development of the waste site. The diversity of political interests controlling the decision process was curtailed, and as a result the optimistic views on uncertainty of scientists and bureaucrats at the Department of Energy were able to prevail over other perspectives and interests. While these events are unlikely to mark an end to the controversy, they can allow action to proceed, and thus mark the beginning of what Schön and Rein (1994, p. xix) have called a "policy drama," where the discussion increasingly focuses on assessing progress toward a particular goal, e.g., the safe storage of nuclear waste, rather than on impossible-to-resolve questions such as whether "safe" storage is possible.

Why Scientize?

The organization of science—its methodological and disciplinary diversity; the multiple institutional settings in which it is conducted—make it a remarkably potent catalyst for political dispute. Recognizing that simple, linear formulations leading from "more science" to "less uncertainty" to "political action" are inherently flawed, others have suggested that society needs to adopt new ways of thinking about the conduct of science, new ways of evaluating how and when science is valid or potentially useful, new institutions for mediating the processes by which science is integrated into political decision making, and new geographic and temporal scales for conducting and using science (e.g., Funtowicz and Ravetz, 1992; Gallopín et al.,

2001; Lee, 1993; Nowotny et al., 2001; Gibbons, 1999; NRC, 1999b; Jasanoff, 1990, 1996b). While accepting the value and salience of all of these suggestions, they focus on the problem of understanding how scientific knowledge, in all its multifarious, social complexity, can best be integrated into contentious decision making processes. In cases where problems are fairly well circumscribed in terms of institutional players and problem definition, such approaches may make particular sense. Yet they do not engage this overarching observation: we have few good examples of science providing sufficient clarification to point the way through politically charged, open-ended environmental controversies,[24] yet innumerable examples of decisive political action in all realms of society taken despite controversy and uncertainty, and with science playing little or no formal role in the debate. One must wonder if it is worth approaching the problem from the opposite direction. So I would like to conclude by briefly exploring the following question: why is it that some political controversies become scientized, while others do not?

To return to the 2000 Presidential election, on its face, the vote count should have been much more amenable to scientific investigation and uncertainty reduction than even the simplest environmental controversy. On the other hand, the election was broadly accepted as a relatively pure process for adjudicating competing values and interests. Moreover, those values and interests had been on public display for months through the election campaign process. Even though the adjudication process itself was a technical one (counting votes), once the authoritativeness of that process was called into question, numerous other mechanisms for adjudicating value disputes were available and mobilized. Because these mechanisms were unabashedly political (or, in the case of the US Supreme Court, perhaps abashedly so), they were not subject to criticism for using junk science or for politicizing scientific results. Contesting sides were overtly seeking to advance their own interests. If there are complaints to be made about this process, they must address the mechanisms by which interests are advanced, such as campaign financing or the process of selecting judges, rather than the mechanism by which the controversy was ended.

What are the interests and values at stake in controversies over global climate change, nuclear waste disposal, or genetically modified foods? While it may not be very hard to arrive at plausible hypotheses about the value preferences of people holding various positions in such controversies, the scientific debate itself conceals those preferences behind technical arguments. This camouflaging process reflects, in part at least, the enduring social commitment to the idea of scientific facts as detached from values, and the consequent desire of everyone on all sides of a given controversy to legitimate their value preferences with an allegedly independent body of facts (e.g., Nelkin, 1995). This commitment is codified through a variety institutions and agreements, for example, the Intergovernmental Panel on Climate Change, which is supposed to provide the scientific basis for making international decisions about climate, and the World Trade Organization, which allows nations to regulate trade in agricultural goods based on risks to human, animal, and plant health only if such regulation is based on accepted scientific principles and standards (World Trade Organization, 1995). What Wynne (1991, p. 120) observed more than a decade ago seems to be no less true today: despite the policy implications of scholarly insight into the contextual origins of scientific knowledge, "the overall trend in the structure and control of science is currently running in the opposite direction."

Any political decision (indeed, any decision) is guided by expectations of the future. Such expectations can in turn be less or more informed by technical knowledge, but the capacity of such knowledge to yield an accurate and coherent picture of future outcomes is very limited indeed. Ultimately, most important decisions in the real world are made with a high degree of uncertainty, but are justified by a high level of commitment to a set of goals and values. Such past political acts as the passage of civil rights legislation, the reform of the US welfare system, or the decision to invade Iraq were not taken on the basis of predictive accuracy or scientific justifications about what the future *would* look like, but on the basis of convictions about what the future *should* look like, informed by plausible expectations of what the future *could* look like. From this perspective it is useful to recall that, when comprehensive environmental laws were enacted in the US during the late 1960s and early 1970s, scientific knowledge about the state of the environment was much less comprehensive and sophisticated than it is today, when almost all environmental laws and regulations are under political attack. The implementation of a broad legal framework for environmental protection in the US was a response to a social and political consensus, not authoritative knowledge (e.g., Kraft and Vig, 1997).

It is difficult to avoid the conclusion that there is no a priori reason why some types of political controversies should be highly scientized, and others should not be. For example, from a purely technical standpoint, the difficulties of predicting future climate outcomes and impacts over the next century, or predicting the behavior of a nuclear waste site over the next 10,000 years, cannot be much less complex, and are likely much more complex, than predicting the future of, say, different immigration policies or medical insurance systems.

Why, then, do some controversies become more scientized than others? Possibilities include:

1. advocates or opponents of action believe that scientific knowledge will advance their value positions or interests;
2. advocates or opponents of action believe that scientific uncertainty will advance their value positions or interests;
3. scientists are involved in the political framing of the controversy; and
4. available policy options for addressing the controversy are insufficiently broad or appealing to attract a political consensus.

In contrast, reasons why some controversies do not become highly scientized might include:

1. value positions are well articulated from the beginning of the controversy;

2. values underlying the controversy are widely viewed as inappropriate for scientific adjudication;
3. effective mechanisms for eliciting and adjudicating value disputes are already in place and well-accepted; and
4. available policy options are broad and appealing enough to attract a political consensus.

Political decision making can fruitfully be understood as a process of adjudicating value disputes (Lasswell, 1977, pp. 184–185; Sandel, 1996, pp. 17–18). This understanding certainly does not imply that facts have no place in political debate. People can only make sense of the world by finding ways to reconcile their beliefs with some set of facts about how reality must operate (e.g., Simon, 1983; Schön and Rein, 1994). So politics can isolate values from facts no more than science can isolate facts from values. The nature of this interaction has been a central subject of science studies scholarship (e.g., Jasanoff, 1987; Jasanoff, 1990).

The problem is that this symmetry does not manifest in political processes. Political debate permits the mobilization of a broad range of weaponry, including scientific facts, religious dogma, cultural norms, and personal experience, in defense of one's values and interests. But scientized debate must suppress the *open* discussion of value preferences; were it not to do so it would have no claim to distinction from politics.[25] This need can strike at the heart of democratic vitality. For example, as mentioned, World Trade Organization rules require that nations can only restrict trade in genetically modified foods on the basis of scientific risk assessments. These rules, of course, are meant to ensure a more open flow of goods across national boundaries, and thus give precedence to economic values over all others. But the well documented opposition in many European countries to GM foods appears to have little to do with scientifically determined levels of risk, and much to do with non-economic values. In many European nations, a majority of people surveyed say that they would not purchase GM foods even if they were known to be safe, environmentally friendly, *and* cheaper than non-GM equivalents. Survey data show that people's concerns are related more to a desire for transparency in decision making about GM foods, a suspicion about the economic motives of multinational companies who sell such foods, a concern about the implications of GM products for the European agricultural system (which in turn connects to concerns about landscape and culture), and worries about the implications of globalization for quality of life (Marris et al., 2001; Gaskell et al., 2003; Rayner, 2003). In the scientized controversy over GM foods, these diverse values have no legitimated part in the debate over levels of risk. Thus, not only are expressions of these values suppressed, but they are suppressed in favor of an alternative value—economic openness—that remains camouflaged behind the commitment to carrying out the debate in scientific terms alone.

Scientization of controversy also undermines the social value of science itself. In the absence of agreed upon values that can inform the articulation of social goals, we cannot recognize the broad range of policy options that might be available to achieve those goals, nor can we possibly know how to prioritize scientific research in support of the goals.[26] Scientific resources end up focused on the meaningless task of reducing uncertainties pertinent to political dispute, rather than addressing societal problems as identified through open political processes. The opportunity cost may be huge.[27] Consider what has taken place in the climate change arena. Certainly the Kyoto Protocol stands as the most significant political achievement related to climate change thus far. The Protocol represents the translation of a set of scientific insights about the relation between greenhouse gas emissions and global temperatures into a political decision to take a first step toward reducing those emissions.[28] But no one can possibly know what the consequences of these emissions reductions will be, either in terms of climate behavior or socio-economic outcomes. Thus, the only coherent value that can be extracted from the decision to adopt such reductions is that reducing greenhouse gas emissions is an inherently good thing to do. But this raises its own set of problems. The Kyoto goals could be achieved, for example, through a variable combination of emissions and sequestration schemes, which might or might not actually result in decreased hydrocarbon consumption. They could be pursued by enhancing global economic equity, for example, through diffusion of new technologies, or by further concentrating global wealth, for example, through policies that fail to spur economic development in poor countries (thus keeping energy consumption low).[29] And the pursuit of the Kyoto goals is not likely to have any discernible effect at all on the impacts of climate on society. If concerns about the negative impacts of climate change are a motivating value behind emissions reductions, those concerns will not be met.

Were such goals and values as, say, absolute reductions in hydrocarbon consumption through greater energy efficiency, more equitable global economic development, and decreased impacts of climate on society openly adopted as worthy of pursuit by society with the help of science, then global scientific priorities would look considerably different than they do now (e.g., Sarewitz and Pielke, 2000; Pielke and Sarewitz, 2003), perhaps corresponding more closely to what some have termed "sustainability science" (NRC, 1999b; Kates et al., 2001).

From these brief discussions I hope to have made clear that there is no reason why environmental controversies must be highly "scientized." Even if science brings such a controversy into focus (for example, by documenting a rise in atmospheric greenhouse gases), the controversy itself exists only because conflict over values and interests also exists. Bringing the value disputes concealed by—and embodied in—science into the foreground of political process is likely to be a crucial factor in turning such controversies into successful democratic action, and perhaps as well for stimulating the evolution of new values that reflect the global environmental context in which humanity now finds itself (Jamieson, 1992). Moreover, the social value of science itself is likely to increase if scientific resources relevant to a particular controversy are allocated after these value disputes have been brought out into the open, their implications for society explored, and suitable goals identified.[30]

A variety of researchers have sought to develop methods for integrating values into environmental research, for example, by developing scenarios of the future evolution of environment

and society that respond to different sets of value preferences (e.g., van Asselt and Rotmans, 1996; Rotmans and De Vries, 1997; Costanza, 2000; UK Climate Impacts Programme, 2001). It is not clear whether or not this is a step in the right direction because these approaches still depend on the ability of mathematical models to yield plausible scenarios of the future, where plausibility will in part be judged within the normative perspectives of those using the models (which themselves embody the normative perspectives of those who build the models). Moreover, the process of articulating concrete future alternatives through models could have the affect of exacerbating political controversy by claiming to make it clearer who future winners and losers are likely to be, given a set of decisions and predicted outcomes (e.g., Glantz, 1995). These approaches also beg the question of how values will actually be elicited and adjudicated in choosing what scenarios society should actually pursue.

What I am suggesting is that progress in addressing environmental controversies will need to come primarily from advances in political process, rather than scientific research. Perhaps such advances will require the formal or informal imposition of a sort of "quiet period" for scientific debate when environmental controversies become highly publicized and gridlocked, to create time and space for underlying value disputes to be brought into the open, explored, and adjudicated *as such* in democratic fora. During such a "quiet period," those who make scientific assertions in fora of public deliberation would have to accompany those claims with a statement of value preferences and private interests relevant to the dispute. This rule would be enforced for scientists as well as lay people. Science does not thereby disappear from the scene, of course, but it takes its rightful place as one among a plurality of cultural factors that help determine how people frame a particular problem or position—it is a part of the cognitive ether, and the claim to special authority vanishes.

If this suggestion seems not just playful but frivolous, consider where my discussion began, with the election and Lomborg controversies. In the former case, the factual dispute was subjugated to the practical necessity of arriving at a resolution, and politics was allowed to do its job. In the latter case, an insistence by all parties that the dispute is about who is in charge of the right environmental facts merely recapitulates in miniature the escalating political gridlock surrounding environmental politics. The technical debate—and the implicit promise that "more research" will tell us what to do—vitiates the will to act. Not only does the value dispute remain unresolved, but the underlying problem remains unaddressed.

The point is not that stripping away the overlay of scientific debate must force politicians to take action. But if they choose not to act they can no longer claim that they are waiting for the results of the next round of research—they must instead explain their allegiance to inaction in terms of their own values and interests, and accountability now lies with them, not with science or scientists. To the extent that our democratic political fora are incapable of enforcing that accountability, the solution must lie in political reform, not more and better scientific information.

Yet one question remains: what, then, becomes of science? One part of the answer is: nothing, it is still there, in the background, along with all the other influences on people's political interests and behavior. But the other part of the answer is that science is liberated to serve society and the environment, for, as I have suggested, it is only after values are clarified and some goals agreed upon that appropriate decisions about science priorities can emerge.

No longer able to hide behind scientific controversy, politics would have to engage in processes of persuasion, reframing, disaggregation, and devolution, to locate areas of value consensus, overlapping interests, or low-stakes options (e.g., "no regrets" strategies) that can enable action in the absence of a comprehensive political solution or scientific understanding (Sarewitz and Pielke, 2000; Pielke and Sarewitz, 2003). In particular, the abandonment of a political quest for definitive, predictive knowledge ought to encourage, or at least be compatible with, more modest, iterative, incremental approaches to decision making that can facilitate consensus and action. Such approaches call upon science not to be a predictive oracle to guide policy choices, but a tool to support, monitor, and assess the implementation of policies that have been selected through the political process (Brunner, 2000; Herrick and Sarewitz, 2000; Lee, 1993). "Sustainability," write Rayner and Malone (1998, p. 132) "is about being nimble, not being right." And being nimble is about taking small steps and keeping one's eyes open. Politics helps us decide the direction to step; science helps the eyes to focus.

Notes

1. This dynamic does not discriminate on the basis of ideology: in the case of the Yucca Mountain nuclear waste repository, the claim that the science is sufficient to justify action (i.e., construction of the site) has most strongly been invoked by those with a politically conservative bent; in the case of climate change, it is political liberals who have been more likely to claim that the science is sufficient to justify a particular line of action (mandated reductions of greenhouse gas emissions).

2. Perhaps paradoxically, the notion persists despite carefully argued claims about the delegitimation of the authority of science in society (e.g., Ezrahi, 1990).

3. In fact, these are the titles of published papers: Leib and Dittmer (2002), US General Accounting Office (2001), and Wand et al. (2001).

4. Studies by various social scientists (e.g., Wand et al., 2001 and Heron and Sekhon, 2003) have sought to show that the intent on many overvoted ballots could be inferred, and have concluded that high prevalence of uncounted overvotes in strongly Democratic counties worked strongly in favor of Bush.

5. I am content to use a conventional definition of "fact," e.g., a statement about the world whose truth or falseness can be tested.

6. The first goal listed in the *Strategic Plan for the US Climate Change Science Program* (Climate Change Science Program, 2003) is to reconcile surface and lower atmosphere temperature records, whose inconsistencies "reduce confidence in understanding of how and why climate has changed."

7. My point is not to excuse conscious manipulation of facts, or to deny that some research simply is not of a very good quality.

But the elimination of these two problems would have little if any effect on the phenomenon I am describing. The problem is not "good" versus "bad" but "ours" versus "theirs" (e.g., see Herrick and Jamieson, 2001 on "junk science").

8. A conspicuous example of such of non-intersecting worldviews is the almost complete lack of cross fertilization between scientists who generate model-based scenarios of future climate behavior, and researchers who study hazards and their reduction. As one indication of this divide, these communities use the word "mitigation" in opposite senses. To the climate modeling community, mitigation means prevention of climate change through greenhouse gas emissions reduction (mitigation of *cause*). To the hazards community, mitigation means protection from climate impacts through, e.g., better land use planning or infrastructure (mitigation of *effects*).

9. Similarly, Barke and Jenkins-Smith (1993) showed that scientists' perception of risk related to nuclear waste disposal was related to disciplinary orientation.

10. The point applies equally to Chapela, a long-time opponent of GM crops.

11. For this reason, he also speculates—correctly, it turns out (Harrison, 2004)—that some or most of the peer reviewers of Lomborg's book must have been economists.

12. The Yucca Mountain story is taken from Metlay (2000).

13. These bodies were the Nuclear Waste Technical Review Board, an independent government agency, and the nongovernmental Board on Radioactive Waste Management of the National Research Council.

14. A similar dynamic has been demonstrated in the area of clinical medical trials, where a number of studies have shown that trials directly or indirectly supported by pharmaceutical companies often yield more favorable assessments of new therapies— greater certainty about positive results—than trials that are not tied to the private sector in any way. The point here is not that scientists are engaging in fraudulent research in an effort to bolster a desired conclusion. The reality seems to be more interesting—that "close and remunerative collaboration with a company naturally creates goodwill [that] can subtly influence scientific judgment in ways that may be difficult to discern" (Angell, 2000, p. 1517). This conclusion is similar to Metlay's suggestion that "organizational norms and culture" influenced uncertainty estimates at Yucca Mountain.

15. Climate feedbacks are processes in the climate system that can either magnify or reduce the system's response to climate forcing such as greenhouse gas emissions.

16. Perhaps most notable among these was the massive *Pathways* report, which states on its opening page (NRC, 1999a, p. 1), first that "we can now focus attention on the critical unanswered scientific questions that must be resolved to fully understand and usefully predict global change," and therefore, that "we need to reduce uncertainties in the projections that shape our decisions for the future."

17. Quotes from a 6 June 2002 draft of the report.

18. Quotes from an 8 August 2002 internal memo to the panel from NRC staff.

19. Memo to the panel, 9 July 2003.

20. As just one example: in a discussion of cloud feedbacks, successive paragraphs refer first to "reduction in the uncertainty

of climate sensitivity," which is generally reported as a model uncertainty, and then states that "[a]nother key uncertainty in cloud–climate interactions is the response of anvil clouds to surface temperature." The second use of the word seems simply to mean "incomplete knowledge" (NRC, 2003, p. 27). Two pages later "uncertainty" is used to mean the difficulty of quantifying a specific value ("the amount of radioactive heating that occurs within the atmosphere versus how much heating occurs at the surface . . .") (NRC, 2003, p. 30).

21. For example, "Research into carbon uptake by the land and ocean as outlined in the US Carbon Cycle Plan . . . should be undertaken to characterize and reduce the uncertainty associated with carbon uptake feedbacks" (NRC, 2003, p. 10).

22. For example, the propensity of experts to display overconfidence in probability estimates has long been recognized (Kahneman et al., 1982). Moreover, as Rosenberg (1994) points out, the very act of estimating probabilities increases the complexity of the system being studied because that system now includes the cognitive states of the experts who are doing the estimating.

23. Jamieson (1995, p. 37) has made a similar point in arguing that "[r]ather than being a cause of controversy, scientific uncertainty is often a consequence of controversy." However Shackley and Wynne (1996, p. 276) suggested the opposite— that "the less a science is tied to policy uses, the more its practitioners would be free to express uncertainty."

24. The successful negotiation of the Montreal Protocol, mandating the phase out of stratospheric-ozone-depleting chlorofluorocarbons (CFCs), is the most obvious candidate for such a success story. The identification of a single chemical culprit (CFCs) as the major cause of ozone depletion suggests that this problem was scientifically "easier" than other high-profile controversies. Yet the incentives for policy action were strongly enhanced by a political and diplomatic climate that was receptive to reductions in CFC use even *before* the resolution of the major scientific questions about ozone depletion (Benedick, 1991; Sarewitz, 1996). Also important was the invention of CFC alternatives by DuPont (Rowlands, 1993), which allowed private sector interests to align with calls for a CFC phase out. The ozone story is less one of controversy resolved by science than of positive feedback among convergent scientific, political, diplomatic, and technological trends.

25. In reality, of course, scientists are constantly engaged in a process of sub rosa, perhaps even subconscious, negotiation that is unavoidably political, whether working in the closed context of a research community (e.g., Fleck, 1979; Collins, 1985), or in the more highly charged atmosphere of a science advisory panel (Jasanoff, 1990).

26. Some environmental controversies, such as climate change, only exist because of scientific research that allowed the problem to be recognized in the first place. However, the role of bringing an environmental problem to public attention is not at all the same as resolving the value-based controversies raised by the problem (e.g., Herrick and Sarewitz, 2000; Alario and Brün, 2001).

27. As one indication of this opportunity cost, consider that the US federal commitment to basic research on climate change has been about four times more than its expenditures on renewable energy research (EIA, 2001).

28. That is, for industrialized nations, by 5% below 1990 levels by the year 2012.

29. In this light it is notable that the Kyoto Protocol's relatively modest commitment to technology development and diffusion is more than offset by a variety of economic and technology policies endorsed by many signatory nations that increasingly undermine the development prospects of poor countries (e.g., Stiglitz, 2003).

30. At this point I must, unfortunately, emphasize that this is neither a brief for nor against "basic research." There is, indeed, every reason to believe that the political determination of value preferences and goals in the realm of the environment would need to be followed by research of many different kinds aimed at helping to pursue those goals.

References

Alario, A., Brün, M., 2001. Uncertainty and controversy in the science and ethics of environmental policy making. Theory Sci. 2 (1), http://theoryandscience.icaap.org/content/vol002.001/02alariobrun.html.

Andronova, N., Schlesinger, M.E., 2001. Objective estimation of the probability distribution for climate sensitivity. J. Geophys. Res. 106 (D19), 22605–22612.

Angell, M., 2000. Is academic medicine for sale? N. Engl. J. Med. 342 (20), 1516–1518.

Ausubel, J., 1991. Energy and environment: the light path. Energy Syst. Policy 15, 181–189.

Bakun, W.H., Lindh, A.G., 1985. The Parkfield, California, earthquake prediction experiment. Science 229, 619–624.

Barke, R., Jenkins-Smith, H., 1993. Politics and scientific expertise: scientists, risk perception, and nuclear waste policy. Risk Anal. 13, 425–439.

Benedick, R., 1991. Protecting the ozone layer: new directions in diplomacy. In: Mathews, J.T. (Ed.), Preserving the Global Environment: The Challenge of Shared Leadership. W.W. Norton, New York.

Bowker, G., Star, S.L., 2001. Pure, real and rational numbers: the American imaginary of countability. Social Stud. Sci. 31 (3), 422–425.

Brunner, R., 2000. Alternatives to prediction. In: Sarewitz, D., Pielke Jr., R.A., Byerly Jr., R. (Eds.), Prediction: Science, Decision Making, and the Future of Nature. Island Press, Covelo, CA, pp. 299–313.

Cartwright, N., 1999. The Dappled World, A Study of the Boundaries of Science. Cambridge University Press, Cambridge, UK.

Changnon, S.A., Pielke Jr., R.A., Changnon, D., Sylves, R.T., Pulwarty, R., 2001. Human factors explain the increased losses from weather and climate extremes. Bull. Am. Meteorol. Soc. 81 (3), 437–442.

Chichilnisky, G., Heal, G., 1995. Markets for tradable CO_2 emission quotas principles and practice. OECD Economics Department Working Papers No. 153, OCDE/GD(95)9, Organization for Economic Co-Operation and Development, Paris.

Climate Change Science Program, 2003. Strategic Plan for the US Climate Change Science Program. Climate Change Science Program, Washington, DC.

Collingridge, D., Reeve, C., 1986. Science Speaks to Power: The Role of Experts in Policy. St. Martin's Press, New York.

Collins, H., 1985. Changing Order: Replication and Induction in Scientific Practices. Sage, London.

Costanza, R., 2000. Visions of alternative (unpredictable) futures and their use in policy analysis. Conserv. Ecol. Online 4 (1), http://www.ecologyandsociety.org/vol4/iss1/art5/index.html.

Dupré, J., 1993. The Disorder of Things, Metaphysical Foundations of the Disunity of Science. Harvard University Press, Cambridge, M.A.

The Economist Print Edition, 2003. Hot potato revisited. Economist November, 76.

EIA (Energy Information Agency), 2001. Renewable energy 2000: issues and trends. Available at www.eia.doe.gov/cneaf/solar. renewables/rea_issues/reatabp2.html.

Ezrahi, Y., 1990. The Descent of Icarus: Science and the Transformation of Contemporary Democracy. Cambridge University Press, Cambridge, MA.

Fleck, L., 1979. Genesis and Development of a Scientific Fact. University of Chicago Press, Chicago (original published in 1935).

Food and Agriculture Organization of the United Nations, 2003a. Weighing the GMO arguments: against. Available at http://www. fao.org/english/newsroom/focus/2003/gmo8.htm.

Food and Agriculture Organization of the United Nations, 2003. Weighing the GMO arguments. Available at http://www.fao. org/english/newsroom/focus/2003/gmo7.htm.

Forest, C.E., Stone, P.H., Sokolov, A.P., Allen, M.R., Webster, M.D., 2002. Quantifying uncertainties in climate system properties with the use of recent climate observations. Science 295, 113–117.

Funtowicz, S.O., Ravetz, J.R., 1992. Three types of risk assessment and the emergence of post-normal science. In: Krimsky, S., Golding, D. (Eds.), Social Theories of Risk. Praeger, Westport, CT, pp. 251–273.

Gallopín, G.C., Funtowicz, S., O'Connor, M., Ravetz, J., 2001. Science for the twenty-first century: from social contract to the scientific core. Int. Social Sci. J. 53, 219–229.

Gaskell, G., Allum, N., Stares, S., 2003. Europeans and biotechnology in 2002. A report to the EC Directorate General for Research from the project 'Life Sciences in European Society', QLG7-CT-1999-00286, http://europa.eu.int/comm/ public_opinion/archives/eb/ebs_177_en.pdf.

Gibbons, M., 1999. Science's new social contract with society. Nature 402 (Suppl.), C81–C84.

Glantz, M.H., 1995. Assessing the impacts of climate: the issue of winners and losers in a global climate change context. In: Zwerver, S., van Rompaey, R.S.A.R., Kok, M.T.J., Berk, M.M. (Eds.), Climate Change Research: Evaluation and Policy Implications. Elsevier, Amsterdam.

Gough, M., 2003. Politicizing Science. Hoover Institution Press, Stanford, CA.

Harrison, C., 2004. Peer review, politics and pluralism. Environ. Sci. Policy 7(5), 357–368.

Heron, M.C., Sekhon, J.S., 2003. Overvoting and representation: and examination of overvoted presidential ballots in Broward and Miami-Dade counties. Electoral Stud. 22 (1), 21–47.

Herrick, C., Jamieson, D., 1995. The social construction of acid rain: some implications for science/policy assessment. Global Environ. Change 5, 105–112.

Herrick, C., Jamieson, D., 2001. Junk science and environmental policy: obscuring public debate with misleading discourse. Philosophy Public Policy Q. 21 (Spring), 11–16.

Herrick, C., Sarewitz, D., 2000. Ex post evaluation: a more effective role for scientific assessments in environmental policy. Sci. Technol. Hum. Values 25 (3), 309–331.

Holling, C.S., 1998. Two cultures of ecology. Conserv. Ecol. 2 (2), http://www.consecol.org/vol2/iss2/art4.

Holt, M., 2003. Civilian nuclear waste disposal. Congressional Research Service, Order Code IB92059, Washington, DC, http:// www.NCSEonline.org/NLE/CRS/abstract.cfm?NLEid=17012.

Houghton, J.T., Ding, Y., Griggs, D.J., Noguer, M., van der Linden, P.J., Dai, X., Maskell, K., Johnson, C.A., 2001. Climate Change 2001: The Scientific Basis. Contribution of Working Group I to the Third Assessment Report of the Intergovernmental Panel on Climate Change. Cambridge University Press, Cambridge, UK.

Hull, D.L., 1988. Science as a Process: An Evolutionary Account of the Social and Conceptual Development of Science. University of Chicago Press, Chicago.

Jamieson, D., 1992. Ethics, public policy, and global warming. Sci. Technol. Hum. Values 17 (2), 139–153.

Jamieson, D., 1995. Scientific uncertainty and the political process, annals. Am. Acad. Political Social Sci. 545, 35–43.

Jasanoff, S., 1987. Contested boundaries in policy-relevant science. Social Stud. Sci. 17 (2), 195–230.

Jasanoff, S., 1990. The Fifth Branch: Science Advisors as Policymakers. Harvard University Press, Cambridge, MA.

Jasanoff, S., 1996a. Beyond epistemology: relativism and engagement in the politics of science. Social Stud. Sci. 26 (2), 393–418.

Jasanoff, S., 1996b. The dilemma of environmental democracy. Issues Sci. Technol. 13 (1), 63–70.

Jasanoff, S., Wynne, B., 1998. Science and decision making. In: Rayner, S., Malone, E. (Eds.), Human Choice and Climate Change, vol. 1: The Societal Framework. Battelle Press, Columbus, OH, pp. 1–87.

Kagan, Y.Y., 1997. Statistical aspects of Parkfield earthquake sequence and Parkfield prediction experiment. Tectonophysics 270 (3–4), 207–219.

Kahneman, D., Slovic, P., Tversky, A., 1982. Judgment Under Uncertainty: Heuristics and Biases. Cambridge University Press, New York.

Kaiser, R.G., 2000. Is this any way to pick a winner. Washington Post 26 May, A1.

Kaplinsky, N., Braun, D., Lisch, D., Hay, A., Hake, S., Freeling, M., 2002. Maize trangene results in Mexico are artefacts. Nature 416, 601–602.

Kates, R.W., Clark, W.C., Corell, R., Hall, J.M., Jaeger, C.C., Lowe, I., McCarthy, J.J., Schellnhuber, H.J., Bolin, B., Dickson, N.M., Faucheux, S., Gallopin, G.C., Grübler, A., Huntley, B., Jäger, J., Jodha, N.S., Kasperson, R.E., Mabogunje, A., Matson, P., Mooney, H., Moore III, B., O'Riordan, T., Svedin, U., 2001. Sustainability science. Science 292 (5517), 641–642.

Kitcher, P., 2001. Science, Truth, and Democracy. Oxford University Press, New York.

Knutti, R., Stocker, T.F., Joos, F., Plattner, G.K., 2002. Constraints on radiative forcing and future climate change from observations and climate model ensembles. Nature 416, 719–723.

Kraft, M.E., Vig, N.J., 1997. Environmental policy from the 1970s to the 1990s: an overview. In: Vig, N.J., Kraft, M.E. (Eds.), Environmental Policy in the 1990s. CQ Press, Washington, DC, pp. 1–30.

Lasswell, H.D., 1977. Psychopathology and Politics. University of Chicago Press, Chicago.

Lee, K., 1993. Compass and Gyroscope, Integrating Science and Politics for the Environment. Island Press, Covelo, CA.

Leib, J.I., Dittmer, J., 2002. Florida's residual votes, voting technology and the 2000 election. Political Geography 21, 91–98.

Lepkowski, W., 2002a. Biotech's OK corral. Sci. Policy Perspect. 13, http://www.cspo.org/s&pp/060902.html.

Lepkowski, W., 2002b. Maize, genes, and peer review. Sci. Policy Perspect. 14, http://www.cspo.org/spp/102502.html.

Lomborg, B., 2001. The Skeptical Environmentalist: Measuring the State of the Real World. Cambridge University Press, Cambridge, UK.

Lovejoy, T., 2002. Biodiversity: dismissing scientific process. Scientific American.com, January, http://www.sciam.com/article.cfm?articleID=000F3D47-C6D2-1CEB-93F6809E C5880000&pageNumber=12&catID=2.

Marland, G., Pielke Sr., R.A., Apps, M., Avissar, R., Betts, R.A., Davis, K.J., Frumhoff, P.C., Jackson, S.T., Joyce, L., Kauppi, P., Katzenberger, J., MacDicken, K.G., Neilson, R., Niles, J.O., Sutta, D., Niyogi, S., Norby, R.J., Pena, N., Sampson, N., Xue, Y., 2003. The climatic impacts of land surface change and carbon management, and the implications for climate-change mitigation policy. Climate Policy 3, 149–157.

Marris, C., Wynne, B., Simmons, P., Weldon, S., 2001. Public perceptions of agricultural biotechnologies in Europe. Final report of the PABE Research project, http://www.lancs.ac.uk/depts/ieppp/pabe/docs.htm.

Merzer, M., 2001. "Overvotes" leaned to Gore. The Miami Herald. com, 11 May, http://www.miami.com/mld/miami/news/2067453.htm.

Metlay, D., 2000. From in roof to torn wet blanket: predicting and observing groundwater movement at a proposed nuclear waste site. In: Sarewitz, D., Pielke Jr., R.A., Byerly Jr., R. (Eds.), Prediction: Science, Decision Making, and the Future of Nature. Island Press, Covelo, CA, pp. 199–228.

Metz, M., Fütterer, J., 2002. Suspect evidence of transgenic contamination. Nature 416, 600–601.

Michael, D.N., 1995. Barriers and bridges to learning in a turbulent human ecology. In: Gunderson, L., Holling, C., Light, S. (Eds.), Barriers and Bridges to the Renewal of Ecosystems and Institutions. Columbia University Press, New York, pp. 461–485.

Morgan, M.G., Keith, D.W., 1995. Subjective Judgments by climate experts. Environ. Sci. Technol. 29, A468–A476.

Nakicenovic, N., 1996. Freeing energy from carbon. Daedalus 125 (3), 95–112.

Nature Publishing Group, 2002. Conflicts around a study of mexican crops. Nature 417, 897–898.

Nelkin, D., 1975. The political impact of technical expertise. Social Stud. Sci. 5 (1), 35–54.

Nelkin, D., 1979. Controversy: Politics of Technical Decisions. Sage, London.

Nelkin, D., 1995. Science controversies: the dynamics of public disputes in the United States. In: Jasanoff, S., Markle, G.E., Petersen, J., Pinch, T. (Eds.), Handbook of Science and Technology Studies. Sage, Thousand Oaks, CA, pp. 444–456.

Nigg, J., 2000. Predicting earthquakes: science, pseudoscience, and public policy paradox. In: Sarewitz, D., Pielke Jr., R.A., Byerly Jr., R. (Eds.), Prediction: Science, Decision Making, and the Future of Nature. Island Press, Covelo, CA, pp. 135–156.

Norgaard, R.B., 2002. Optimists, pessimists, and science. BioScience 52 (3), 287–292.

Nowotny, H., Scott, P., Gibbons, M., 2001. Rethinking Science: Knowledge and the Public in an Age of Uncertainty. Polity, Cambridge.

NRC, 1999a. Global Environmental Change: Research Pathways for the Next Decade. National Academy Press, Washington, DC.

NRC, 1999b. Our Common Journey. National Academy Press, Washington, DC.

NRC, 2000. Reconciling Observations of Global Temperature Change. National Academy Press, Washington, DC.

NRC, 2001. Climate Change Science, An Analysis of Some Key Questions. National Academy Press, Washington, DC.

NRC, 2003. Understanding Climate Change Feedbacks. National Academy Press, Washington, DC.

Nuclear Waste Technical Review Board, 2000. Spring 2000 board meeting transcript. Nuclear Waste Technical Review Board, Arlington, VA, http://www.nwtrb.gov/meetings/meetings.html.

Nuclear Waste Technical Review Board, 2003. An evaluation of key elements in the US Department of Energy's Proposed System for Isolating and Containing Radioactive Waste. Nuclear Waste Technical Review Board, Arlington, VA, http://www.nwtrb.gov/reports/reports.html.

Oreskes, N., 2004. Science and public policy: what's proof got to do with it? Environ. Sci. Policy 7 (5), 369–383.

Parrot Lab, 2003. Parrot Lab, http://www.cropsoil.uga.edu/~parrottlab/.

Pielke Jr., R.A., Rubiera, J., Landsea, C., Fernandez, M.R., Klein, R., 2003. Hurricane vulnerability in Latin America and the Caribbean: normalized damage and loss potential. Nat. Hazards Rev. 4 (3), 101–114.

Pielke Jr., R.A., Sarewitz, D., 2003. Wanted: scientific leadership on climate. Issues Sci. Technol. Winter, 27–30.

Pimentel, D., 2002. Exposition on skepticism. BioScience 52 (3), 295–298.

Quist, D., Chapela, I.N., 2001. Transgenic DNA introgressed into traditional maize landraces in Oaxaca, Mexico. Nature 414, 541–543.

Rayner, S., 2000. Prediction and other approaches to climate change policy. In: Sarewitz, D., Pielke Jr., R.A., Byerly Jr., R. (Eds.), Prediction: Science, Decision Making, and the Future of Nature. Island Press, Covelo, CA, pp. 269–296.

Rayner, S., 2003. Democracy in the age of assessment: reflections on the roles of expertise and democracy in public-sector decision making. Sci. Public Policy 30 (3), 163–170.

Rayner, S., Malone, E., 1998. Ten suggestions for policymakers. In: Rayner, S., Malone, E. (Eds.), Human Choice and Climate Change, vol. 4: What Have We Learned. Battelle Press, Columbus, OH, pp. 109–138.

Roeloffs, E.A., Langbein, J., 1994. The earthquake prediction experiment at Parkfield, California. Rev. Geophys. 32 (3), 315–336.

Rosenberg, A., 1994. Instrumental Biology or the Disunity of Science. University of Chicago Press, Chicago.

Rotmans, J., De Vries, B., 1997. Perspectives on Global Change. Cambridge University Press, Cambridge, UK.

Rowlands, I.H., 1993. The fourth meeting of the parties to the Montreal protocol: report and reflection. Environment 35 (6), 25–34, http://www.ciesin.org/docs/003-077/003-077.html.

Sandel, M.J., 1996. Democracy's Discontent: America in Search of a Public Philosophy. Harvard University Press, Cambridge, MA.

Sarewitz, D., 1996. Frontiers of Illusion: Science, Technology, and the Politics of Progress. Temple University Press, Philadelphia.

Sarewitz, D., 2000. Science and environmental policy: an excess of objectivity. In: Frodeman, R. (Ed.), Earth Matters: The Earth Sciences, Philosophy, and the Claims of Community. Prentice-Hall, Upper Saddle River, NJ, pp. 79–98.

Sarewitz, D., Pielke Jr., R.A., 2000. Breaking the global warming gridlock. Atlantic Monthly July, 54–64.

Schneider, S., 2002. Global warming: neglecting the complexities. Sci Am. January, 62–65.

Schön, D.A., Rein, M., 1994. Frame Reflection: Toward the Resolution of Intractable Policy Controversies. Basic Books, New York.

Schwarz, M., Thompson, M., 1990. Divided We Stand: Redefining Politics, Technology and Social Choice. University of Pennsylvania Press, Philadelphia.

Shackley, S., Wynne, B., 1996. Representing Uncertainty in Global Climate Change Science and Policy: Boundary-Ordering Devices and Authority. Sci. Technol. Hum. Values 21 (3), 275–302.

Simon, H.A., 1983. Reason in Human Affairs. Stanford University Press, Stanford, CA.

Simon, H.A., 1997. Administrative Behavior, fourth ed. Free Press, New York.

Snow, A., 2003. Research interests. The Department of Evolution, Ecology, and Organismal Biology, The Ohio State University, http://www.biosci.ohio-state.edu/~eeob/resints/res_snow.html.

Stiglitz, J.E., 2003. Globalization and its Discontents. Norton, New York.

Supreme Court of the United States, 2000. 531 US 98.

US General Accounting Office, 2001. Elections: statistical analysis of factors that affected uncounted votes in the 2000 Presidential election. GAO-02-122, 15 October, http://www.gao/gov/cgi-bin/getrpt?gao-02-122.

United States House of Representatives, 2003. Politics and Science in the Bush Administration, Committee on Government Reform—Minority Staff. Washington, DC, http://www.house.gov/reform/min/politicsan-dscience/.

UK Climate Impacts Programme, 2001. Socio-economic scenarios for climate change impact assessment: a guide to their use in the UK. UK Climate Impacts Programme, Oxford, UK, http://www.ukcip.org.uk/research_tools/research_tools.html.

van Asselt, M.B.A., Rotmans, J., 1996. Uncertainty in perspective. Global Environ. Change 6 (2), 121–157.

van der Sluijs, J., van Eijndhoven, J., Shackley, S., Wynne, B., 1998. Anchoring devices in science for policy: the case of consensus around climate sensitivity. Social Stud. Sci. 28 (2), 291–323.

Wand, J., Shotts, K., Sekhon, J., Mebane Jr., W., Herron, M., Brady, H., 2001. The butterfly did it: aberrant vote for Buchanan in Palm Beach County, Florida. Am. Political Sci. Rev. 95 (4), 793–810.

Wilson, E.O., 1998. Consilience: The Unity of Knowledge. Alfred A. Knopf, New York.

Wolfenbarger, L.L., Phifer, P.R., 2000. The ecological risks and benefits of genetically engineered plants. Science 290, 2088–2093.

World Trade Organization, 1995. Agreement on the application of sanitary and phytosanitary measures. Article 2.2.

Wynne, B., 1989. Sheepfarming after chernobyl: a case study in communicating scientific information. Environment 31 (2), 11–15, 33–39.

Wynne, B., 1991. Knowledges in context. Sci. Technol. Hum. Values 16 (1), 111–121.

DANIEL SAREWITZ is Professor of Science and Society and Director of the Consortium for Science, Policy, and Outcomes (CSPO) at Arizona State University. Recent publications include *Living with the Genie: Essays on Technology and the Quest for Human Mastery* (co-edited with Alan Lightman and Christina Desser, Island Press, 2003); *Prediction: Science, Decision making, and the Future of Nature* (co-edited with Roger Pielke Jr. and Radford Byerly Jr, Island Press, 2000), and *Frontiers of Illusion: Science, Technology, and the Politics of Progress* (Temple University Press, 1996).

Acknowledgements—Roger Pielke Jr., Beth Raps, Richard Nelson, Charles Herrick, and Naomi Oreskes provided valuable comments on this paper. I also thank three anonymous reviewers for their extraordinarily penetrating critiques of the penultimate version (to the numerous arguments against anonymity in peer review, add this: one is deprived of the pleasure and benefit of continued engagement with thoughtful reviewers). Discussions with Daniel Metlay have contributed greatly to my understanding of the Yucca Mountain story, as well as to the broader question of how institutions confront uncertainty.

From *Environmental Science & Policy,* Issue 7, 2004, pp. 385–403. Copyright © 2004 by Elsevier Ltd. Reprinted by permission via Rightslink.

Test-Your-Knowledge Form

We encourage you to photocopy and use this page as a tool to assess how the articles in *Annual Editions* expand on the information in your textbook. By reflecting on the articles you will gain enhanced text information. You can also access this useful form on a product's book support Web site at *http://www.mhcls.com/online/*.

NAME: DATE:

_____

TITLE AND NUMBER OF ARTICLE:

_____

BRIEFLY STATE THE MAIN IDEA OF THIS ARTICLE:

_____

LIST THREE IMPORTANT FACTS THAT THE AUTHOR USES TO SUPPORT THE MAIN IDEA:

_____

WHAT INFORMATION OR IDEAS DISCUSSED IN THIS ARTICLE ARE ALSO DISCUSSED IN YOUR TEXTBOOK OR OTHER READINGS THAT YOU HAVE DONE? LIST THE TEXTBOOK CHAPTERS AND PAGE NUMBERS:

_____

LIST ANY EXAMPLES OF BIAS OR FAULTY REASONING THAT YOU FOUND IN THE ARTICLE:

_____

LIST ANY NEW TERMS/CONCEPTS THAT WERE DISCUSSED IN THE ARTICLE, AND WRITE A SHORT DEFINITION:

We Want Your Advice

ANNUAL EDITIONS revisions depend on two major opinion sources: one is our Advisory Board, listed in the front of this volume, which works with us in scanning the thousands of articles published in the public press each year; the other is you—the person actually using the book. Please help us and the users of the next edition by completing the prepaid article rating form on this page and returning it to us. Thank you for your help!

ANNUAL EDITIONS: Environment 08/09

ARTICLE RATING FORM

Here is an opportunity for you to have direct input into the next revision of this volume.
We would like you to rate each of the articles listed below, using the following scale:

1. **Excellent: should definitely be retained**
2. **Above average: should probably be retained**
3. **Below average: should probably be deleted**
4. **Poor: should definitely be deleted**

Your ratings will play a vital part in the next revision.
Please mail this prepaid form to us as soon as possible.
Thanks for your help!

RATING	ARTICLE	RATING	ARTICLE
_____	1. Climate Change 2007	_____	16. The Future of Nuclear Power
_____	2. How Many Planets?	_____	17. Personalized Energy
_____	3. Five Meta-Trends Changing the World	_____	18. Hydrogen: Waiting for the Revolution
_____	4. Globalization's Effects on the Environment	_____	19. Strangers in Our Midst
_____	5. Do Global Attitudes and Behaviors Support Sustainable Development?	_____	20. America's Coral Reefs: Awash with Problems
_____	6. Population and Consumption	_____	21. Markets for Biodiversity Services
_____	7. Can You Buy a Greener Conscience?	_____	22. Tracking U.S.Groundwater: Reserves for the Future?
_____	8. A New Security Paradigm	_____	23. How Much Is Clean Water Worth?
_____	9. Where Oil and Water Do Mix	_____	24. Searching for Sustainability
_____	10. The Irony of Climate	_____	25. Swift Boating, Stealth Budgeting, and Unitary Executives
_____	11. Avoiding Green Marketing Myopia	_____	26. The Truth About Denial
_____	12. Gassing Up with Hydrogen	_____	27. How Science Makes Environmental Controversies Worse
_____	13. Wind Power		
_____	14. Whither Wind?		
_____	15. The Rise of Renewable Energy		

BUSINESS REPLY MAIL
FIRST CLASS MAIL PERMIT NO. 551 DUBUQUE IA

POSTAGE WILL BE PAID BY ADDRESSEE

McGraw-Hill Contemporary Learning Series
501 BELL STREET
DUBUQUE, IA 52001

NO POSTAGE
NECESSARY
IF MAILED
IN THE
UNITED STATES

ABOUT YOU

Name   Date

_____

Are you a teacher? ☐ A student? ☐
Your school's name

_____

Department

_____

Address   City   State   Zip

_____

School telephone #

_____

YOUR COMMENTS ARE IMPORTANT TO US!

Please fill in the following information:
For which course did you use this book?

_____

Did you use a text with this ANNUAL EDITION? ☐ yes ☐ no
What was the title of the text?

_____

What are your general reactions to the Annual Editions concept?

_____

Have you read any pertinent articles recently that you think should be included in the next edition? Explain.

_____

Are there any articles that you feel should be replaced in the next edition? Why?

_____

Are there any World Wide Web sites that you feel should be included in the next edition? Please annotate.

_____

May we contact you for editorial input? ☐ yes ☐ no
May we quote your comments? ☐ yes ☐ no